Traffic Safety

Research for Innovative Transports Set

coordinated by
Bernard Jacob

Volume 4

Traffic Safety

Edited by
George Yannis
Simon Cohen

iSTE

WILEY

First published 2016 in Great Britain and the United States by ISTE Ltd and John Wiley & Sons, Inc.

ISTE Ltd
27-37 St George's Road
London SW19 4EU
UK

www.iste.co.uk

John Wiley & Sons, Inc.
111 River Street
Hoboken, NJ 07030
USA

www.wiley.com

Library of Congress Control Number: 2016936178

British Library Cataloguing-in-Publication Data
A CIP record for this book is available from the British Library
ISBN 978-1-78630-030-0

Contents

Acknowledgments

The European Commission, DG MOVE and RTD, the Conference of European Road Directors (CEDR), the European Road Transport Research Advisory Council (ERTRAC), the European Rail Research Advisory Council (ERRAC) the European technology platform WATERBORNE-TP are acknowledged for their support and active contribution to the Programme Committee of the TRA2014, in charge of reviewing and selecting the papers presented at the conference, which forms the main input of this volume.

The French Institute for science and technology for transport, development and network (Ifsttar) is aknowledged for the successful organisaation organization of the conference TRA2014, in which 600 high quality papers were presented.

Joëlle Labarrère, former secretary of the Programme Committee of TRA2014, and executive assitant of the department COSYS with Ifsttar, is aknowledged for her valuable help to the editors and for this volume making.

Preface

The transport sector is very much concerned about environmental adaptation and mitigation issues. Most of these are related to the objective of curbing GHG emission by 20% by 2020, alternative energy and energy savings, sustainable mobility and infrastructures, safety and security, etc. These objectives require the implementation of advanced research work to develop new policies, and to adjust education and industrial innovations.

The theme and slogan of the Transport Research Arena held in Paris (TRA2014) were respectively: "Transport Solutions: From Research to Deployment" and "Innovate Mobility, Mobilise Innovation". Top researchers and engineers, as well as private and public policy and decision-makers, were mobilized to identify and take the relevant steps to implement innovative solutions in transport. All surface modes were included, including walking and cycling, as well as cross modal aspects.

Policies, technologies and behaviors must be continually adapted to new constraints, such as climate change, the diminishing supply of fossil fuels, the economic crisis, the increased demand for mobility, safety and security, i.e. all the societal issues of the 21st Century. Transport infrastructures and materials, modal share, co-modality, urban planning, public transportation and mobility, safety and security, freight, logistics, ITS, energy and environment issues are the subject of extensive studies, research work and industrial innovations that are reported in this series of books.

This book is a part of a set of six volumes called the *Research for Innovative Transports* set. This collection presents an update of the latest academic and applied research, case studies, best practices and user perspectives on transport carried out in Europe and worldwide. The presentations made during TRA2014 reflect on them. The TRAs are supported by the European Commission (DG-MOVE and DG-RTD), the Conference of European Road Directors (CEDR) and the modal European platforms, ERRAC (rail), ERTRAC (road), WATERBORNE, and ALICE (freight), and also by the European Construction Technology Platform (ECTP) and the European Transport Research Alliance (ETRA).

The volumes are made up of a selection of the best papers presented at the TRA2014. All papers were peer reviewed before being accepted at the conference, and they were then selected by the editors for the purpose of the present collection. Each volume contains complementary academic and applied inputs provided by highly qualified researchers, experts and professionals from all around the world.

Each volume of the series covers a strategic theme of TRA2014.

Volume 1, *Energy and Environment*, presents recent research work around the triptych “transports, energy and environment” that demonstrate that vehicle technologies and fuels can still improve, but it is necessary to prepare their implementation (electromobility), think about new services and involve enterprises. Mitigation strategies and policies are examined under different prospective scenarios, to develop and promote alternative fuels and technologies, multi-modality and services, and optimized transport chains while preserving climate and the environment. Evaluation and certification methodologies are key elements for assessing air pollution, noise and vibration from road, rail and maritime transports, and their impacts on the environment. Different depollution technologies and mitigation strategies are also presented.

Volume 2, *Towards Innovative Freight and Logistics*, analyzes how to optimize freight movements and logistics; it introduces new vehicle concepts, points out the governance and organization issues, and proposes an assessment framework.

Volumes 3 and 4 are complementary books covering the topic of traffic management and safety.

Volume 3, *Traffic Management*, starts with a survey of data collection processes and policies and then shows how traffic modeling and simulation may resolve major problems. Traffic management, monitoring and routing tools and experience are reported and the role of traffic information is highlighted. Impact assessments are presented.

Volume 4, *Traffic Safety*, describes the main road safety policies, accident analysis and modeling. Special focus is placed on the safety of vulnerable road users. The roles of infrastructure and ITS in safety are analyzed. Finally railway safety is focused upon.

Volume 5, *Materials and Infrastructures*, is split into two sub-volumes, investigating geotechnical issues and pavement materials' characterization, innovative materials, technologies and processes and introducing new techniques and approaches for auscultation and monitoring. Solutions to increase the durability of infrastructures and to improve maintenance and repair are presented, for recycling as well as for ensuring the sustainability of the infrastructures. Specific railways and inland navigation issues are addressed. A focus is put on climate resilient roads.

Volume 6, *Urban Mobility and Public Transport*, highlights possible innovations in order to improve transports and the quality of life in urban areas. Buses and two-wheelers could be a viable alternative in cities if they are safe and reliable. New methodologies are needed to assess urban mobility through new survey protocols, a better knowledge of user behavior or taking into account the value of travel for public transport. The interactions between urban transport and land planning are a key issue. However, these interactions have to be better assessed in order to propose scenarios for new policies.

Bernard JACOB, Chair of the TRA2014 Programme Committee

Jean-Bernard KOVARIK, Chair of the TRA2014 Management Committee

March 2016

Introduction

Advances in telecommunications and information technologies are changing the practices used in both everyday life and in professional life. The transport world, sensitive to innovation, does not escape this movement.

Our daily environment demonstrates successful mutations. New equipment is deployed along the roads or on board vehicles. Variable message signs display real-time travel times. Cameras detect incidents and trigger alerts. Information terminals provide service schedules and waiting times for buses or trains. Other technologies facilitate the management of daily travel, making it more reliable, safer and more comfortable.

These developments highlight various aspects of advanced traffic management as well as transport safety. Behind, there is transport research. Its role is to imagine, assess and support the emergence of new approaches and innovative systems. Multi-disciplinary by essence, transport research is well adapted to deal with these issues. This is the purpose of this volume resulting from the international TRA2014 Conference, held in Paris in April 2014. The Conference was organized under the sign of the transition from *Research to deployment in Transport solutions*.

The topic of traffic is organized in two separate but complementary volumes: Volume 3 on Traffic Management and Volume 4 on Traffic Safety; both presenting a selection of papers in the aforementioned fields. As a major event on transport in Europe, the conference covered a broad range of issues linked to Traffic Management and Safety. Naturally, the shortlist presented in these two volumes does not cover the wide spectrum of these areas. It aims to highlight its diversity through a choice of updated papers from the conference. Selection is primarily based

---

Introduction written by George YANNIS and Simon COHEN.

on a quality criterion, also taking into account the geographical diversity of papers in order to restore the originality and richness of the current research.

Main findings

The selected 23 chapters that are included in this volume on traffic safety demonstrate how technological innovations as well as new methodologies applied to transport safety can modify usual practices and offer efficient solutions to the ongoing challenges of safety considerations, needs of vulnerable road users, environmental issues and economical constraints. Both theoretical papers and practical case studies explore topics such as road safety management and policies, accident analysis and modeling, vulnerable road users' safety, road infrastructure safety, ITS and railway safety.

Nowadays, the issue of road safety plays an increasingly important role in traffic and mobility planning and management. In the European Union, systematic efforts for gathering and harmonizing road safety data at the European level have led to a significant upgrade and enhancement of the EU databases, supporting decision-making of both national and international authorities and stakeholders. The availability of detailed, high-quality road safety data is a prerequisite for accident analysis and modeling that can investigate the evolution of road fatalities and casualties, identify the risk of road injuries or allow the development of accident prediction models.

Vulnerable road users require special attention as far as safety is concerned. Innovative research methodologies, such as the use of scanners to track the trajectories of multiple pedestrians in a crowd open new fields of research that can eventually assist in the improvement of pedestrians' safety. Furthermore, the rising popularity of cycling as a means of transportation in urban environments necessitates the re-evaluation of aspects of road design and operation, e.g. at intersections, in order to properly accommodate cyclists' needs.

Speed management is an important issue in road safety, with a direct link to accident severity. Relevant research plays an important role in the evaluation of speed related measures, based on the availability of data for reliable statistical analyses.

The concepts of the safe system approach and the vision of zero accidents are becoming increasingly accepted by researchers, road safety practitioners and stakeholders internationally, and, within these concepts, the challenge to combine road safety with efficient traffic operations and capacity is investigated and promoted to decision makers.

The expansion of ITS applications to vulnerable road users (pedestrians, children travelling to/from school etc.), security and privacy issues related to ITS and the exploitation of further applications such as spreading public awareness of traffic safety seem to be some of the challenges that are currently investigated.

Research in railways safety also indicates that there is a significant potential for improvement, considering the latest technical innovations and developments. Innovative systems are being developed to assist railway management with regards to safety (e.g. evacuation of passengers) and new risk assessment methodologies are presented to help in risk identification and setting priorities.

This fourth volume, extracted from the TRA Conference 2014, will interest both the research and higher education communities, professionals in the management of road and rail traffic, economic and institutional decision-makers increasingly solicited on new forms of transport management. They will find both the state of the art of some key issues, chapters on various methods and illustrative case studies.

This volume on traffic safety includes six parts, covering aspects such as road safety management and policies, accident analysis and modelling, vulnerable road users' safety, road infrastructure safety, ITS and safety as well as railway safety.

Part 1 deals with safety management in general, road safety policy and strategy and development of road safety knowledge systems. Researchers and decision makers can find a thorough investigation and analysis of road safety management in European countries, as well as a presentation of the European Road Safety Knowledge System that includes a wealth of data on road safety and various analyses results. Furthermore, decision makers may be interested in user-friendly tools allowing to integrate traffic safety in urban mobility plans.

Part 2 deals with detailed statistical analysis of accident data, in order to identify or understand road safety critical issues and develop accident models. The issue of the evolution of the number of road fatalities in Poland, in relation to economic factors, is presented, along with an analysis aiming to identify the risk of road traffic injuries for pedestrians, cyclists, car occupants and PTW riders in Rhône, France, based on a road trauma registry and travel surveys. Furthermore, interesting accident prediction models for main rural roads in Hungary are developed, with imminent and obvious practical applications.

Part 3 discusses road safety issues of vulnerable road users: pedestrians, cyclists, young drivers and PTWs. Decision makers will find the analysis of PTW mobility and safety in the OECD countries useful, which concludes in a number of measures integrated with the development of a safe system approach. Research methodologies are proposed to track multiple people in crowds of pedestrians. Finally, analysis of the results of two interesting surveys is presented: one on the patterns of drink driving processes for car drivers and motorcyclists and another on the perceived risk of urban cycling.

Part 4 refers to road infrastructure safety, with particular focus on speed limits, road restraint systems, infrastructure safety management, and various design issues. The part includes both theoretical and practical issues; a comprehensive review on the application of Vehicle Restraint Systems, evaluations of implemented safety measures, such as speed related measures in Sweden, and investigation of traffic signalization issues. Also, a presentation of a data-driven approach to assess the safety effects of a speed limit reducing operation before its implementation can be of assistance to road safety practitioners and decision makers, and the development of software tools for Road Safety Inspections can assist road agencies in the selection of road infrastructure rehabilitation and maintenance projects.

Part 5 explores the use, effectiveness and acceptability of Intelligent Transportation Systems (ITS) technologies in road safety. It focuses on safety and mobility impacts of ITS applications for vulnerable road users, on security and privacy enhancing technologies, on the development of routing algorithms and on the development of a web application to increase public awareness of the state of traffic safety.

Finally, Part 6 discusses railway safety, and includes a comprehensive overview of recent mainline freight train derailments in Europe, the proposal of a risk assessment methodology, and a discussion on the application of Interpretive Structural Modelling (ISM) to security systems in Indian Railways. Furthermore, a system conceived to provide additional clarification and guidance for the evacuation of large numbers of persons within a railway station during an emergency may prove useful to railway safety managers.

Conclusion

The work gathered in this volume provides an insight into research, best practices and transport policies with focus on state-of the-art advances in the field of traffic safety. They demonstrate the progress made in the various processes of data collection, modeling, management, information and assessment, assisting academics, transport professionals, practitioners and decision makers to a better understanding of the current and future trends. The crucial and increasing role of

ITS applications becomes evident, and more frequently researchers and practitioners are applying a universal approach and interdisciplinary methodologies to address transport related issues, including global approaches in modeling. Furthermore, special focus is given to sustainability of presented traffic and safety solutions with special emphasis to the needs of vulnerable road users and to new concepts such as the safe system approach.

PART 1

Road Safety Policy

1

Analysis of Road Safety Management Systems in Europe

The objective of this chapter is the analysis of road safety management in European countries and the identification of "good practice". A road safety management investigation model was created, based on several "good practice" criteria. Road safety management systems have been thoroughly investigated in 14 European countries on 2010, by means of interviews with both governmental representatives and independent experts, who filled in an extensive questionnaire. A reliable and accurate picture ("profile") was created for each country, allowing for country comparisons. Then, statistical methods were used to make rankings of the countries, and analyze the relationship between road safety management and road safety performance. The results of the analyses suggest that it is not possible to identify one single "good practice". Nevertheless, there were several elements that emerged as "good practice" criteria. On the basis of the results, recommendations are proposed at national and European level.

1.1. Introduction

In Muhlrad *et al.* [MUH 11] a road safety management system is defined as "a complex institutional structure involving cooperating and interacting bodies, which supports the tasks and processes necessary to the prevention and reduction of road traffic injuries". By definition, a road safety management system should meet a number of "good practice" criteria spanning the entire policy-making cycle, from agenda setting to policy formulation, adoption, implementation and evaluation and including efficient structure and smooth processes, to enable evidence-based policy-making.

Chapter written by Nicole MUHLRAD, Gilles VALLET, Ilona BUTLER, Victoria GITELMAN, Etti DOVEH, Emmanuelle DUPONT, Heike MARTENSEN, Pete THOMAS, Rachel TALBOT, Eleonora PAPADIMITRIOU, George YANNIS, Luca PERSIA, Gabriele GIUSTINIANI, Klaus MACHATA and Charlotte BAX.

Effective organization of road safety management is assumed to be one of the conditions for obtaining good road safety results at the country level [DAC 12, ELV 12]. Moreover, as road safety is becoming more and more integrated into broader scoped transport or environment policies, and given the effects of the current economic recession on road safety resources, the need for optimization of road safety management systems becomes even more pronounced.

Within the DaCoTA research project, a road safety management investigation model proposed by Muhlrad *et al.* [MUH 11] is based on several "good practice" criteria, defined by an exhaustive literature review, to address the need for optimized road safety management systems, leading to better road safety performance in a changing environment.

The objective of this chapter is to present the analysis of a road safety management framework in European countries and the identification of "good practice" for the optimization of road safety management processes, carried out within the DaCoTA research project.

For that purpose, road safety management systems have been thoroughly investigated in 14 European countries in 2010, by interviews with governmental representatives and independent experts in each country, who filled in an extensive questionnaire on the degree to which the various road safety management systems meet the "good practice" criteria. A shorter version of the DaCoTA questionnaire has also been prepared in collaboration with the European Transport Safety Council (ETSC) and dispatched to the ETSC-PIN panel of experts. The data was then analyzed by means of both quantitative and qualitative analysis.

This chapter is structured as follows: in section 1.2, the road safety management investigation model is presented, and the data collection and handling procedures are described. In section 1.3, the results of qualitative analysis of the data are presented, while section 1.4 concerns the results of quantitative analysis. Section 1.5 presents the conclusions of the research in terms of road safety management "good practice" in Europe. Finally, section 1.6 summarizes the DaCoTA key messages and recommendations for the improvement of road safety management systems in Europe.

1.2. Methodology

1.2.1. *Road safety management investigation model*

The investigation model of [MUH 11] describes road safety management structures and outputs according to the policy-making cycle (agenda setting, policy formulation, adoption, implementation and evaluation) set against the background of a typical hierarchical national government organization (Figure 1.1). The most complete RS management system, which would have been obtained for a country fulfilling all the "good practice" criteria that was identified and was used as a reference (Figure 1.2). For each country, "good practice" elements, a lack of such elements and peculiarities can be then summarized in a "diagnosis" including structures, processes, policy-making tasks and outputs according to the investigation model.

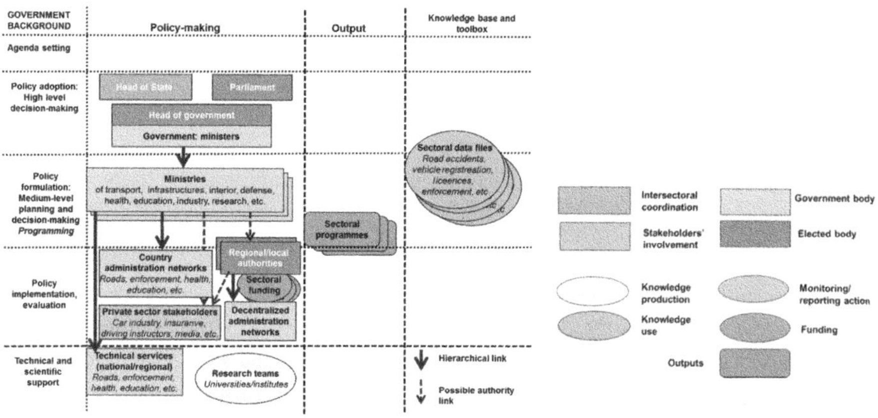

Figure 1.1. *Government organization background. For a color version of the figure, see www.iste.co.uk/jacob/safety.zip*

1.2.2. *Data collection and handling*

On the basis of the investigation model, an extensive DaCoTA questionnaire was developed, by which various road safety management systems meet the "good practice" criteria. The questions related to the five main areas of Road Safety Management:

– institutional organization, coordination and stakeholders' involvement;

– policy formulation and adoption;

– policy implementation and funding;

– monitoring and evaluation;

– scientific support and information, capacity building.

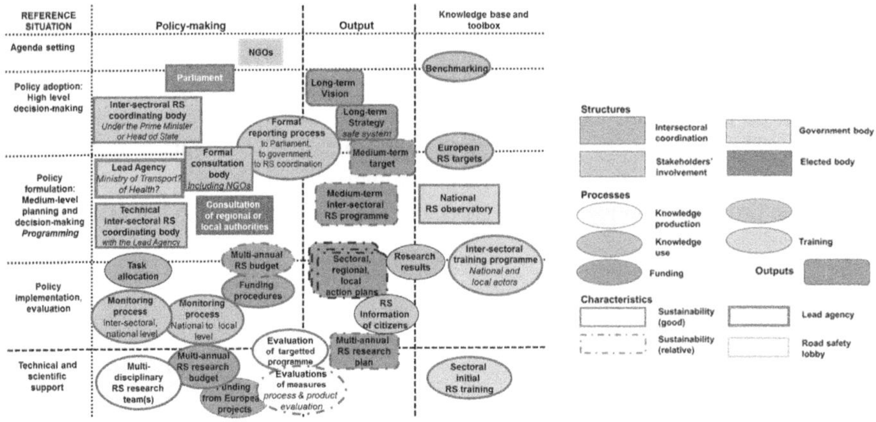

Figure 1.2. *"Reference" country meeting all the "good practice" criteria. For a color version of the figure, see www.iste.co.uk/jacob/safety.zip*

The questionnaire was filled in for 14 countries. More specifically, the DaCoTA partners represented the 12 countries: Austria, Belgium, Finland, France, Greece, Israel, Italy, Latvia, the Netherlands, Poland, Spain and the United Kingdom, and were able to collect data in the native language of a further two: Ireland and Switzerland. To maximize the representativeness of the sample, questionnaires were sent by email to road safety actors in Latvia and Spain to fill in independently without an interview. However, clarifications were sought when necessary.

Two groups of road safety professionals were targeted:

– government representatives: road safety practitioners who are or have been directly involved in policy and decision making over a long enough period of time for them to have acquired wide-ranging experience in road safety;

– independent experts: road safety researchers or scientists who may contribute to policy but do not have a decision making role and could offer a non-partisan view of the Road Safety Management systems in place.

A shorter version of the DaCoTA questionnaire was been prepared in collaboration with the European Transport Safety Council (ETSC). This questionnaire includes 11 key questions similar to those of the original DaCoTA questionnaire and was dispatched to the PIN panel of the ETSC, i.e. the 30 high level national experts from ETSC network of member organizations. This gave a general overview of the Road Safety Management system in 30 countries, although in much less detail than the DaCoTA data.

The combined use of the two questionnaires allowed on the one hand the coverage of basic road safety management elements for all European countries (DaCoTA/ETSC-PIN questionnaire), and on the other hand the full in-depth analysis for a subset of European countries (DaCoTA questionnaire).

1.3. Qualitative analyses of road safety management systems in Europe

Within the qualitative analysis of the DaCoTA research project, a thorough analysis and cross-checking of the questionnaire responses and related comments was carried out, for both the governmental representatives and the independent experts, in order to draw a reliable and accurate picture or "profile" for each country, and allowing for in-depth country comparisons for selected key items. For further details, the reader is referred to Papadimitriou *et al.* [PAP 12].

1.3.1. *Road safety management profiles*

Country profiles of the road safety management systems in the 14 European countries were analyzed and compared to the reference "good practice" system (Figure 1.2). Road safety management structures and outputs are described according to the policy-making cycle (agenda setting, policy formulation, adoption, implementation and evaluation) and set against the background of a typical hierarchical national government organization. Because such a typical organization is not suited to managing road safety policies, which involve most government sectors, specific structures have been set up in most countries, modifying or short-circuiting the typical hierarchical administration.

For each country, these structures as well as the working processes were charted to provide a graphic picture of the road safety management situation ("country profile"), such as the one presented in Figure 1.3 for Belgium, and the identification of "good practice" elements, such as those presented in Figure 1.4 for Belgium. Focus was on the national organization and the relationships between national and regional/local structures and not on road safety management at the decentralized level, as it was agreed at an earlier stage of methodology building that this aspect could not be tackled in the timeframe of the DaCoTA project.

The thorough analysis of the country profiles, together with additional information from the DaCoTA/ETSC-PIN data, allowed for an in-depth analysis and comparison of countries, leading to several observations and conclusions. These are summarized in section 1.3.2.

1.3.2. *Summary of country analyses*

1.3.2.1. *Institutional organization, coordination and stakeholders' involvement*

A large variation was observed in the structures and processes at the higher level of road safety management. The component "Lead Agency formally appointed to take responsibility for road safety" had a higher availability level among the countries. However, different types of Lead Agencies (from strong departments of ministries, to interministerial committees and road safety councils) and with different specific roles were identified. In several cases, it is not easy to identify the "lead agency".

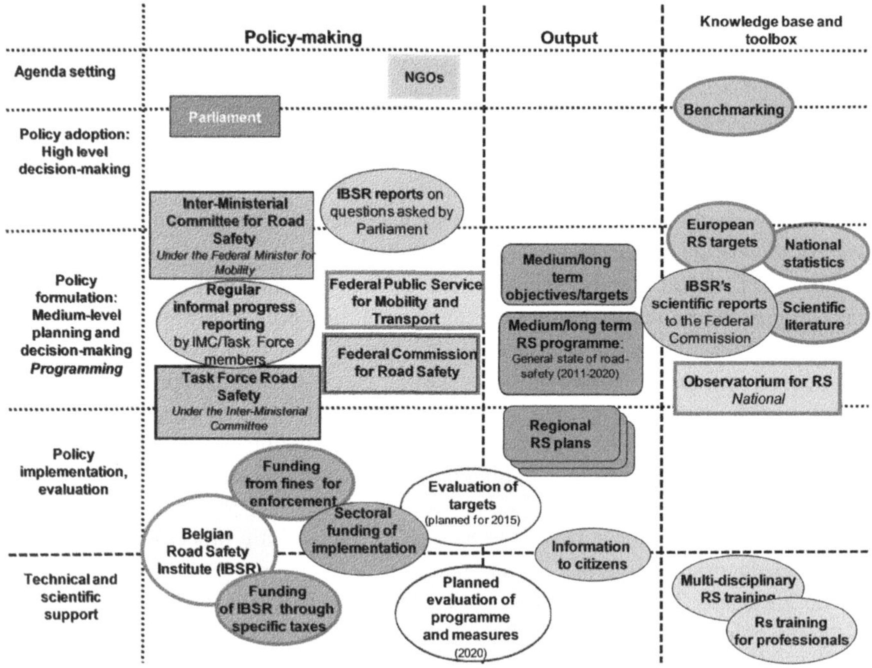

Figure 1.3. *Structures, processes and outputs in Belgium, 2010. For a color version of the figure, see www.iste.co.uk/jacob/safety.zip*

Although it is widely acknowledged that effective road safety management can be achieved with lead agencies of various structural and procedural forms (BLI 09), the results of DaCoTA suggest that road safety management systems based on strong departments of ministries, or that use government agencies specifically established for this purpose, with clear responsibility for the government's road safety policy, are more effective.

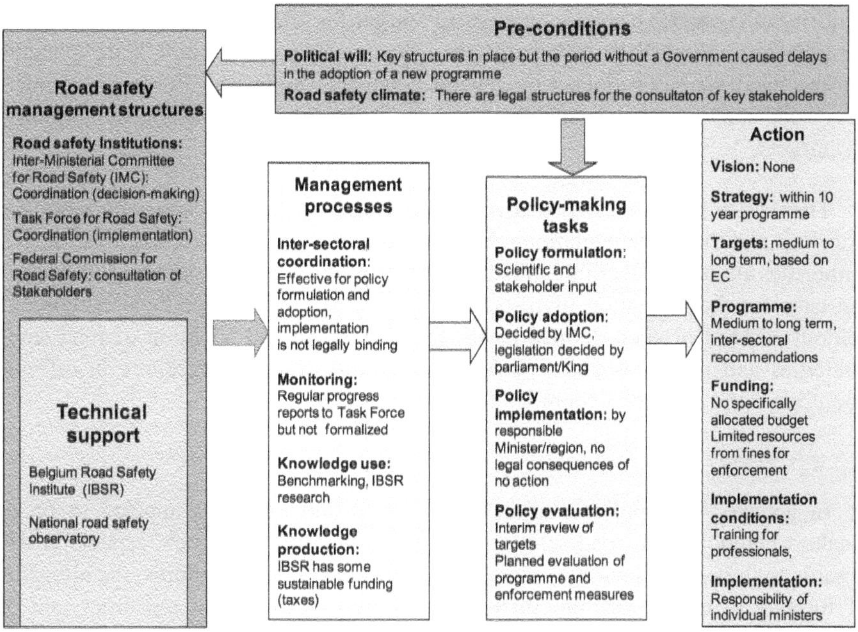

Figure 1.4. *Overview of road safety management good practice elements in Belgium, 2010*

The DaCoTA results clearly indicate that the establishment of a structure and process alone is not sufficient for effective road safety management. In several countries coordination and budget are the most critical links for setting the processes in motion. The effectiveness of road safety management systems can also be largely affected by the degree to which regional authorities, NGOs, stakeholders or the public at large are involved via systematic consultation at all stages of the policy-making process. Very few countries demonstrate such routines and fruitful consultation processes.

1.3.2.2. *Policy formulation and adoption*

Road safety policy formulation showed the largest degree of "consensus" between countries, especially with regards to the presence of a road safety strategy with specific quantitative targets for fatality reduction. Nevertheless, several inconsistencies and uncertainties are involved in the adoption of road safety programs and the participation or consultation of regional and local authorities.

Road safety visions and targets appear to be strongly influenced by either European Union proposals or road safety "leader" countries in Europe. The vast majority of countries have adopted the EU target for 2020, as they had also adopted the previous one of 2010. "Vision Zero", "Sustainable Safety" and "Safe Systems" are the main visions endorsed by several countries. Almost all European countries have road safety strategies and programs, with the majority boasting the ambitious EU targets.

There is a lot of inconsistency in the design of the programs, the setting of priorities and the implementation schedule. Proposals coming from regional or local authorities are hardly ever integrated into national road safety programs. The same is the case for the allocation of resources, so that the regional or local budgets are seldom ensured or even defined at all. Finally, the formal adoption of road safety strategies and programs takes place under quite different procedures in different countries – and in several countries it remains pending.

1.3.2.3. *Policy implementation and funding*

In general, the implementation of programs and measures appears to be the weakest component of road safety management systems in Europe, especially with regards to the establishment of formal resource allocation procedures, the allocation of funding to evaluation, the sufficiency of funds and human resources and the drafting of plans to support implementation.

The problem of providing stable economic foundations for implementing and managing road safety programs is the key to improved effectiveness and efficiency of road safety work. A decision is seldom taken to ensure the availability of a budget for road safety activities from the national budget. Moreover, the lack of information on measures implementation costs at national and international level, combined with a lack of knowledge on the methods appropriate to calculate these costs, makes the evaluation of the actual implementation expenses an estimation by itself.

Moreover, formal procedures for budget allocation to the various actors are seldom in place, especially for the regional or local authorities. As a consequence, the agency responsible for implementation has to rely on its own budget, and the implementation itself depends on the resources available in this agency as well as on the priority it assigns to road safety.

In countries with a clearly designated "lead agency", this agency takes over the majority of program management duties, otherwise it is not always clear who is responsible for what part of the implementation. A lack of coordination at the operational level is clearly identifiable, resulting in some sectors being more efficient than others in performing the road safety interventions that they have been assigned.

1.3.2.4. *Monitoring and evaluation*

A satisfactory level of availability was identified with respect to "benchmarking" for monitoring progress in the road safety situation in relation to the other countries. Nevertheless, most elements related to monitoring and evaluation had a medium or lower level of availability across the countries. In the majority of cases it involves collecting information when a program ends; only a couple of countries monitor programs while they are still in progress.

Only in a few countries is the evaluation of safety measures part of the culture and routine within the road safety program, with a dedicated budget. In several countries, evaluation is very rare and adjusted to the available budget. Even when evaluation is consistently performed, it is usually limited to infrastructure and enforcement measures, or to specific behaviors targeted by specific measures. Formal efficiency assessment techniques are not always implemented. As regards the evaluation of the overall road safety program, it is mostly limited to a "checklist" of the specific measures foreseen, rather than an actual evaluation.

1.3.2.5. *Scientific support and information, capacity building*

In most countries, a higher than medium level of availability is observed for a number of elements related to scientific support and information, such as the use of research results for formulating road safety policies, the systematic information of citizens on the national road safety policy and interventions and their effects, and the presence of articles or programs in the media, which review, criticize or challenge current road safety policies.

Moreover, in most countries, there is at least one research institute or university department performing multi-disciplinary road safety research. While national road safety observatories exist in most countries, there is great variation in their type, role and operation. Only in a few countries are the road safety observatories a part of the lead agencies, while in most cases this role is taken over by research centers, statistical offices or the police. Capacity building and training of road safety actors is seldom a systematic procedure with a dedicated budget. A better use of the scientific capacity appears to be one of the major challenges for evidence-based road safety policy-making in the European countries.

1.4. Quantitative analyses

The quantitative analyses carried out within DaCoTA used statistical methods to identify patterns and rankings of countries, as regards both the road safety management characteristics, and the relationship between road safety management and road safety performance. For further details, the reader is referred to Papadimitriou *et al.* [PAP 12].

1.4.1. *Clustering of countries on the basis of road safety management components*

Statistical clustering techniques were used to group and rank the 14 European countries on the basis of their level of availability of the various road safety management "good practice" elements, separately for each one of the five areas of the DaCoTA questionnaire (see section 1.2.2). Figure 1.5 presents, as an example, the clustering results concerning the first area of the questionnaire, namely "institutional organization". A ranking of countries in terms of their road safety management system, per area and as a whole, was also presented.

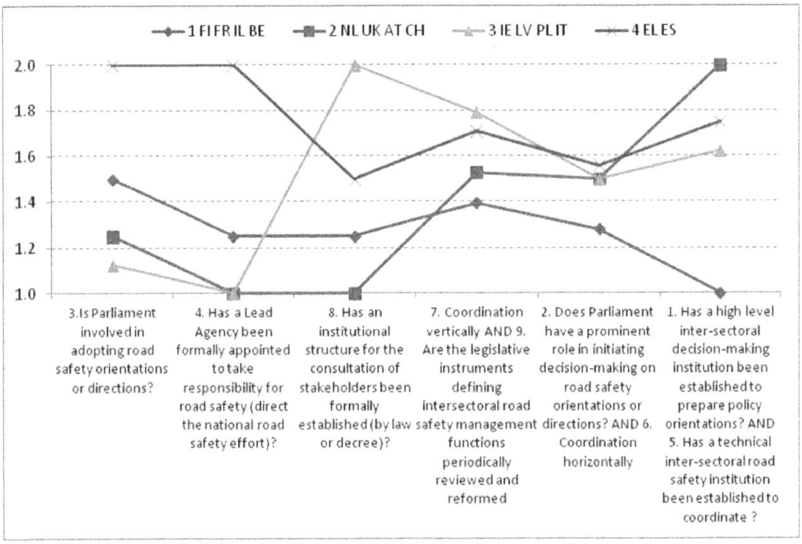

Figure 1.5. *"Institutional organization": mean values of availability of road safety management elements, by clusters of countries. For a color version of the figure, see www.iste.co.uk/jacob/safety.zip*

The analyses confirmed that the complexity and variability of road safety management systems is such that the task of ranking the countries in terms of road safety management is very demanding. It was revealed that all the countries are completely different when road safety management systems are considered as a whole, making it impossible to propose a single overall ranking of countries' road safety management systems. However, it is possible to compare the countries when parts of the road safety management system are considered separately.

However, even when examining the various parts of road safety management systems separately, no two countries were found to belong to the exact same ranking. Across all the analyses, a number of countries with a consistently higher level of availability of some road safety management components could be identified, and others with a consistently lower level of the same features. Interestingly, however, the countries that were ranked systematically at the top of road safety management components were not always those that are known to be the best road safety performing countries.

For example, according to the clustering results, Switzerland, Israel, Finland were found to be consistently ranked best in the various stages of the road safety policy-making cycle, whereas, the UK and the Netherlands were not always ranked best in road safety management components. On the other hand, the countries that were found to be consistently ranked at the lowest of the scale were Poland and Greece, which is in accordance with these countries' ranking on the basis of fatality rates.

The inconsistencies that emerged when comparing the rankings of road safety management with road safety performance, especially for the "good" performing countries, brought forward the need for a dedicated analysis on the potential links between these two.

1.4.2. *Statistical models linking road safety management with road safety performance*

In this analysis, regression models were developed to test whether road safety management is associated with road safety performance. Different road safety outcomes (fatalities, reduction in fatalities, safety performance indicators – the intermediate outcomes) were tested against road safety management indicators and other background variables. Composite indicators were also used both as dependent and explanatory variables [BAX 12].

The dedicated analysis of road safety management and road safety performance was based on the SUNflower pyramid [KOO 02] tackling the entire hierarchy from structure and culture, to programs and measures, to safety performance indicators (intermediate outcomes) and to road safety final outcomes (i.e. fatalities and injuries). Due to the complexity of road safety management systems, as identified by the analysis of the DaCoTA questionnaire responses, this analysis was based on a shorter version of the questionnaire, namely the common DaCoTA/ETSC-PIN questions. The various models developed are summarized in Table 1.1.

	Model 1	Model 2	Model 3	Model 4	Model 5
Dependent variable	Fatalities per million inhabitants	Fatalities per million passenger-kilometers	% reduction in fatalities 2001-2010	Composite index of road safety outcomes	Composite index on Safety Performance Indicators (SPI)
Dependent variable type	Rate	rate	Percentage	Values within [0,1]	Values within [0,1]
Model	Quasi-Poisson	Quasi-Poisson	Beta regression	Beta regression	Beta regression
Explanatory variables	– Background indicator – Composite SPI – RSM	– Background indicator – Composite SPI – RSM	– Background indicator – Composite SPI – RSM	– Background indicator – Composite SPI – RSM	– Background indicator – RSM
Significant effects	– Background indicator – Composite SPI	– Background indicator		– Background indicator	– Background indicator – RSM

Table 1.1. *Summary of models linking road safety management (RSM) with road safety performance*

The results suggest no direct relationship between road safety management and the final outcomes of the road safety systems. However, they do suggest a relationship between road safety management and road safety performance indicators [PAP 13]. This is what is in fact suggested by the SUNflower pyramid, namely that the policy context and input will first affect the intermediate outcomes, i.e. the operational level of road safety, which corresponds to the level of road infrastructure, the maturity of road user behavior, the protection offered by vehicles, etc. These operational conditions are thought to be the result of policies and interventions, and the final outcomes are results of these operational conditions.

Of course, the fact that European countries constitute a small sample, does not allow for the identification of strong relationships, but rather to the indication of the presence of relationships. Moreover, there are some confounding factors that could not be accounted for, such as mobility, economy, weather, long traditions, etc.

1.5. Conclusion

The results of the DaCoTA analyses on road safety management systems suggest that although a number of "good practice" elements can be established as regards road safety management structures, processes and outputs, it is not possible to identify one single "good practice" model at national level. The best performing countries are not always ranked best in terms of road safety management components. On the other hand, the proposed "good practice" criteria seem to work as regards the worst performing countries.

The differences in European road safety management systems notwithstanding, there have been several elements that emerged as more critical "good practice" criteria, such as the presence of a strong lead agency, the efficiency of the implementation – monitoring – evaluation part of the policy-making cycle, the embedding of programs in sustainable and result-focused structures and processes and the distribution and coordination of responsibilities between federal, regional and local levels. In particular, the implementation, funding, monitoring and evaluation elements showed the lowest level of availability in the European countries and appear to be the most problematic sections of the road safety management systems.

When examining the relationship between road safety performance and road safety management in the different countries, road safety management was found to be associated with safety performance indicators (SPIs), reflecting the operational level of road safety in each country.

However, a factor that should be taken into account is the time of observation. In some countries, road safety management components may be so recent that they have not yet had the time to deploy their full potential; or they may have been around for such a long time that their impact has already gradually faded away.

Finally, it was indicated that expert responses may reflect an independent and more objective view and that future analysis might be better to use experts' opinion as a prime source.

1.6. Key messages and recommendations

The improvement of road safety management systems is a key component for the improvement of road safety at national and European level, through improved structures, processes and training. The present research revealed that there is no single ideal road safety management model, but there are several good practice criteria which may be adapted to the national conditions in each case. One clear finding is that a similar performance in road safety management can be achieved by means of differing structures and implementation processes. Similarly, similar road safety performance in terms of final outcomes (i.e. fatalities) may be the "result" of substantially different road safety management systems. On the basis of the DaCoTA analysis of road safety management systems in Europe, a number of key messages and recommendations are outlined in Table 1.2.

Recommendations at national/regional level	Recommendations at European level
Develop objective knowledge of RSM within countries Decentralization with care Establishment of an Independent Lead Agency Inter-sectoral and vertical coordination Continuous stakeholders consultation Focus on implementation rather than vision and strategy Strengthen the link from policy formulation to policy adoption Regular monitoring and evaluation Resources and funding Knowledge-based policies Capacity building and training Handle road safety management in times of recession	Adopting the safe systems approach Exploiting the synergies of road safety and environmental policies Adoption of serious injury reduction targets Focusing on the essentials, leaving the details to the individual countries Strengthening the role of the European Road Safety Observatory Publication of a Road Safety Management Good Practice Manual Building on the existing framework and improving where necessary Political will and commitment from all stakeholders

Table 1.2. *Key messages and recommendations for the improvement of road safety management in Europe*

1.7. Acknowledgments

The DaCoTA WP1 group would like to address special thanks to the governmental representatives and the independent experts of the 14 European countries analyzed for filling in the questionnaire and providing useful and insightful comments on the road safety management systems in their countries. The group is also grateful to Professor Richard Alsop, Chairman of the ETSC PIN panel for his useful comments and remarks on the results of this research.

1.8. Bibliography

[BAX 12] BAX C. (ed.), Developing a Successful Composite Index; End report, Deliverable 4.9 of the Dacota project, European Commission, Brussels, 2012.

[BLI 09] BLISS T., BREEN J., *Implementing the Recommendations of the World Report on Road Traffic Injury Prevention. Country Guidelines for the Conduct of Road Safety Capacity Reviews and the Related Specification of Lead Agency Reforms, Investment Strategies and Safety Projects*, World Bank Global Road Safety Facility, Washington, DC, 2009.

[ELV 12] ELVIK R., "Does the use of formal tools for road safety management improve safety performance?", *The Proceedings of the 2012 TRB Annual Meeting*, Transportation Research Board, Washington DC, 2012.

[DAC 12] DACOTA, Road Safety Management, Deliverable 4.8p of the EC FP7 project DaCoTA, Available at: http://safetyknowsys.swov.nl/Safety_issues/pdf/Road%20Safety%20Management.pdf (accessed 18/9/2013), 2012.

[JOS 12] JOST G., ALLSOP R., STERIU M., A Challenging Start towards the EU 2020 Road Safety Target, 6th Road Safety PIN Report. ETSC Brussels, 2012.

[KOO 02] KOORNSTRA M., LYNAM D., NILSSON G. *et al.*, *SUNflower: a comparative study of the development of road safety in Sweden, the United Kingdom, and the Netherlands.* SWOV Institute for Road Safety Research, Leidschendam, 2002.

[MUH 11] MUHLRAD N, GITELMAN V, BUTTLER I. *et al.* (eds), Road safety management investigation model and questionnaire, Deliverable 1.2 of the EC FP7 project DaCoTA, 2011.

[PAP 12] PAPADIMITRIOU E., YANNIS G., MUHLRAD N. *et al.* (eds), Analysis of road safety management in the European countries, Deliverable 1.5 vol. 2 of the EC FP7 project DaCoTA, 2012.

[PAP 13] PAPADIMITRIOU E., YANNIS G., "Is road safety management linked to road safety performance?" *Accident Analysis and Prevention*, vol. 59, pp. 593–603, 2013.

2

Conceptualizing Road Safety Management through a Territorialized Complex System: Context and Goals

Efforts to improve road safety notwithstanding, the road accident statistics remain a concern throughout the world. This work, after considering France, in particular, we can say that the possible reasons for this proposes to extend to the issue of Road Safety to the land planning problematic, which responds to the characteristics of a spatialized complex system. The methodology used in the ANR program CRITERE mixes interdisciplinary and multi-stakeholder purpose. It combines representations structured by explicit data models, and the more empirical concepts expressed by experts in charge of local public action. The tool provides a user-friendly mapping formulation of partners' opinions and conclusions, highlighting the possible correlations between forms and characteristics of the area, road safety measures, accident and infrastructure observed. It gives the possibility to identify, spatialize, qualify and compare areas for treatment. This application is placed in the perspective of hoisting the issue of road safety among the strategic concerns of territorial management.

2.1. Introduction

For many years, in France, public stakeholding has been preoccupied with displaying road safety as an issue of major importance. Moreover, in 2002 (President J. Chirac official decision), road safety was announced as a "national concern". However, even today, this engagement has not evolved into a clear strategic action at the metropolitan level, as opposed to other issues that emerged over the same period such as those related to sustainable development (pollution, noise, etc.).

Chapter written by Thierry SAINT-GÉRAND, Mohand MEDJKANE, Abdelkrim BENSAID, Dominique FLEURY, Jean-François PEYTAVIN, Eliane PROPECK-ZIMMERMANN and Maroua BOUZID.

In road risk literature, safe travel is primarily addressed in terms of behavior and technical organization of road networks. Indeed, targeting these two groups of factors in the evolution of traffic regulation and control has undoubtedly contributed to the reduction of road deaths observed over the last 15 years in particular. However, the present level (around 3,000 killed people per year) does seem to be decreasing significantly any longer, suggesting that different aspects of this problem operate, and need to be analyzed as well.

One of them is the territorial dimension of road risk and road safety problems. Geographers and accident specialists strongly suspect that factors resulting from the configuration of the territory on which mobility occurs need to be highlighted. Indeed, many of these factors significantly influence traffic conditions: morphology and structure of urbanized spaces, the organization of buildings [ENG 86], meshing the shapes of networks, the visual characteristics of the user's environment [FLE 07], the distribution of functional areas that control flows and their itineraries, spatial ergonomics [SAI 02], etc. An increasing hypothesis is progressively admitted, not only among geographers but also among accidentologists and even some stakeholders, like many types of risks (health, industrial, social, etc.), road risk contains a leading socio-technical component. Thus, it takes part in a vast transdisciplinary reflection on the link between development and safety within territories, and on consistency-related problems in urban design relative to the risks and regulations made by the actors (users or managers) at different levels.

The study of relationships between the structure and the surface of a territory involves questioning the management of road safety through public policy not only in behavioral (the competence and civility of users) or technical terms (safety planning), but at the strategic level as well, across various levels, based on the territory itself at a macro-scale. So, it is necessary to have access to territorial information that is both relevant for the problem and for the scales of analysis, and that is in line with its operational objectives. However, there exists one major challenge which hinders the constitution of such information: the undoubtedly reductionistic and deterministic nature of classical spatial analysis tools, such as those ordinarily used in this domain by both researchers and actors. The ANR (French Research National Agency) CRITERE program aims to overcome this major difficulty, trying to create an approach that is able to also model possible road risk and road safety in their spatial complexity.

2.2. Methodological challenge: integration of different road safety concepts into territorial complex system modeling

Two approaches can be mobilized to develop an approach that is able to model as broadly as possible the complexity of road risk/road safety problems:

– the theoretical construction by geographers of accident-prone spaces by combining relevant variables based on given assumptions,

– the compilation of local actors' (experts) know-how which is based on observed concrete case studies.

Thus, the challenge is in the building of a relevant method suitable to associate these two different approaches.

In direct collaboration with LMCU ("Lille Métropole" Urban Community in Northern France, 1,100,000 inhabitants), the ANR CRITERE program seeks to develop an interactive researchers/experts dialogue tool. This tool must articulate the spatial generalizations of the experts' know-how acquired empirically and in an "ad hoc" manner on the ground, and relate them to other spatial generalizations resulting from the models developed by the researchers. The tool is designed to disseminate this knowledge at the metropolitan level, then at different levels of disaggregation that are significant from the road risk and road safety perspective (sector individualized by experts, the neighborhood, interchange, intersection, route, accident scenes, road safety planning, etc.). A Geographic Information System (GIS developed using ArcGIS 10.1 ESRI) constitutes the core of this application. The geodatabase is based on a Hypergraph Based Data Structure (HBDS) conceptual model. It collects space-time and semantic scales (infrastructures, transport, flows, speeds, buildings, land use, population, activities, habitat, recreational and services' spaces, environment, etc.) and links them. This provides a structured basis of localized data, to model the territory. It integrates these data, combines them, weighs them and develops cartographic representations which give rise to interpretations, or even to new explorations that are discussed through consultation.

To implement this protocol, data must have been designed, developed and calibrated in advance. This chapter presents the approach that guides the construction of such a tool. The objective of this information system that is both scientific and participative is to position road risk and its management through road safety at a strategic level in relation to urban action. We will present a few results as demonstrative examples.

First, we will present the methodological approach: the first principle, using intermediate GIS, consists of placing at the center of the research, a face to face encounter between the researcher and the expert and the analysis tools and spatial complexity. In a second phase, we will present how the GIS was developed to map potential accident-prone locations identified by the experts and analyze their territorial configuration at different levels. This information is likely to help us understand their location within the territory, and grasp the associated factors relevant to our problem.

The beginning phase involves collecting the discourse of actors (developers, researchers or specialized technicians on the operational issues of road safety) and encoding them into GIS entities. The second phase involves generalizing the statements across the territory, mapping them and proposing them anew to experts to develop a systemic understanding jointly that is action-oriented in relation to risk at different territorial levels. An example on the accident-prone zones involving vulnerable road users (pedestrians, cyclists) will further illustrate this approach.

Any territory, such as a geographical area driven by a system of interactions that are primarily internally regulated, is inherently complex. This complexity results from reciprocal, diversified and volatile influences which come between all key elements, notably human, material, immaterial and organizational, even environmental. For their part, risk systems are mainly socio-technical systems that are complex as well, and are, therefore, dynamic and non-linear. Numerous interactions and regulations operate at different levels between the components and the actors of the system. Addressing this double complexity using classic analysis tools highlights a major methodological problem. In the field of road risk/road safety where control by the user and the planner constantly interacts and evolves, this tensioning is very hazardous, producing disappointing results at the macro level of territorial analysis.

Road risk systems arise from dysfunctions within transportation systems. So, three forms of regulation are at work inside and are interrelated:

– the socio-economic planning that determines the morphology of urban (and suburban) networks, flows and the resulting itineraries;

– specific planning by road safety managers;

– the adaptive spatial behavior spontaneously adopted by users (pedestrians and drivers).

On the ground, mobile users – motorists, two-wheeler users, pedestrians – adapt their behavior very efficiently in almost all the encountered circumstances (speed, circumventing, etc.) in real time; planners, observing behavior, redesign the space in non-real time in order to improve security. However, the capacity of adaptation of the users is highly varied, and this implies that the results of a public policy can never be totally predictable across the entire territory [ASM 84, FLE 98]. From a scientific point of view, such a process occurs with an operational constraint: the simple causality is replaced by a complex nonlinear causality that is rather relational and multifactorial, where effects can provide feedback on the causes [MOR 77]. This, indeed, raises questions on the scope of the study which is always difficult to define *a priori*, owing to the fact that configurations evolve, itineraries change, accident locations "migrate". This indicates the need to examine spatial representations across several levels that enable us to adjust the scope and the focus of observation, instantly modifying the boundaries of the study area as well as the scale of analysis.

Temporal activities and adjustments carried out to tackle these rhythms also raise questions: some land planning decisions can lead to unexpected or even dire consequences on risk. For example, a wide road designed to prevent traffic congestion during peak hours attracts significantly higher flows and increased speeds notably during off-peak periods [MIL 03]. This dynamic reality as well as the resulting incertitude projects actors into a prospective dimension rather than promoting reliable forecasts [TRE 06]. This calls for taking risk into account at the strategic level of urban action, the level where the global coherence ought to be considered. All this points out the crucial question of the tools that are suitable for that purpose.

Considered today as the most convenient spatial analysis tools, geographic information systems (GIS) are largely used within the geographer community because they make it possible to model spatial system structures and underlying interactions across different scales [HUG 00, STE 04]. Among administrators, despite increasing usage of GIS, applications remain primarily sectorized and managed for internal use. In their institutions, developing a global, integrated and shared approach among the concerned actors is not identified as an objective. The major challenge to overcome today concerns the integration of knowledge of "those who know" within the spatial analysis tool designed by researchers in such a way that it becomes, not a simple analysis and visualization tool of the risks on a territory, but indeed an agent of the decision-making process.

The originality of this GIS-based complex system modeling lies in the way it is used to tackle the territorial complexity of road risk/road safety: extraction and confrontation of a wealth of experience coming from two sources:

– the wealth of experts (local risk managers (technicians of regional authorities, the State, the private sector, etc.) who possess substantial knowledge obtained from concrete accident cases, action and the observation of effects;

– the wealth of researchers who possess spatial theories, methods and systemic analysis and generalization tools.

The experts on the ground develop an empirical and inductive approach based on their experience with concrete cases; the researchers develop global methods, concepts based, formalized in a rather hypothetico-deductive manner. Although, fundamentally different, these two visions are today to be put into perspective as they are likely to complement each other and even constitute appropriate means of mutual validation of their logics and the results that they produce. Through

CRITERE, we undertake the construction of a platform that enables this enrichment and mutual validation.

The platform's architecture is of modular design (Figure 2.1). It includes:

– the spatial analysis module (GIS): this expresses the territorial problem of the road risk/road security interface at different scales of time, space and representations (points of view). On the one hand, the "actual territory" comprising physical and socio-economic road risks (buildings, infrastructure, population, mobility, itineraries, transport systems, etc.) and on the other, road accidents that occur there;

– local expertise extraction module: this expresses at different levels the "cognitive territories" of road safety stakeholders who have their own experience, visions and priorities based on their activities and objectives within political and/or technical structures;

– integrating the two previous modules and redefining dynamic areas of road safety interest. It is necessary to reformulate discussions in order to identify two types of spatial representations: the "sensitive territories" for road risks and "prioritized territories" for road safety action. The objective is to create the conditions for consistency across the different development and control measures;

– the user interface (GUI under Human Machine Interface) for visualization and interactive dialogue: the tool must be able to visualize and develop the territorial representations instantaneously in order to keep up with the discussions with the actor.

Loop iterations compare the discourse level by level, the concepts of expert partners/geographers and the resulting spatial entities. Surveys make it possible to map the potential accident-prone locations based on the information according to the experts, and to analyze their territorial configurations. In parallel, formalized processing of spatial analysis using GIS is carried out by researchers (the typology of multifactorial spaces, computing network accessibility, analyzing the road environment, visibility indices) in order to provide complementary elements for dialogue.

By discussing the representations via the interface module, the confrontation process is initiated. The loop stops, in other words, the progressive reformulation of the empirical spatial entities by the experts and the theoretical spatial entities by the geographers comes to an end, when convergence (agreement) is reached: the experts accept the GIS formulation of their entities as reconstructed by the geographers, while the geographers validate the methodological consistency of the experts. In this way, the expert broadens his knowledge on the territorial articulation of road risk/

road safety and the geographer validates his assumptions and the tested cognitive schema from the field.

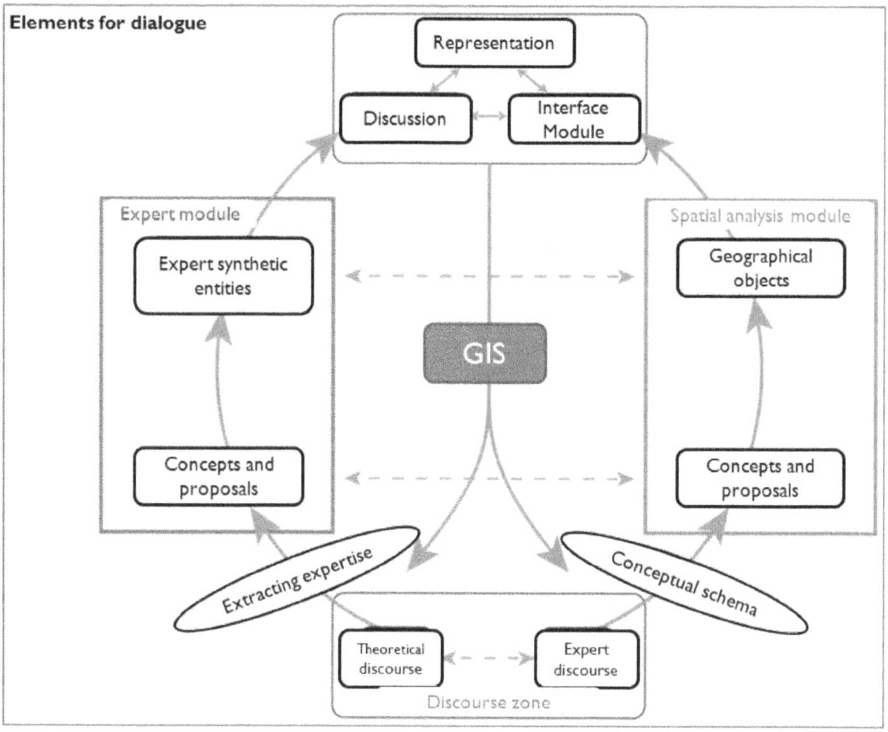

Figure 2.1. *CRITERE Platform's architecture design*

As the approach is based on dialogue between the experts and the spatial analysis tool, it is necessary to:

– collect the discourse of actors (planners, researchers or technicians specialized in operationalization issues of road safety);

– distinguish the key notions that emerge from the relationship between development and road safety in order to transform them into GIS components (layers, spatial entities or data attributes);

– insert them inside the GIS, in coherence with the large amount of factual data previously collected by geographers.

The way used by experts to express their own territorial representations through job-oriented terms needs attention. In his profession, the expert has his own language, which must be deciphered. His examples are based on real accidents from specific situations observed. By describing a concrete case, he activates a logic, a syntax, a vocabulary that is different from that of the researcher. In this way, he uses relative and synthetic notions whose meanings are more or less nuanced, but also terms that are ambiguous, vague, pictorial, even intuitive within the semantic (strong local life, intense city, commercial atmosphere, high traffic routes disconnected from their urban environment, etc.) as well as the localization domain ("near", "next to", "above," etc.) and which must be put back into context and placed within the territory. These experts' concepts are entities of high-level synthesis. Transcribing them into GIS entities implies a methodology that restructures variables of lower levels.

A series of interviews among experts, completed by technical reading (CERTU documents, development documents, etc.), made it possible to identify and define the key notions or expressions prominent in the technical and operational discourse on road security. We analyzed the discourse in order to extract spatial entity concepts (singular spatial objects – a train station – or aggregates – a neighborhood, a residential area, etc.) that we could reconstitute in the spatial analysis tool using query syntax. For example, the "generators/attractors" of population (and subsequently of users' mobility) can be defined by the weighted combination of spatial representations relative to where consumption areas are located, transit stops, train stations, hospitals, schools, etc. Obviously, this does not mean explaining all useful technical knowledge (concerning traffic signal cycle or nature of a road surface) but rather focusing on the elements that make it possible to describe the space for preventive measures identified at the strategic level. We grouped the variety of notions into five items based on convenience ("network", "traffic and speed", "environment", "mobility-travel", "socio-spatial characteristics"). The nature of these items clearly shows that road safety is a global issue that falls within the strategic level.

In order to translate them into cartographical representations, these notions must be measureable. Based on how the terms are defined – and the issues involved – indicators are proposed in order to adapt them into criteria that could be used within GIS. This phase is reflected on by the LMCU experts who also forward proposals in order to align each viewpoint concerning the possibility of translating the notions, in view of the technical capacities of GIS and the available data. The development of a semantic and syntax structure made it possible to construct discourse on the state of insecurity.

2.3. A practical example: ZIVAG

A series of interviews among experts, completed by technical reading (CERTU documents, development documents, etc.), made it possible to identify and define the key notions or expressions prominent in the technical and operational discourse on road security such as the areas that "generate/attract" population (and subsequently user mobility). In GIS, these areas can be identified and located by using the weighted combination of spatial representations relative to where consumption areas are located, transit stops, train stations, hospitals, schools, etc.

The study consists primarily in analyzing the discourse of experts to extract spatial entity concepts (singular spatial objects – a train station – or aggregates – a neighborhood, a residential area, etc.) that can be reconstituted in the spatial analysis tool using query syntax. A semantic network is obtained. This is in the form of formal language (dictionary) of spatial entities and of their relationships. This semantic network makes it possible to simultaneously reconcile quantitative and/or qualitative spatial data attached to accurate geometrical shapes of GIS entities (buildings and road sections) and data that is imprecise but relevant from the experts' perspective, in the form of qualitative spatial relationships (neighborhood relations, distance, orientation).

Based on how the terms are defined – and the issues involved – we proposed indicators in order to adapt them into "spatial variables" useable within spatial analysis tools. This phase was subjected to the reflection and the propositions of the project partners as well as to the LMCU experts. It was indeed the occasion for numerous discussions between the partners in order to align each vision on how to translate the notions, based on the technical capacities of GIS and the available data.

The "ZIVAG" (Vulnerable Severely Damaged Involved Zones) concept was created following the discussions with the stakeholders: this is a priority zone for security action which targets vulnerable road users (pedestrians and cyclists) and the severity of their injuries. Consequently, to address this issue, a zoning action must be defined in which specific security arrangements can be elaborated. ZIVAG is the result of crossing zones of high urban density and those of high traffic flows [MED 11].

In view of our problem, urban density can be defined as a measure of the density and interferences in a specific area, of elements, factors and sets of forces that generate potentially hazardous movements. It regroups:

– "local life": elements of urban living orientated towards leisure and consumption, generators of pedestrian flows, for instance points of activity (restaurants, post offices, pharmacies, museums, bars, shopping facilities, etc.);

– “road users main generators”: particular elements of high urban activity that generate important mobility flows (large hospitals, train stations, universities);

– “public transport” (“PT”): common public transits (bus, tram, metro (subways)).

The mobility vectors are comprised of elements that describe traffic flows: elements of the network with traffic of over 6,000 vehicles a day.

Each layer of vector objects from the two groups is first converted into a regular network (raster) based on the density of points (kernel method, Silverman [SIL 86], p. 76, equation 4.5). Each resulting layer of raster data contains a single value per cell, representing the local value of the criteria for which the layer is intended. The sides of the cell measure 25 meters, chosen following control tests in order to adjust effectively to the variations of urban areas. For each criterion, we obtain a surface of intensity with zones of varying concentration relative to the criterion considered. Each criterion is then standardized to obtain the same representation but with intensity values varying from 0 to 100.

Criteria combination functions through sums (Figure 2.2(a)) that can be weighted (Figure 2.2(b)).

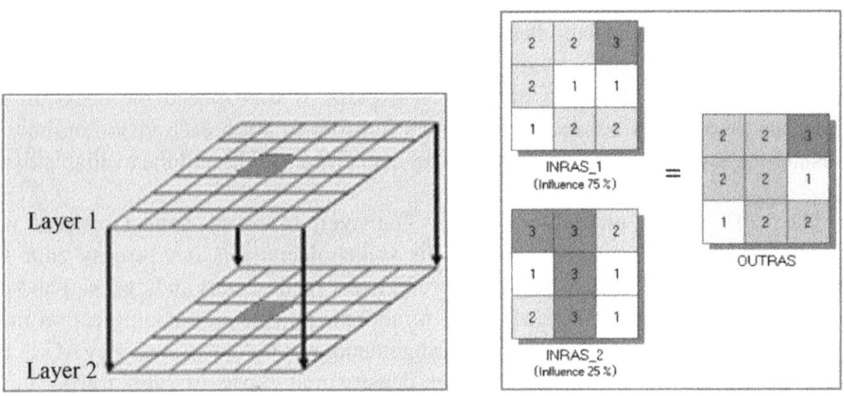

Figure 2.2. *a) Sum of rasters; b) weighted sum of rasters*

Subsequently, we identified four zones:

– “10”: zones with the highest density representing 10% of the total LMCU zone;

– “25”: zones of average density;

– “50”: zones of low density;

– “100”: zones with the lowest density.

A percentage of accidents involving the most vulnerable users was calculated for each zone. The criterion representing “urban intensity” was tested using several types of weights (Figure 2.3).

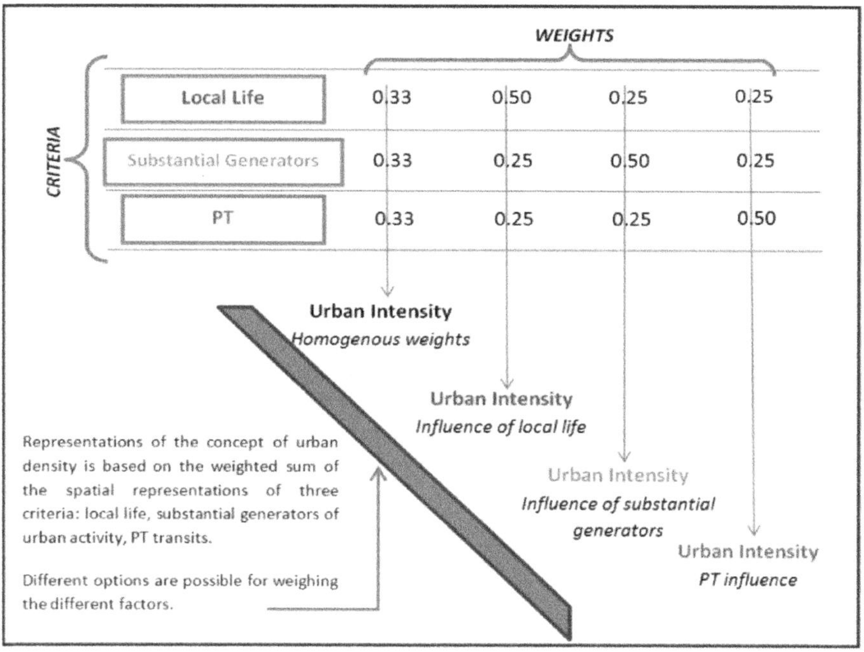

Figure 2.3. *“Urban intensity” modeling*

There are four possible representations of the urban intensity concept that we then combined to the representation of traffic with two possibilities as well: high values in relation to traffic, or to urban density. Both the maps below therefore represent ZIVAG based on different weighting choices. The first test map (Figure 2.4(a)) represents ZIVAG by crossing:

– the “urban intensity” criterion based on a simple sum of the elements of “local life”, “substantial generators” and “PT” (homogenous weights are shown on the map legend by the triangle, square and circle of similar size);

– the “traffic flow” criterion where weighting values are high (the focus placed on traffic is represented on the map legend by a black arrow).

The second test map (Figure 2.4(b)) represents ZIVAG by crossing:

– the “urban intensity” criterion based on the weighted sum of elements of “local life” , “substantial generators” and “PT” (focus is on the factor “local life,” which is represented on the map legend by the triangle which is larger than the square and the circle);

– the “traffic flow” criterion where weighting values are low (the focus placed on urban intensity is represented on the map legend by a black arrow).

2.4. Conclusion and followings

The objective of the research presented here was to design a tool able to position road safety strategically, at all necessary levels, in line with the development of metropolitan areas.

A study that seeks to better account for risk in socio-economic planning policies must be placed in the complexity domain. Indeed, integrating risk (and notably road risk that is specifically covered here) at the local level comes up against interactions and controls by both the users and those in charge of spatial planning across varying scales. The proposed approach juxtaposes two forms of distinct conceptions: that of researchers, theoretical, elaborated in a conceptual model of road risk and road safety environment, and that of field experts expressed in implicit vocabulary that is meaningful to insiders (local technicians, the State).

The entire approach is considered to be a “circuit” of an iterative task which emerges between the spatial analysis tool and the expert: the experts’ knowledge and their own language are translated in the GIS, the multifactorial analyses and generalizations are re-examined depending on different dimensions. The tool is therefore involved in the decision making process.

Our objective is, therefore, to show the significance of new forms of local practices of cooperation between researchers/experts in order to better take into account road risk within urban action strategies.

However, to ensure dialogue between researchers, experts and decision makers, information must also be easily and rapidly accessible, via user-friendly and flexible interfaces which enable the users to instantaneously exploit their content depending on the current needs.

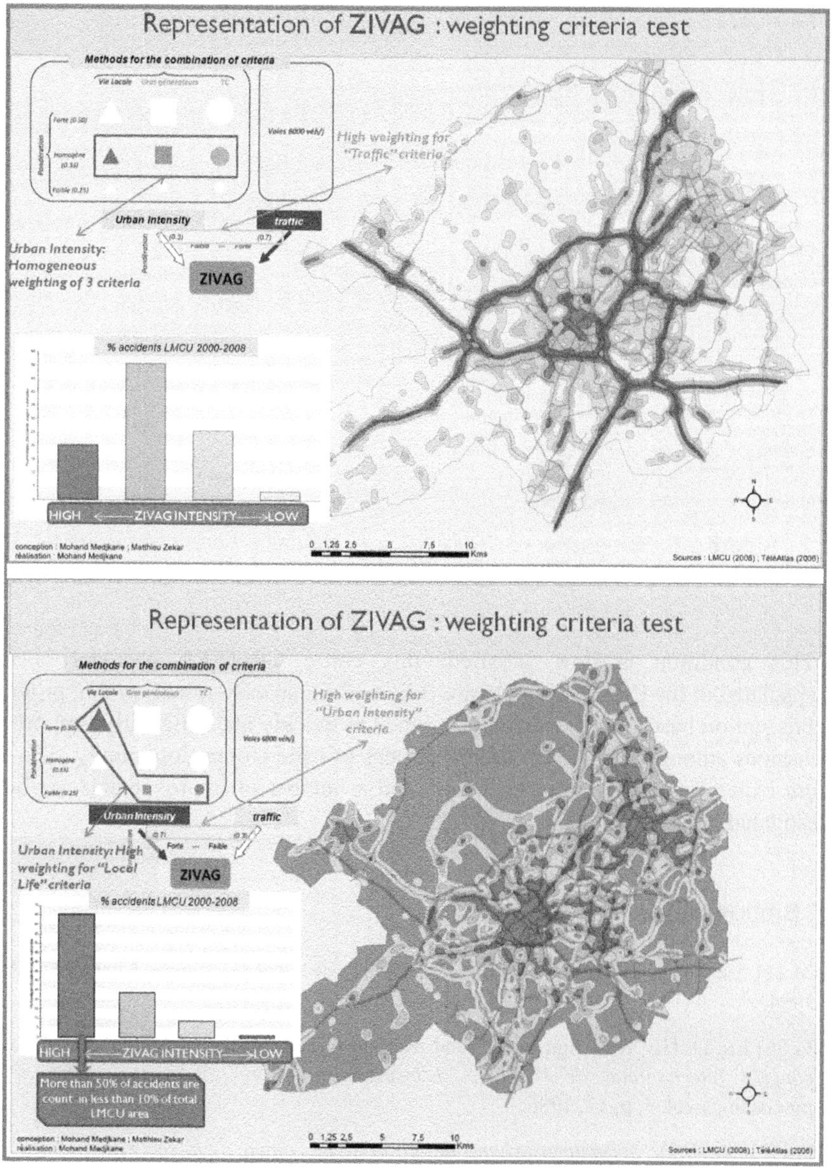

Figure 2.4. *a) First test map; b) second test map. For a color version of the figure, see www.iste.co.uk/jacob/safety.zip*

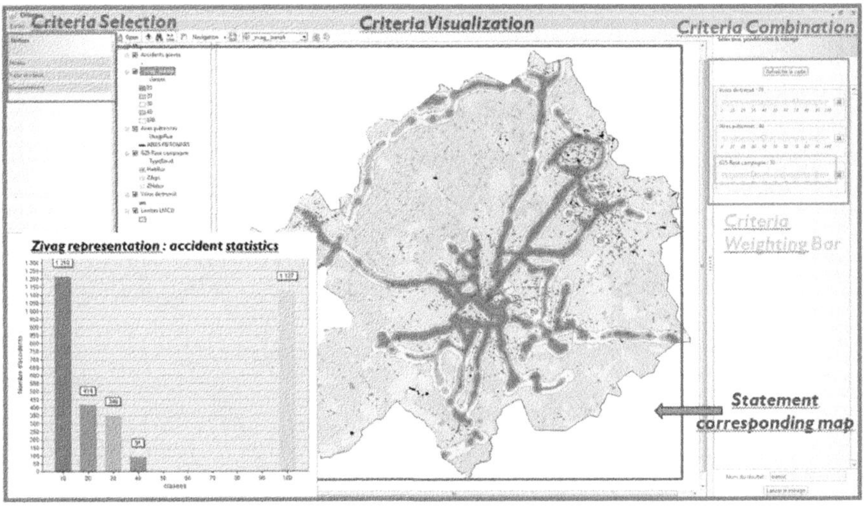

Figure 2.5. *Screenshot of CRITERE user interface. For a color version of the figure, see www.iste.co.uk/jacob/safety.zip*

This condition is now satisfied: this entire CRITERE protocol is now encapsulated in the CRITERE software (Figure 2.5), an easy-to-use *ad-hoc* program, ArcGis station based, and especially designed to quickly and efficiently support live discussions among the different urban services of Lille Urban Community. The first *in situ* experimentations are convincing, and a number of improvements are being investigated.

2.5. Bibliography

[ASM 84] ASMUSSEN E., *OCDE programmes intégrés de sécurité routière*, Chapter III, Paris, 1984.

[ENG 86] ENGEL U., "Risk figures for road users in different urban street categories", *5ème congrès international de l'ATEC, "L'insécurité routière"*, 9-13 June 1986, Paris, proceedings vol. 3, p. 17, 1986.

[FLE 98] FLEURY D., *Sécurité et urbanisme, La prise en compte de la sécurité routière dans l'aménagement*, Presses de l'ENPC, p. 299, Paris, 1998.

[FLE 07] FLEURY D., "Urban Road Safety, How to deal with complexity, Physics and the City", GIORGINI B., BAZZANI A., RAMBALDI S. (eds), *International Review Advances in Complex Systems*, vol. 10, no. 2, Laboratorio di Fisica della Città and CIG-Bologna University, Italy, 2007.

[HUG 00] HUGUENIN-RICHARD F., Approche géographique des accidents de la circulation: propositions de modes opératoires de diagnostic: application au territoire de la métropole lilloise, PhD Thesis, University of Franche-Comté, p. 322, 2000.

[LEC 06] LECOZE JC., TRÉMOLIÈRES A., "Complexité, interdisciplinarité et gestion des risques industriels", *Colloque "Conduites, Organisations et Systèmes complexes"*, University of Caen – MRSH, 9–10 March, 2006.

[MED 11] MEDJKANE M., L'espace des risques routiers: Apports conceptuels et méthodologiques de la géographie aux problématiques territoriales de sécurité routière. PhD Thesis, University of Caen, p. 188, 2011.

[MIL 03] MILLOT M., Développement urbain et insécurité routière: l'influence complexe des formes urbaines, PhD Thesis, Ecole Nationale des Ponts et Chaussées, Paris, 2003.

[MOR 77] MORIN E., *La nature de la nature, La méthode 1*, Seuil, Points essais, p. 399, 1977.

[SAI 02] SAINT-GERAND TH., SIG: structures conceptuelles pour l'analyse spatiale, Thesis, University of Rouen, 2002.

[SIL 86] SILVERMAN B.W., *Density Estimation for Statistics and Data Analysis*, Chapman and Hall, 1986.

[STE 04] STEENBERGHEN T., DUFAYS T., THOMAS I. *et al.*, "Intra-urban location and clustering of road accidents using GIS: a Belgian example", *International Journal of Geographical Information Science*, vol. 18, no. 2, 2004.

3

Development of the European Road Safety Knowledge System

The objective of this chapter is the presentation of the European Road Safety Knowledge System, which was developed within the DaCoTA research project of the 7th Framework Program of the European Commission. This knowledge system includes a number of components, concerning data and tools, road safety issues and countries. A wealth of data was gathered, on road fatalities, exposure, safety performance indicators, socioeconomic indicators, health and causation indicators, road user attitudes and behaviors, traffic laws and regulations, road safety management, etc., for 30 European countries. This data was used in road safety analyses leading to the Basic Fact Sheets, and the Annual Statistical report. Comprehensive summaries of the literature on important road safety issues and methodologies were carried out, in the form of web-texts. Country analyses were also carried out, including road safety management "profiles," "country overviews" and country forecasts. These were integrated through a web-based Road Safety Knowledge System. This system can be a very useful support for road safety research in Europe, and ideal to link research and policy-making.

3.1. Introduction

Even though some European countries present a remarkable road safety level, being among the countries with the best road safety performance globally, there are several constraints that do not allow for accurate road safety analyses at a European level. The lack of specific data and of related details (accidents, injuries, exposure, performance indicators, etc.), the data compatibility and comparability issues among

---

Chapter written by George YANNIS, Pete THOMAS, Nicole MUHLRAD, Heike MARTENSEN, Emmanuelle DUPONT, Letty AARTS, Petros EVGENIKOS and Eleonora PAPADIMITRIOU.

the European countries and the low reliability of data in several cases are the most common problems that need to be confronted. Additionally, the absence of standard methodologies for data/information collection and analysis is observed. Through existing analyses, the correlations between various parameters are identified, but not the causation of the accidents, thus analyses are not solution-oriented. On that purpose, the necessity for systematic collection of road safety data and knowledge through a comprehensive tool is now more urgent than ever, allowing us to consolidate and organize existing data and information and making data and information easily available as an one-stop service. Moreover, such a system should provide a complete tool-kit (analyses, methodologies, benchmarking tools) and support road safety decision making at all levels.

The objective of this study is the presentation of the European Road Safety Knowledge System, which was developed within the DaCoTA research project of the 7th Framework Program of the European Commission and includes a number of components, concerning data and tools, road safety issues and countries. A three-step methodology was adopted for the development of this European Road Safety Knowledge System. Initially, a wealth of data and information was gathered and stored in an organized way, on road fatalities, exposure, safety performance indicators, socioeconomic indicators, health and causation indicators, road user attitudes and behaviors, traffic laws and regulations, road safety management structure and processes etc., for 30 European countries.

These data were subsequently used in road safety analyses leading to specific outputs on important road safety issues, namely the Basic Fact Sheets (on e.g. motorcycles, older drivers, rural areas etc.) and the Annual Statistical report. At the same time, comprehensive summaries of the literature on important road safety issues and methodologies were carried out, in the form of web-texts (e.g. on motorcycles, cost-benefit analysis, etc.). Country analyses were also carried out, including a number of outputs: road safety management "profiles" were developed for each country, including the related structures and processes, the "good practice" elements, as well as the elements needing improvement. Moreover, "country overviews" were created, with country-specific data and information. Finally, country forecasts were developed, on the basis of a robust and specially developed methodology, in which the developments in road fatalities and exposure over the period 1970–2010 are used to forecast road fatalities on 2020, according to different mobility scenarios.

Finally, the structure of a web-based Integrated Road Safety Knowledge System was designed and developed as a comprehensive and integrated road safety information system containing in a structured way all above mentioned outputs (statistics, interactive data, knowledge and tools). In this system different types of road safety data and knowledge were included and respective structures were tested, allowing their future exploitation into the EC European Road Safety Observatory system by giving easy access to data, information and tools and thus supporting the road safety policy-making in Europe.

3.2. Data/knowledge collecting and processing

3.2.1. *Assembly of road safety data*

As a first step, various types of existing road safety data and information were gathered from various national and international sources, initially directly from the sources and at a second phase through national experts of the CARE/RSPI Experts Groups of the European Commission. For that purpose, appropriate comprehensive Master Tables, as in Table 3.1 below, with all types of national data were developed and were gradually filled-in for 30 European countries.

Regarding road accident data, as they are already harmonized at the European level through CARE, the community database with road accident data at disaggregated level, a list of 73 road accident elements (variables and values) collected from all EU countries using a uniform protocol was established. The main criteria for the selection of these basic figures were that the combined variables and values must be useful for macroscopic road accident analysis at EU level, but also that they are available and reliable in all EU countries. This set comprises basic figures regarding number of persons killed as total figures, but also at a disaggregated form, i.e. total vehicle occupants killed by vehicle age group, passenger car occupants killed by vehicle age group, motorcyclists killed by vehicle age group, moped riders killed, etc.

With reference to risk exposure data, a first assembly through EUROSTAT and IRTAD took place and a list of 97 risk-exposure elements was developed and included in the Master Tables: population by age group, vehicle fleet by vehicle type and vehicle age, person-kilometers by vehicle type, vehicle-kilometers by vehicle type, vehicle-kilometers by road class, ton-kilometers, road length by road type, traffic per road type, economic and social indicators such as GDP, unemployment rate, fuel and alcohol consumption, etc.

	Time series road safety data (1975–2010)																				
		1991	1992	1993	1994	1995	1996	1997	1998	1999	2000	2001	2002	2003	2004	2005	2006	2007	2008	2009	2010
	Number of persons killed																				
1	Total figures	2.112	2.158	2.159	2.253	2.411	2.157	2.105	2.182	2.116	2.037	1.880	1.634	1.605	1.670	1.658	1.657	1.612	1.553	1.456	1.281
2	Drivers killed	1.114	1.186	1.212	1.256	1.361	1.207	1.199	1.261	1.228	1.193	1.131	979	1.010	1.017	1.053	1.077	1.013	1.020	964	838
3	Passengers killed	525	509	450	518	569	528	497	504	489	469	411	376	338	360	371	313	344	285	290	241
4	Pedestrians killed	473	464	498	479	481	422	409	417	399	375	338	279	257	293	234	267	255	248	202	179
5	Age group 0–14 (children)	30	31	30	38	24	23	25	25	18	14	19	16	9	15	11	11	13	12	10	4
6	Age group 15–17	11	11	9	4	11	5	5	9	5	5	4	5	6	3	3	2	3	1	1	1
7	Age group 18–24	28	15	27	30	19	20	15	18	13	16	9	12	9	11	8	14	5	3	7	4
8	Age group 25–49	58	78	70	77	99	62	73	58	68	61	61	38	41	60	36	56	47	40	40	32
9	Age group 50–64	105	90	103	83	80	63	75	68	67	58	59	48	36	46	38	33	39	36	31	28
10	Age group 65+	242	238	256	248	243	237	204	223	209	210	173	152	148	143	126	141	140	142	98	100
11	Unknown	0	1	4	1	6	12	12	16	19	11	13	88	15	12	10	8	14	15	10	
12	Total vehicle occupants killed	1.639	1.694	1.661	1.774	1.929	1.735	1.696	1.765	1.717	1.662	1.542	1.355	1.348	1.377	1.424	1.390	1.357	1.305	1.254	1.079
13	Vehicle age <1 year	110	144	94	101	97	80	76	88	83	104	108	92	73	87	100	104	92	93	46	–
14	Vehicle age 1–2 years	245	251	284	250	221	257	245	273	305	310	345	262	251	242	244	242	248	207	179	–
15	Vehicle age 3–5 years	412	430	510	512	558	303	298	271	280	270	230	267	259	305	264	246	245	236	231	–
16	Vehicle age 6–10 years	473	452	407	454	512	320	323	358	335	318	292	239	233	222	272	313	282	276	265	–
17	Vehicle age 11–15 years	399	418	366	455	542	208	242	215	206	202	163	143	187	202	237	187	176	158	167	–
18	Vehicle age >15 years	–	–	–	–	–	187	177	230	217	177	168	163	132	117	131	116	151	155	192	–
19	Unknown	–	–	–	0	–	380	335	330	291	281	236	189	213	202	176	182	163	180	174	1.079

Table 3.1. *Snapshot of Master Table*

Furthermore, data on selected safety performance indicators (SPI), reflecting the operational conditions of the road traffic system, which influence the system's safety performance, have been gathered through the outputs of the SafetyNet project (http://ec.europa.eu/transport/wcm/road_safety/erso/safetynet/content/safetynet.htm) but also through the National Experts using the Master Tables. The SPIs for which data were collected concerned alcohol and drug use, such as the percentage of fatalities resulting from crashes involving at least one driver impaired by alcohol, percentage of drivers above legal limit for alcohol in roadside checks, amount of roadside checks by the police, etc. Similarly, regarding speeding, the SPIs for which the related data were collected concerned average speed, percentage of speed limit offenders, and the percentage of vehicles over speed limit by road type, etc. Regarding protection systems for seat belt wearing, the SPIs concern: passenger cars – front seat (separated by driver and front seat passenger if available), passenger cars – rear seats, passenger cars – correct protection of children < 12 years, etc. For helmet use, the SPIs concern: cyclist helmets, moped helmets and motorcycle helmets. Regarding daytime running lights (DRL), the SPIs are: total usage of DRL, usage rate of DRL per road type, usage rate of DRL per vehicle type, DRL usage by road type and total DRL usage. Regarding vehicles, the SPIs are: vehicle fleet distribution by age, and the percentage of vehicle fleet tested by EuroNCAP, average EuroNCAP score for the vehicle fleet, vehicle fleet composition by vehicle type, crash worthiness, etc. Regarding enforcement, the collected data concerned the following SPIs: number of speeding tickets by the police, amount of alcohol tickets by the police, amount of seatbelt wearing tickets by the police and amount of helmet use tickets by the police. In total, 39 elements related to SPIs are available through the Master Tables.

Moreover, harmonized data and information regarding accident causation, across a number of European countries, was used, exploiting the in-depth Safety Net Accident Causation Database collected from six countries following a common methodology and, importantly, a detailed process for recording causation is called the SafetyNet Accident Causation System (SNACS). This resource includes 1,006 cases split between Germany, Italy, The Netherlands, Finland, Sweden and the UK and was used to produce some basic causation factors.

As combining road accident data with data on road accidents derived from the health sector can provide a better insight on the severity of the road accidents, an assembly of health data took place. Information on medical environment from EUROSTAT was exploited and the following health indicators were defined: health personnel by the type of personnel, hospital facilities and main causes of deaths.

Additionally, several other pieces of data useful for road safety analyses were gathered through the Master Tables. More specifically, data on underreporting of casualties and data on basic country characteristics (area – km^2, amount of unused land – the percentage of total area, average winter temperature of the capital city, average summer temperature of the capital city, annual precipitation level for the capital city (mm), population density and population living in urban areas). In total, 263 data elements are available through the Master Tables, including 37 elements related to traffic laws and road safety measures, 27 elements related to road safety management and 14 elements related to under-reporting, social costs and country characteristics. These elements were selected as the most interesting among many others gathered, as described in section 3.2.2.

3.2.2. *Assembly of road safety knowledge*

For the selection of the information, the Basic Principles for the DaCoTA Data Warehouse were applied, namely:

– quality: data and information are made public only after thorough quality control (availability, reliability, comparability, etc.);

– transparency: all data and information available to everybody, accompanied with the related meta-data (sources, definitions, etc.);

– independence: data, information and especially analysis results should be checked for their consistency and any bias should be properly highlighted;

– usability and accessibility: an advanced user interface should guarantee easy access to all data and information.

Appropriate templates were developed as checklists for every type of information to be collected and the data collection was carried out in three levels. First, all international and national sources, research projects and any other available sources and links identified were explored and exploited. Second, the CARE/RSPI Experts Groups were consulted to validate and add any further information. Finally, in some cases missing information was collected through direct contact with national contacts.

Data on basic road safety programs in 30 European countries were gathered and examined, and several elements such as the existence of a broad national road safety strategy with measurable targets, a specific national road safety plan with quantitative goals, the progress achieved, the responsible organization for implementing the safety strategy plans, etc. are considered. Additionally, information on Road Safety Management for the various countries was gathered through the Master Tables. More specifically, 27 relevant elements were gathered regarding key functions in road safety policy-making, road safety strategy or vision of the country, national plans and targets, availability of road safety management components and enforcement and remarkable road safety policy issues.

Moreover, data on road safety measures in European and other countries were gathered and organized in respective categories, covering different road safety areas and geographical levels. Various data sources were used concerning mainly results from research projects (PROMISING, ROSEBUD, SUPREME, RIPCORD-ISEREST) and final reports/studies of CEDR, COWI and IRTAD. These measures were categorized into four main categories concerning Road User Behavior, Road Environment, Vehicle and Road Safety Management consisting of more detailed sub-categories. In total, 655 safety measures were identified for the 34 different sub-categories and for each of them an exhaustive description and related information were recorded.

Information about traffic rules in the EU Member States was also gathered. Several rules were examined and a list of the most appropriate in terms of importance and EU availability was selected. In accordance with those arrangements, the data was divided into four groups: drivers, pedestrians, vehicles and emergency phone number. The scope of data collection was defined for each group and overall for the 4 categories 54 variables were defined and gathered for 27 Member States and Switzerland in an appropriate Table. Several websites were reviewed (95) to find out the necessary information, such as the European Commission (DG MOVE), World Health Organization, International organizations (e.g. ETSC – European Transport Safety Council, International Transport Forum), Research Institutes and National sources as Ministry or road safety organizations. Each source and its data were evaluated and from each one the most reliable data were selected.

As the information on how road users perceive rules, measures and behavior in traffic can give additional insight in the public support for certain measures taken or to be taken and the self-reported behavior also gives some additional insight in road user behavior, related data on road user attitude and behavior were selected and gathered. The SARTRE studies provided an appropriate source for this information. The studies span a number of years (1996, 1999, 2003 and 2011); the data are harmonized between European countries and are updated. From the SARTRE studies, the following issues were selected because they are relevant for road safety: driver behavior (self-reported) and attitudes towards risk taking.

Regarding driver behavior, data on the following issues were gathered: speeding frequency by road network type, drink driving frequency during last week (over the legal limit and driving with some alcohol), protective system usage frequency by area type and road network type, red light (amber) running frequency, overtaking frequency in situations where it can just be made, frequency tailgating too close to the vehicle in front, giving way to pedestrians frequency.

Regarding attitudes towards risk taking, data on the following issues were gathered:

– alcohol and drugs: (1) agreement on freedom for people to decide for themselves how much they can drink and drive and (2) agreement on more severe penalties for drunk-driving offences. This information is also available by age group and gender;

– speeding: agreement on more severe penalties for speeding. Information is also available by age group and gender;

– protective system usage: (1) feeling of comfort when not wearing a seat belt and (2) attitude towards the need of wearing a safety belt.

Moreover, as the costing of road accidents can also be envisaged as a tool of improvement of decision-making and a mean of classifying the politics, the projects, and the research regarding road safety, several existing studies and reports on accident costs calculation were reviewed and a synthesis report has been prepared, providing also recommendations for the harmonization of the calculation methodology between the countries.

3.3. Key road safety analyses and summaries

3.3.1. *Annual Statistical Report and Basic Road Safety Fact Sheets*

Several editions of the Annual Statistical Report were delivered (2010, 2011 and 2012) including 52 Tables and 26 Figures with the most interesting combination of selected road accident data related to: person class, person killed, area type, motorway, junction type, weather conditions, modes of transport, month, day of the week, hour of day, from 27 European countries for a decade [YAN 12a].

As access to the CARE database is only permitted to a restricted range of users, it has been important to develop a comprehensive range of publications based on these data that are accessible to the general public. The concept of the Basic Fact Sheet (BFS) with disaggregated road accident data for a decade on selected road safety topics, with worth-noticing comments outlined in the "highlight boxes" was developed, and progressively more Basic Fact Sheets are prepared and published annually. Within the framework of DaCoTA, three new editions were developed. In the most recent edition, which was released in 2012, a separate Basic Fact Sheet on causation was included. The set of eighteen Fact Sheets is listed in Table 3.2 below. One part of the development comprised adding details of accident causation to Fact Sheets where appropriate, based on in-depth accident data collected during the SafetyNet project, health indicators by the EU Injury Database, but also maps on specific road safety topics derived from the CARE system.

Basic Fact Sheet	Health indicators section	Causation section
Main figures	Yes	No
Children (aged <15)	Yes	No
Young people (aged 18–24)	No	Yes
The Elderly (aged>64)	Yes	Yes
Pedestrians	No	Yes
Cyclists	Yes	Yes
Motorcycles and mopeds	No	Yes
Car occupants	Yes	Yes
Heavy Goods Vehicles and Buses	No	Yes
Motorways	No	No
Junctions	No	Yes
Urban areas	No	No
Youngsters (age 15–17)	Yes	No
Roads outside urban areas	No	No
Seasonality	No	No
Single vehicle accidents	No	Yes
Gender	Yes	Yes
Accident causation	No	–

Table 3.2. *2012 Basic Fact Sheets*

Both the Annual Statistical Report and Basic Road Safety Fact Sheets were prepared according to the guidelines (design principles) that are to be followed when such outputs are drafted or redrafted in the future, as more uniform style was desirable. The guidelines concern the formatting, the design of tables and figures, the use of specific data (i.e. fatality data), the calculation of summaries and the statistical principles when presenting data. All these are based as far as possible on scientific principles, even though in order to achieve consistency it has been necessary to make rather arbitrary choices.

3.3.2. *Country overviews*

To facilitate road safety comparisons between countries, Country Overviews were developed in DaCoTA for each country [AAR 12], not only presenting the current state of road safety in terms of annual number of crashes or traffic victims, but also containing information on precursors for crashes, such as behavior and policy in a country. This information is organized by using the Road Safety Pyramid [WEG 05] which covers all layers related to: structure and culture, programs and measures, road safety performance, indicators, road safety outcomes and social cost.

The overviews start with a presentation of basic facts of a country, the organization in relation to road safety and attitudes of the drivers, presenting the structure and culture layer. Next, the road safety goals, vision, actions and programs are mentioned following a fixed format. The data for each country are provided along with a European reference (European average or mode) whenever the information is available. Road-user behavior and other system-quality characteristics of the country are described in the safety performance indicators part. It contains information on speed, drink driving, vehicle safety and use of protective systems. The next part contains a description of the annual number of road deaths and their characteristics, such as road transport mode, age and gender, location, lighting and weather conditions and crash type. Numbers are provided for 2001 and the last year available (2009, 2010 or 2011). The average annual change and the share of the number or fatalities in the last year available are also provided. Furthermore, some risk figures can be found, as well as information on under-reporting of fatalities and severely injured road users. The country overview proceeds with information on road safety costs, which constitute the top layer of the road safety pyramid. The most prominent characteristics of the country in relation to road safety are finally summarized in a summary, where the safety position of the country is recorded, the scope of the main problem is noted and any recent progress and any remarkable road safety policy issues are presented.

3.3.3. *Road safety management profiles*

The data and information gathered has allowed for the preparation of the Road Safety Management Profile for each European country. In particular, a road safety management investigation model and questionnaire was used [MUH 11], describing road safety management structures and outputs according to the policy-making cycle (agenda setting, policy formulation, adoption, implementation and evaluation) set against the background of a typical hierarchical national government organization. The most complete RS management system which would be obtained for a country fulfilling all the "good practice" criteria identified was used as a reference. For each country, "good practice" elements, a lack of such elements and peculiarities were then summarized in a "diagnosis" including structures, processes, policy-making tasks and outputs according to the investigation model.

For each country, the structures and the working processes were charted to provide a graphic picture of the road safety management situation ("country profile"), based on coded answers to the questionnaire and comments of governmental and independent Experts, interviewed within the framework of DaCoTA in the first quarter of 2010. Focus was on the national organization and the relationships between national and regional/local structures and not on road safety management at the decentralized level, as it was agreed at an earlier stage of methodology building that this aspect could not be tackled in the timeframe of the DaCoTA project [PAP 12].

3.3.4. *Forecast fact sheets*

Data and information gathered was exploited within DaCoTA for the estimation of road traffic fatalities based on time-series analysis, as it is important to know in what direction the annual casualties are developing, and how fast this development is expected to go. The methods applied to achieve the forecasts are sophisticated statistical tools, not easily understood by non-experts [THO 13]. The forecast results, however, are of direct interest for road safety practitioners with all levels of statistical expertise, therefore it was decided not only to develop a technical description of the forecasting model and of the process that led to its selection for each country, but also the Country Forecast Fact Sheets [DUP 12]. The forecast factsheets are meant to give a relatively non-technical description of the past development of the fatalities (and of the exposure if available). The road traffic fatalities, the traffic volume and the fatality risks are forecasted to 2020 and also forecasts according to mobility scenarios are carried out for all 30 European countries, with exposure as most important explaining variable. If known, the (possible) reasons for the developments are shortly described. Forecasts of the road safety situation in every country include a description of the method adopted to produce these forecasts.

3.3.5. *Summaries on key road safety issues/web-texts*

Finally, summaries on key road safety issues were prepared in the form of 22 web texts, containing high quality information on important road safety topics. The information is scientifically based, easy to read and ready to use and for each of the subjects discussed, includes an overview of the magnitude of the problem, prevalence and countermeasures. The topics that are covered by the web texts are:

– age groups: children, novice drivers or older drivers;

– road users: pedestrians and cyclists, powered two-wheelers;

– hazardous behavior: driver distraction, cellphone use while driving, fatigue; alcohol/drugs, speed and speed management, work-related road safety;

– post-crash: post impact care, eSafety; road safety measures: roads, speed enforcement, vehicle safety;

– policy issues: quantitative targets, cost-benefit analysis, safety ratings, road safety management, integration of road safety in other policy areas.

All these web texts were initially developed in SafetyNet and updated in DaCoTA, under the supervision of an editorial group (SafetyNet) and editorial board (DaCoTA), both consisting of international road safety experts, who were responsible for producing the information about a specific road safety subject. Both editions of the web texts (SafetyNet and DaCoTA) are available in the European Road Safety Knowledge System.

3.3.6. *Integrated road safety knowledge system*

The structure of a web-based Integrated Road Safety Knowledge System (see http://safetyknowsys.swov.nl) was designed and developed as a comprehensive and integrated road safety information system containing in a structured way all gathered road safety data and knowledge, as well as the outputs of their analyses (Figure 3.1).

Figure 3.1. *Integrated Road Safety Knowledge System*

In this system, different types content were included and respective structures were tested, allowing their future exploitation into the EC European Road Safety Observatory system by giving easy access to data, information and tools to anyone interested in road safety issues and thus supporting the road safety policy-making in Europe. More specifically, persons using the system will be those interested in road safety related issues, in conducting their own analysis on basis of this data, or in comparing the performance of countries to determine what can be done to improve road safety. For those users who want to do their own analysis, it means that a certain level of knowledge of the quality and analyses of road safety data and tools is required. It is therefore expected that the users of the European Road Safety Knowledge System will consist of policy-makers, researchers and press. Based on the stated objectives and the target groups, the system was designed to meet certain specifications: the data should be easily accessible and also be as interactive as possible. In terms of content, the System consists of five main components [YAN 12b]: safety issues, countries, statistics, methods and links.

The safety issues part is the knowledge base meant for the Integrated Road Safety Observatory of the European Commission, where the user can find high quality information on important road safety issues in the form of web texts. The information is scientifically based, easy to read and ready to use. For each of the subject treated, the information consists of an overview of the magnitude of the problem, prevalence and countermeasures. The subjects are broadly related to age groups, road users, hazardous behavior, post-crash, road safety measures and policy issues.

Moreover, in order to help policy-makers and researchers to have a good view of the road safety state of European countries, a number of country tools have been included in the System, including the Country Overviews and Forecasts.

The statistics part contains road safety related data and important information on what to do with crash data. The data are organized either as part of an interactive data browsing tool (e.g. crash and exposure data) or as static data (Annual Statistical Report, Basic Fact Sheets, data about safety performance indicators and driver behavior/attitude).

To get high quality data, information and well-structured tools, for each road safety product included in the System a respective methodology has been developed and is described. The procedures of gathering safety issues and collecting information for the country tools are presented, as well as the procedure for collecting statistics and the related information (meta-data) and in-depth accident investigation.

Additionally, a complete set of more than 400 links to external files is developed, gathering, linking and standardizing road safety data as well as other sources, by providing reciprocal web links. The type of information that can be disclosed includes National data files, International data files, Research project links and Stakeholder links. The list includes the link to the website, a brief description of the organization, project or database. Also the country and the type of link are specified and can be used to search. In total, more than 400 links are organized in several user-friendly ways, allowing the users to search for the information/data they need by alphabetic order, Country, Focus (each divided by sub-categories) and Organization.

Regarding the functional specifications of the System, these have been defined within the DaCoTa partnership, in consultation with the CARE Experts Group. The

Cognos PowerPlay was used as software allowing the development of a fully-fledged data browser tool within the budget and the time available in the project.

3.4. Conclusion and next steps

During the last two decades, the systematic efforts for gathering and harmonizing road accident data at the European level have led to a significant upgrade and enhancement of the EU databases. Moreover, important data collection and harmonization efforts have provided very useful results as regards exposure data and safety performance indicators; however, the availability, completeness and level of harmonization of this data varies significantly. At the same time, additional data and related information sources have been established at EU level, including in-depth data, behavior/attitudes data, programs and measures data, social cost data etc., mainly in the framework of European research projects. However, these data sources are still not of sufficient comparable quality, they are still not sufficiently linked and the aggregate data are not always accessible. Finally, an important amount of national data remains unexploited at the European level.

The European Road Safety Knowledge System presented in this paper is a solid but easily accessible, integrated road safety system that allows not only for better integration of the various data assembly processes, but also for the provision of a complete set of data services, with full documentation of the data and information and their sources, in order to support road safety knowledge and the related decision making. This system is developed as a complete working tool for road safety stakeholders, for which not only a full description of the structure, format and content is designed and specified, but also the related content is provided, consisting of a complete set of aggregate road safety related data (road accident data, risk exposure data, safety performance indicators, in-depth data, health indicators/data), knowledge (programs, measures, legislation, social cost, behaviors/attitudes, regulations) and the related key road safety analyses and syntheses outputs (Annual Statistical Report, Basic Fact Sheets, Country Overviews, web-texts, road safety management profiles, Forecast Fact Sheets, etc.).

The European Road Safety Knowledge System can also serve as example for the further enhancement of the European Road Safety Observatory (ERSO), as new structures and features have been tested during the development procedure and new data, knowledge and analysis outputs have been assembled and have become available for incorporation into the existing ERSO.

The next steps for the improvement of this road safety data and knowledge tool concern carrying out more surveys for collection of exposure data, development of additional performance indicators and detailed recording of driver behavior. These should be supplemented by more large-scale experiments on in-depth accident investigation, naturalistic driving and driving simulator, more research and analysis, enabling the identification of more solutions to real-life problems, thus leading to a more rigid European Road Safety Observatory.

Moreover, it is noted that decision making of national and international authorities and stakeholders will significantly benefit from the operation of a powerful European Road Safety Observatory making available complete data sets and targeted road safety analyses and syntheses, since more data and more knowledge widely available will progressively lead to the continuing reduction of casualties at the European roads.

3.5. Acknowledgments

This chapter is based on work carried out within the scope of the DaCoTA (Data Collection Transfer and Analysis) project of the 7th Framework RTD Program of the European Commission. Additionally, the members of the EC CARE Experts Group, representing the 27 Member States, have given considerable support to the project.

3.6. Bibliography

[AAR 12] AARTS L., "Implementation of Country Overviews: Meta-Data. Deliverable 4.5. of the EU project DaCoTA", Available at: http://www.dacotaproject.eu/ Deliverables/ Country%20overviews/Country%20overview%20meta%20data_def.pdf, 2012

[BRO 12] BROUGHTON J., BRANDSTAETTER C., YANNIS G. *et al.*, Assembly of Annual Statistical Report and Basic Fact Sheets – 2012, Deliverable D3.9 of the EC FP7 project DaCoTA, 2012.

[DUP 12] DUPONT & MARTENSEN (eds), Forecasting road traffic fatalities in European countries, Deliverable 4.4 of the EC FP7 project DaCoTA, Available at:: http://www.dacota-project.eu/Deliverables/DaCoTA_D4_4%20Final2.pdf, 2012.

[MUH 11] MUHLRAD N., GITELMAN V., BUTTLER I. (eds), Road safety management investigation model and questionnaire, Deliverable 1.2 of the EC FP7 project DaCoTA, 2011.

[PAP 12] PAPADIMITRIOU E., YANNIS G., MUHLRAD N. *et al.* (eds), Analysis of road safety management in the European countries, Deliverable 1.5 Vol.II of the EC FP7 project DaCoTA, 2012.

[THO 13] THOMAS P., MUHLRAD N., HILL J. *et al.*, Final Project Report, Deliverable 0.1 of the EC FP7 project DaCoTA, 2013.

[WEG 05] WEGMAN F., EKSLER V., HAYES S. *et al.*, SUNflower: a comparative study of the development of road safety in the SUNflower+6 countries: Final Report, SWOV Institute for Road Safety Research, Leidschendam, The Netherlands, 2005.

[YAN 12a] YANNIS G., EVGENIKOS P., AARTS L. *et al.*, Data Warehouse – Final Report, Deliverable 3.8 of the EC FP7 project DaCoTA, 2012.

[YAN 12b] YANNIS G., EVGENIKOS P., AARTS L.,. Design and development of the road safety data warehouse – Final Report, Deliverable 3.7 of the EC FP7 project DaCoTA, 2012.

PART 2

Accident Analysis and Modeling

4

Structural Time Series Modeling of the Number of Fatalities in Poland in Relation to Economic Factors

In comparison with other European countries Poland's traffic fatality risk rates still remain on very high levels. These last few years, the number of fatalities was around 3,500, and the mortality rate reached about nine killed per 100,000 of the population, which places Poland in a very bad position in Europe. This situation shows a strong need for intensified systemic and preventive work, which also prevails in the field of road safety analysis. The objective of this paper is to describe the models that have been developed for analyzing the short-term changes in the aggregate number of fatalities in Poland, measured on a monthly basis between 1998 and 2012. The models account for the influence of economic conditions on the level of road mortality.

4.1. Introduction

4.1.1. *Context*

Poland's traffic fatality risk rates still remain on very high levels, although in 2012 the death rate decreased by 15% to 3,571, and the fatality rate decreased from 110 per million of the population to 93, still well above the EU average in 2011 of 60. The EU average reduction in road deaths over the period 2001–2011 was 45%, but the reduction in Poland was only 24%. This relatively poor performance put Poland at the bottom of the European league. In 2001 Poland's fatality rate was the same as that of Belgium and Estonia, and lower than the rates in Greece, Latvia, Lithuania, Luxembourg and Portugal, but by 2011 all these countries had improved their positions and Poland had become the country with the highest rate [WOR 13].

Chapter written by Ruth BERGEL-HAYAT and Joanna ZUKOWSKA.

The road toll is not only a human tragedy, but also a substantial burden on the entire economy. The costs of rescue, emergency care, rehabilitation, lifetime care of permanently disabled victims, property damage and the work time lost through the injury and death of productive members of society are not well researched for Poland, but can be expected to be as high as 2% of GNP [WHO 13].

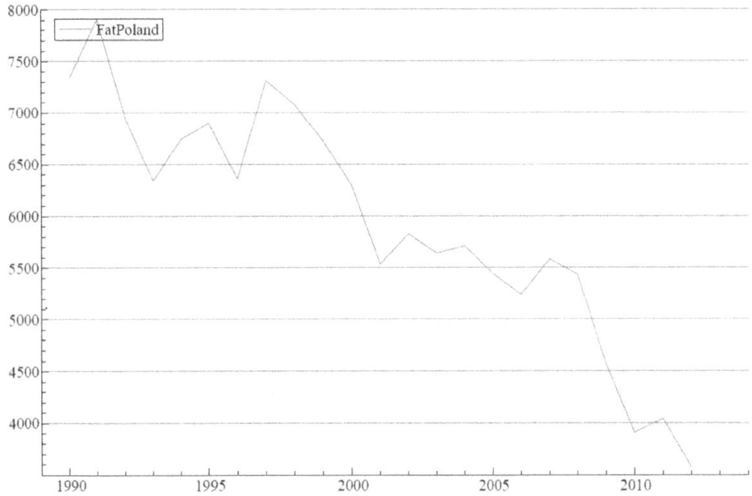

Figure 4.1. *Annual number of fatalities in Poland in the period 1990–2012*

Poland is a signatory to the UN Declaration on road safety. As a member of the EU, Poland is obliged to follow EU specific policies and the EU road safety policy that is described in the document entitled "Towards a European road safety area: policy orientations on road safety 2011–2020". It also proposes to continue with the target of halving the overall number of road deaths in the European Union by 2020, starting from the baseline of 2010. The recently announced target for Poland for 2020 is consistent with this EU target and the UN Decade of Action target, and is no more than 2,000 fatalities.

This situation shows a strong need for intensified systemic and preventive work, which also prevails in the field of road safety analysis. To monitor progress in reaching the long-term targets, the road safety situation must be evaluated and its processes understood. This kind of work had been already done in Poland [HOL 10] although there is still a need to develop the methods used, especially for short- or medium-term trend analysis.

4.1.2. *Research question and objective of the study*

The burden of road accidents and victims was initially the direct consequence of the strong increase in automobility that prevailed after the end of WWII, and the need to analyze road accidents and their severity in relation to risk factors was clearly brought to the attention of the research field by national authorities [OEC 70]. Since then, road safety trend analysis at national level in Europe but also worldwide has been the topic of several researchers' or practitioners' networks, whether formally constituted or resulting from research bodies' initiatives [BER 15]. The turndown in mortality trends for countries in Europe occurred from the 70s onwards, at different points according to the country's development and welfare level; although there is no doubt that this downturn is related to the introduction of basic road safety measures (speed limitation, mandatory seat-belt and helmet wearing, drink-driving limitation), the question of the role of traffic volume[1] and its growth, which reduced significantly since the first oil crisis, in the long-term decreasing trend of road fatalities has been raised. This question of the role of exposure to risk in road safety trend development is central not only for road safety trend analysis at national level, but also for trends benchmarking at international level [JAM 12]. However, exposure to risk measures at national level is neither systematically available, nor fully reliable when available, and they are hardly comparable as they are rarely provided with similar methodologies. Although noticeable progress has been made for allowing accident data to be harmonized at the national level among countries in Europe [BRO 08], it is not the case as regards exposure to risk [YAN 05]. Taking this into account, the focus now turns to a measure of economic activity, such as the economic growth for controlling for exposure at national level; in addition, the role of the influence of a country's economic health on its road safety level is a topic of renewed interest, as the recent financial and economic crisis and the resulting recession that was observed in some countries should logically have contributed to decreasing trends in these recent years [BER 13]. However although economic data are easier to access, the economic factor is to be considered cautiously as its influence on road risk differs, whether in the short or the long term, and this is also the case for the risk exposure factor. Those considerations are also of great interest to the authors and are discussed later in this chapter.

In the former analysis of road safety trends in Poland – due to the unavailability of traffic data, and to the occurrence of two economic breakdowns in 2001 and 2007 onwards – researchers turned to economic factors to explain the simultaneous decrease that occurred in the trend of the number of fatalities in Poland – as in other countries in Europe. Variables such as the GDP and unemployment rate have been used for modeling changes in the number of fatalities in the short- or medium term

1 The traffic volume is often used as the best estimate of risk exposure at national level, although it does not account for the exposure of non-motorized road users.

[ZUK 04]: this approach is investigated further in this chapter using another measure of the economy condition – industrial production.

The following part of this chapter is organized as follows: the current state of knowledge on the short-term relationship between economic conditions and road mortality at national level is first recalled in section 4.2; the econometric approach adopted for estimating that link is then described in section 4.3; an analysis of the dataset used for Poland is given in section 4.4, the modeling results are given and discussed in sections 4.5 and 4.6; and finally, section 4.6 provides a brief conclusion and recommendations for future work.

4.2. Current state of knowledge

The state of knowledge covers both the mechanisms of the process that links the economic conditions to road mortality at national level in the short or medium term, and their quantification by means of appropriate modeling.

The relation between the economic conditions and the aggregate road mortality is not a direct link, which does not make it straightforward to identify its mechanisms. The economic conditions of a country directly influence the aggregate level of transport activity – the largest part of it being taken by road traffic – and consequently influence the national numbers of road injury accidents and casualties. In periods of economic growth, the short-term link to road traffic volumes is positive, as well as to accident and severity numbers. In periods of lower growth, the lack of resources not only induces a reduction of traffic volumes, but also influences the behavior of the driver, who adopts a slower and more cautious type of driving under unchanged traffic volume. Finally, the unemployment level and in particular the unemployment rate, which logically evolves opposite to the economic activity and is commonly chosen to "feel" the economic climate in the country, may as well be used as an appropriate indicator for measuring the influence of economic growth on the road safety level.

All these mechanisms have been investigated and most often quantified. We shall limit ourselves to recalling some quantified results available in the literature as regards the short- or medium-term relation between economic growth and road mortality at national level. This option excludes consideration the results quantified on annual data over long periods of time, and therefore those remaining are not numerous. On the one hand, a number of international studies state that there is a correlation between the total number of traffic fatalities and the degree of activity in the country [THO 92]. On the other hand, some studies also mention the unemployment rate to support that argument [BRÜ 95]: as unemployment grows miles travelled fall – vkt being the most widely used measure of exposure to risk, a factor known to affect directly the accident data – and vice versa. As regards the recent publications on the topic of the relation between the economic conditions and road safety at national level, some relevant results were found using time series analysis techniques on a monthly basis. For Spain, it was demonstrated using state space methods that the industrial production and the road safety indicators – numbers of injury accidents, injured and fatalities – had common cycles of recession and expansion [GAR 07]. However, the same result could not be found for France [AIS 11], and this may be explained by the fact that the type of economic development is not similar in both countries. On the contrary for Poland it was demonstrated, also using state space methods, that the number of fatalities was negatively linked to the unemployment rate [ZUK 04].

4.3. Methodology

The structural time series analysis methods also referred to as state-space methods [HAR 86, COM 07] have been used more and more for modeling the aggregate number of fatalities at national level [DUP 07]. The approach that the authors adopt is innovative, as usually such analyses are led on an annual basis – in order to explain and forecast long-term changes in the aggregate number of fatalities at national level [LAS 01]). On the contrary, short-term changes can only be modeled on an infra-annual basis: similar but uncompleted approaches were taken on a quarterly basis – without the inclusion of exogenous variables [COM 2007] and on a monthly basis – without the inclusion of economic variables [BER 13].

A structural model of periodic (monthly) discrete time series consists of three basic components: the seasonal component, the trend and the irregular interference. It can then be extended according to the needs of the situation: in our case, explanatory variables and interventions are added in order to determine the effect of the selected factors on the number of fatalities in Poland over the analyzed period. Among the selected potential factors which are known to impact the number of fatalities are the industry production and the unemployment rate. Also, the influences weather and road safety measures (or countermeasures) are accounted for when the possible interventions of the model are analyzed (factors or specific events which impact the road safety level).

The general formulas we used for modeling the monthly number of fatalities with the so-called structural model are as follows:

$$\log F_t = \mu_t + \gamma_t + \beta x_t + \sum_{k=1}^{K} \lambda_k w_{kt} + \varepsilon_t \qquad \varepsilon_t = N(0, \sigma_\varepsilon^2)$$

$$\mu_t = \mu_{t-1} + \sum_{l=1}^{L} \lambda_l w_{lt} + \eta_t \qquad \eta_t = N(0, \sigma_\eta^2)$$

$$b_t = b_{t-1} + \zeta_t \qquad \zeta_t = N(0, \sigma_\zeta^2)$$

$$\gamma_t = -\sum_{j=1}^{s-1} \gamma_{t-j} + \omega_t \qquad \omega_t = N(0, \sigma_\omega^2),$$

where

F_t is the monthly number of fatalities;

x_t is the variable measuring the economic factor in the month, with coefficient β;

w_{kt}, k=1,..,K, and w_{lt}, l=1,..,L, are K + L intervention variables, with coefficients λ_k and λ_l;

μ_t and b_t are the level and the slope of the local linear trend;

γ_t is the seasonal component written under a dummy form;

ε_t, η_t, ζ_t and ω_{it}, i=1,..,I, are the error terms, with the variances σ_ε^2, σ_η^2, σ_ζ^2 and $\sigma_{\omega_{it}}^2$ which are not mutually correlated, for $t = 1,...,n$.

Depending on the risk factor, it is assumed in that formulation that the relation of the number of fatalities to a measure x of that risk factor is either multiplicative or semi-multiplicative, at the level of the month.

Thus, $logF_t$ will in all cases be the dependent variable whereas x_t or $logx_t$ will be used as the independent one.

4.4. The data

4.4.1. *Mortality and economic indicators*

The number of fatalities (Police source), the unemployment rate and the industrial production index (Eurostat source) were collected on a monthly basis for the 15-year period 1998–2012. Figure 4.1 represents their changes over that period. It should be noted that the unemployment rate varied between the range of 3–11 over the period but was multiplied by 20 in Figure 4.1 for reasons of commodity. As can be seen, the fatality data represent some outliers – unusually low or high values – which cannot be included in the periodic seasonal pattern. The general trend is decreasing, but several subperiods of time with different rates of decrease can be detected visually.

The correlation between the trend of the fatality data and each of the economic datasets is less visible, and the graphical analysis that is presented below is intended to identify these correlations. However the trends of the two economic datasets seem to evolve in opposition.

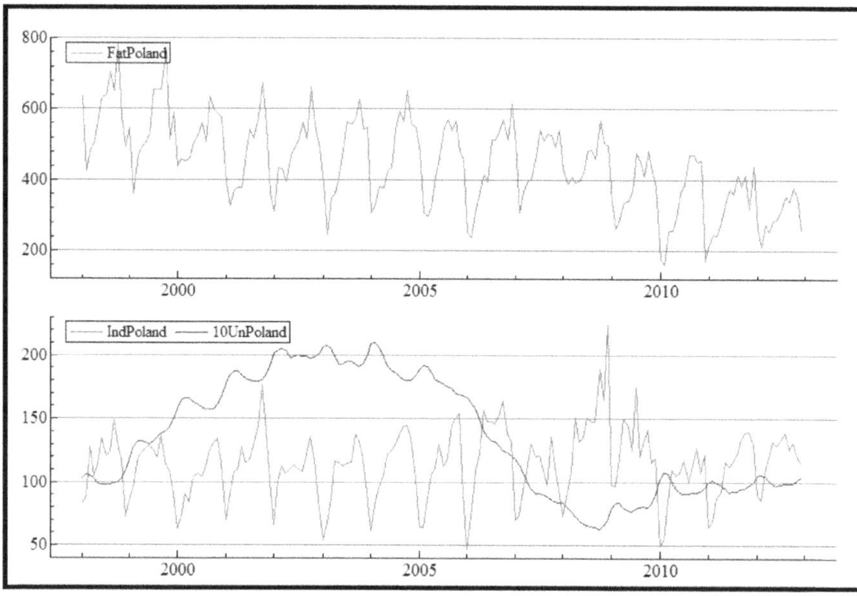

Figure 4.2. *Number of fatalities (Fat), unemployment rate (Un) and industrial production index (Ind) in Poland in the period 1998–2012. For a color version of the figure, see www.iste.co.uk/jacob/safety.zip*

4.4.2. *Weather for explaining the outliers*

Unusual weather in any month is known to affect road safety statistics of that month [BER 13]. In order to explain the outliers with low values, averaged monthly temperature data were collected and compared with the seasonal averages for the month. Thus, six periods of cold spells were detected, and the three most severe were retained for explaining the low outliers of January 2006, January 2010 and December 2010.

4.4.3. *Graphical analysis*

Cross analysis was performed for detecting the relation between the number of fatalities (F) and the industrial production index (IPI), and the relation between the number of fatalities and the unemployment rate (UR) respectively. Figures 4.3 and 4.4 are cross plots of the log number of fatalities and the IPI, of the log number of fatalities and the UR respectively. Six distinct periods with different trends were considered: January 1998–December 2001, January 2002–May 2003, June 2003–December 2004, January 2005–June 2008, July 2008–December 2009, January 2010–December 2012, and the related points represented differently on the cross plots.

A positive correlation appears between the number of fatalities and the IPI for each subperiod (which may be visually represented by a line with a positive slope for each subperiod). This suggests a log-linear relationship between the two variables. Similarly but inversely, the negative correlation which appears between the number of fatalities and the UR for each subperiod suggests a log-linear relationship between the variables. However, log–log relationships were also tested additionally for quantifying these correlations.

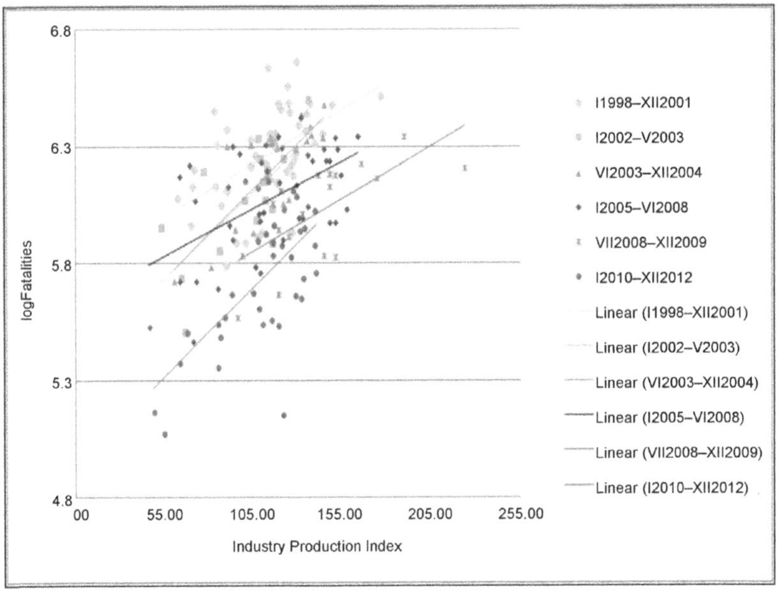

Figure 4.3. *Cross-plot of the log number of fatalities and the industrial production index in Poland in the period 1998–2012. For a color version of the figure, see www.iste.co.uk/jacob/safety.zip*

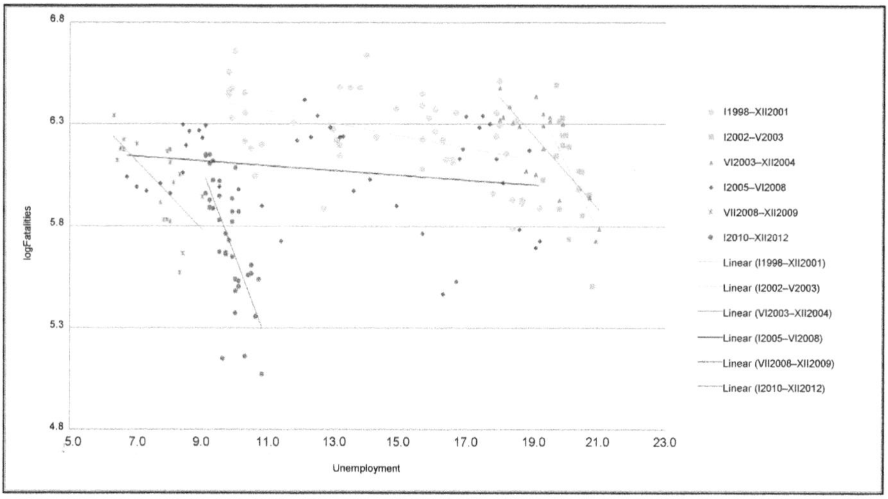

Figure 4.4. *Cross-plot of the log number of fatalities and the unemployment rate in Poland in the period 1998–2012. For a color version of the figure, see www.iste.co.uk/jacob/safety.zip*

As will be detailed in the next section, several state space models with explanatory variables (industrial production index and/or unemployment rate) and interventions were estimated for adjusting the monthly numbers of fatalities over the period, with log–lin and log–log specifications. They all confirm the positive correlation between the number of fatalities and the IPI, and the negative correlation between the number of fatalities and the UR.

Figure 4.5 shows the values of the three components of the adjusted data, provided with Model 5. Figure 4.5 shows a method to approximate the observed data (the log number of fatalities) with the sum: level + regression effects + interventions effects, after the seasonal and the residual are subtracted from the adjusted data.

Figure 4.5 (top) clearly shows that there has been an upwards shift in the level at the beginning of 2002, which lasted until the middle of 2008, after which a serious decrease in the number of fatalities occurred in relation with the crisis. However, the number of fatalities still reached high values during the second half of 2010 and the year 2011.

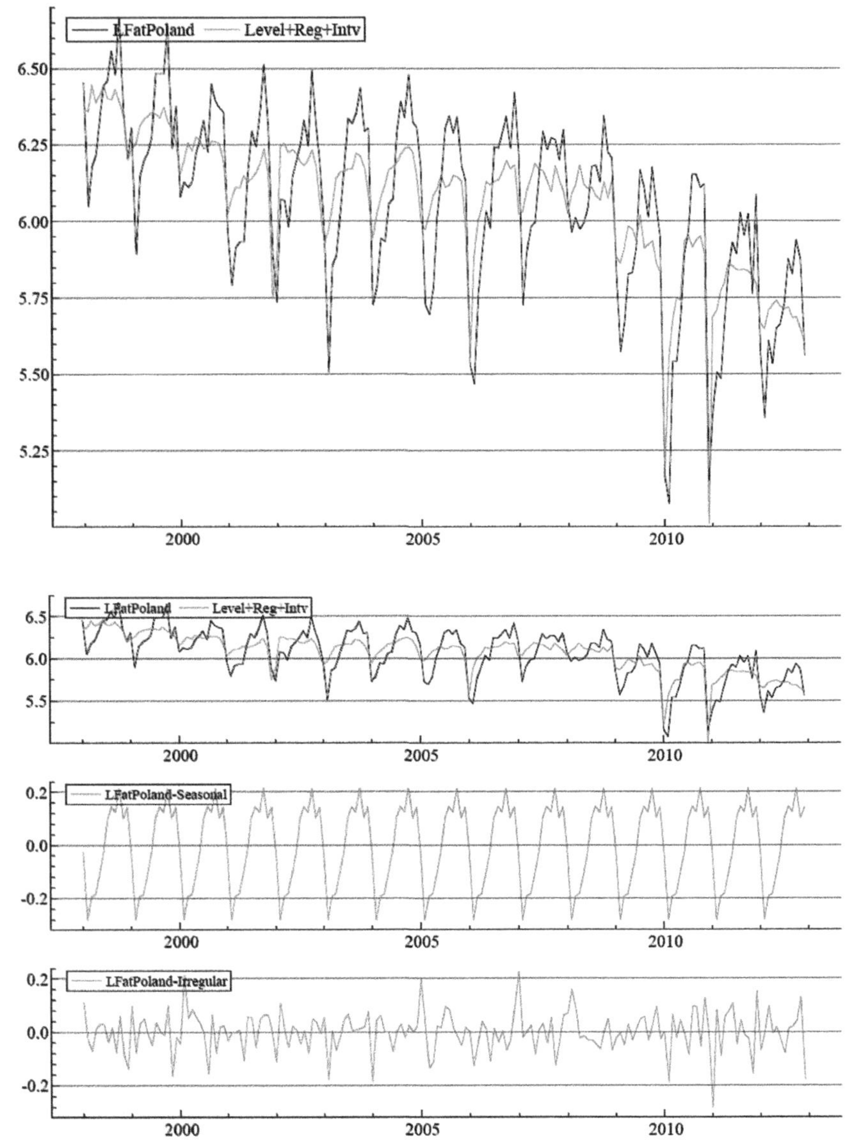

Figure 4.5. *Log number of fatalities: observed values (in black) and modeled as the sum of three components (in red): level + regression + interventions, seasonal, irregular. Outputs from Model 5, for January 1998–December 2012 (outputs Oxmetrics 6.1). For a color version of the figure, see www.iste.co.uk/jacob/safety.zip*

Parameters	**Model 1**	**Model 2**	**Model 3**	**Model 4**	**Model 5**
σ_{ε}^2	0.00846644	0.00852400	0.00782268	0.00746172	0.00752721
σ_{η}^2	0.000793293	0.000810436	0.000956116	0.000923596	0.000799411
σ_{ζ}^2	8.43746e-008	3.21920e-008	2.62767e-008	2.70926e-008	6.03428e-008
σ^2	8.06781e-008	6.61644e-008	4.45049e-008	5.13822e-008	5.56476e-008
μT	5.42498 [0.00000]	5.82728 [0.00000]	4.92173 [0.00000]	3.77896 [0.00000]	4.06372 [0.00000]
b_T	-0.00831 [0.00771]	-0.00764 [0.00432]	-0.00762 [0.00594]	-0.00753 [0.00600]	-0.00803 [0.00600]
Seasonal chi2 test	418.91258 [0.00000]	438.90024 [0.00000]	296.53911 [0.00000]	302.97611 [0.00000]	274.46206 [0.00000]
Seasonal effects					
1	-0.14716 [0.00000]	-0.15026 [0.00000]	-0.08036 [0.02186]	-0.03165 [0.40970]	-0.02775 [0.46615]
2	-0.37274 [0.00000]	-0.37606 [0.00000]	-0.32669 [0.00000]	-0.29251 [0.00000]	-0.28287 [0.00000]
3	-0.21592 [0.00000]	-0.21910 [0.00000]	-0.21028 [0.00000]	-0.20638 [0.00000]	-0.19384 [0.00000]
4	-0.17933 [0.00000]	-0.18091 [0.00000]	-0.18547 [0.00000]	-0.18999 [0.00000]	-0.18371 [0.00000]
5	-0.08463 [0.00078]	-0.08408 [0.00083]	-0.09709 [0.00009]	-0.10795 [0.00001]	0.10859 [0.00001]
6	-0.00672 [0.78744]	-0.00571 [0.81886]	-0.01317 [0.58628]	-0.02437 [0.31158]	-0.02928 [0.22577]
7	0.12781 [0.00000]	0.12922 [0.00000]	0.11566 [0.00001]	0.10430 [0.00003]	0.09928 [0.00007]
8	0.16818 [0.00000]	0.16981 [0.00000]	0.16240 [0.00000]	0.15125 [0.00000]	0.14502 [0.00000]
9	0.16818 [0.00000]	0.17019 [0.00000]	0.14321 [0.00000]	0.12851 [0.00000]	0.12355 [0.00000]
10	0.27544 [0.00000]	0.27776 [0.00000]	0.23545 [0.00000]	0.21951 [0.00000]	0.21389 [0.00000]
11	0.13570 [0.00000]	0.13741 [0.00000]	0.11863 [0.00000]	0.10572 [0.00003]	0.10159 [0.00007]
12	0.13119 [0.00000]	0.13173 [0.00000]	0.13771 [0.00000]	0.14355 [0.00000]	0.14270 [0.00000]

Table 4.1. *Results and performance criteria of the five models fitted on the log of the monthly numbers of fatalities in Poland in the period January 1998–December 2012 (outputs Oxmetrics 6.1)*

NOTE.– For each model, the parameters are given as follows:

– the estimated variances of the level, slope, seasonal component and residue;

– the estimated level and slope at the end of the period (12.2012).

Parameters	**Model 1**	**Model 2**	**Model 3**	**Model 4**	**Model 5**
UR	-0.02702	–	–	–	-0.02033
t-value	-2.46904 [0.01460]	–	–	–	-1.86542 [0.06397]
logUR		-0.28340	–	–	–
t-value		-2.20839 [0.02864]	–	–	–
IPI			0.00193	–	–
t-value			3.57327 [0.00047]	–	–
logIPI	–	–	–	0.29147	0.27241
t-value	–			4.58520 [0.00001]	4.29815 [0.00003]
pulse_12.2001	-0.30653	-0.31299	-0.28412	-0.28886	-0.28270
t-value	-2.84173 [0.00507]	-2.89371 [0.00434]	-2.69786 [0.00773]	-2.81225 [0.00554]	-2.75932 [0.00647]
shift_2.2002	0.26978	0.25566	0.26283	0.25106	0.26558
t-value	3.38701 [0.00089]	3.22675 [0.00152]	3.26131 [0.00136]	3.17852 [0.00178]	3.42989 [0.00077]
pulse_1.2006	-0.34209	-0.34338	-0.28092	-0.20650	-0.21679
t-value	-3.29645 [0.00121]	-3.29610 [0.00121]	-2.74793 [0.00669]	-2.00997 [0.04612]	-2.11691 [0.03582]
pulse_1.2010	-0.43472	-0.42786	-0.41196	-0.35786	-0.35460
t-value	-4.15322 [0.00005]	-4.06078 [0.00008]	-4.04241 [0.00008]	-3.54465 [0.00052]	-3.51789 [0.00057]
shift_6.2010	0.18159	0.17617	0.18809	0.18439	0.18699
t-value	2.37136 [0.01891]	2.29839 [0.02283]	2.41128 [0.01703]	2.41331 [0.01694]	2.51928 [0.01275]
pulse_12.2010	-0.81530	-0.81494	-0.86049	-0.88776	-0.88287
t-value	-7.83931 [0.00000]	-7.80664 [0.00000]	-8.43953 [0.00000]	-8.86506 [0.00000]	-8.83982 [0.00000]
AIC	**-4.3087**	**-4.3061**	**-4.3565**	**-4.4012**	**-4.4090**
BIC	**-3.9362**	**-3.9336**	**-3.9839**	**-4.0287**	**-4.0188**

Table 4.2. *Results and performance criteria of the models fitted on the log of the monthly numbers of fatalities in Poland in the period January 1998–December 2012 (outputs Oxmetrics 6.1) (contin.)*

NOTE.– For each model, the parameters are given:

– the coefficients of the explanatory and intervention variables (in bold), the value of the t-test and the related probability (in italics);

– the performance criteria: Akaike Information Criterion (AIC), Bayesian Information Criterion BIC).

4.5. Results

Five models were estimated for adjusting the log number of monthly fatalities. Two models used the unemployment rate as an explanatory variable, whether UR or logUR (Models 1 and 2), two models used the industrial production index as an explanatory variable, whether IPI or logIPI (Models 3 and 4) and finally a model used the two variables UR and logIPI (Model 5). In all cases, the intervention analysis was also considered.

The first comment is that the structure of the locally linear model is a locally linear trend with stochastic level, fixed slope and fixed seasonal, and this is the case for all five models. This results from the fact that the variances of the disturbances associated with the slopes and the seasonal component are close to zero. It should be noted that the values of the level, measured at the end of the period (December 2012), differ according to the model as the locally linear structure applies to a different variable in each case. The correlation to the explanatory variable is in all cases significant, but the estimated coefficients are to be interpreted differently according to the model[2].

Finally, in addition to the three outliers due to cold spells in January 2006, January 2010 and December 2010 (see section 4.4.2), two intervention analyses are of interest. First, a break in the level, which increased by 25% from February 2002 onwards: the economy was back on track after the sudden economic downturn that was observed in late 2001 – with an exceptionally low value in the number of fatalities in December 2001, considered as another outlier. Second, another break in the level which increased by 18% from June 2010 onwards: an uncontrolled message was given to society regarding new traffic rules (the tolerance to drive

2 As regards Models 1 and 2, an increase of 1 point in the unemployment rate in the month is associated with a decrease of 2.7% in the number of fatalities in that month, whereas an increase of 1% of the unemployment rate in the month is associated with a decrease of 0.3% of the number of fatalities in that month.

As regards Models 3 and 4, an increase of 1 point in the IPI in the month is associated with an increase of 0.2% in the number of fatalities in that month, whereas an increase of 1% of the IPI in the month is associated with an increase of 0.3% of the number of fatalities in that month.

faster on motorways and express roads from 1st January 2011 was in fact anticipated by road users on the entire interurban network).

In consideration of the performance criteria, Models 1 and 4 are to be chosen as the best models with a sole explanatory variable: UR or log IPI. And Model 5 with two explanatory variables UR and log IPI is still better from a statistical point of view.

4.6. Discussion

These results are to be compared with similar results already obtained using the same approach in other countries. In France in particular, similar results were found as regards the relation between the number of fatalities and the unemployment rate, and this indicator will be used for monitoring mortality in the short-term using the Giboulee short-term trend analysis model [ONI 12]. However, although similar results were also found for Spain for the long-term period 1986–2006 – ending before 2007–2008 when the recent economic crisis started and since which unemployment rates have increased very significantly – no significant relation could be found for Greece, even by ending the period of the model's calibration before the start of the crisis. These differences in the results call for forming groups of countries in Europe with similar characteristics as regards their economic conditions in response to the recent crisis.

4.7. Conclusion and outlook

The results confirm that a structural time series model with explanatory and intervention variables is an appropriate tool for explaining the changes in the monthly number of fatalities in Poland for the period 1998–2012, in relation to economic factors such as the industrial production index and/or the unemployment rate. A preliminary graphical analysis was conducted, which confirmed that the correlation between the number of fatalities and the industrial production index (and the unemployment rate respectively) was positive (and negative respectively) on average. Log–log and log–lin specifications were then tested for accounting for these correlations, and three models which confirm this average relation were finally retained as statistically satisfactory and interpretable.

The stability of these models is to be tested by considering several subperiods and, most importantly, the models are to be checked for analyzing the recent crisis period. Scenarios of future short- or medium-term developments of the number of fatalities could also be built, on the basis of scenarios of possible short- or medium-term developments of both industrial production and unemployment.

4.8. Bibliography

[AIS 11] AISSAT O., Modélisation des tendances d'indicateurs de nationaux de sécurité routière, Masters thesis, University of Reims (in French), 2011.

[BER 13a] BERGEL-HAYAT R., CHRISTOFOROU Z., FERRIERE S., The impact of the economic crisis on road mortality an exploratory approach for some countries in Europe, *Proceedings of ETC 2013*, Frankfurt, September 30th–October 2nd, 2013.

[BER 13b] BERGEL-HAYAT R., DEBBARH M., ANTONIOU C., Explaining the road accident risk: weather effects, *Accident Analysis & Prevention*, available at: http://dx.doi.org/10.1016/j.aap.2013.03.006, 2013.

[BER 15] BERGEL-HAYAT R., ZUKOWSKA J., Road safety trends at national level in Europe: a review of time-series analysis performed during the period 2000–12, *Transport Reviews*, vol. 35, pp. 650–671, 2015.

[BRO 08] BROUGHTON J., AMOROS E., BOS N. *et al.*, Final Report on Task 1.5. Deliverable 1.15. of the EU FP6 project SafetyNet, available at: http://www.erso.eu, 2008.

[BRÜ 95] BRÜDE U., What is happening to the number of fatalities in road accidents? A model for forecasts and continuous monitoring of development up to the year 2000, *Accident Analysis & Prevention*, vol. 27, pp. 405–410, 1995.

[COM 07a] COMMANDEUR J., BJLEVELD F., BERGEL R., A multivariate time series analysis applied to SafetyNet data, SafetyNet Deliverable D7.7 of the EU FP6 project SafetyNet, 2007.

[COM 07b] COMMANDEUR J.J.F., KOOPMAN S.J., *An Introduction to State Space Time Series Analysis*, Oxford University Press, 2007.

[DUP 07] DUPONT E., MARTENSEN H. (ed.), Multilevel modelling and time series analysis in traffic safety research – methodology and manual; Deliverables D7.4 and D7.5 of the EU FP6 project SafetyNet, available at: http://www.erso.eu, 2007.

[GAR 07] GARCÍA-FERRER A. *et al.*, The relationship between road traffic accidents and real economic activity in Spain: common cycles and health issues, *health economics*, vol. 16, pp. 603–626, 2007.

[HAR 86] HARVEY A.C., DURBIN J., The effects of seat belt legislation on British road casualties: a case study in structural time series modeling, *Journal of the Royal Statistical Society*, vol. 149, part 3, pp. 187–227, 1986.

[HOL 10] HOLLÓ P., EKSLER V., ZUKOWSKA J., Road safety performance indicators and their explanatory value: a critical view based on the experience of Central European countries, *Safety Science*, vol. 48, pp. 1142–1150, 2010.

[ITF 13] ITF, *Road Safety Annual Report 2013*, OECD, available at: http://www.internationaltransportforum.org/irtadpublic/pdf/13IrtadReport.pdf, 2013.

[JAM 12] JAMROZ K., "The impact of road network structure and mobility on the national traffic fatality rate", *Procedia – Social and Behavioral Sciences*, vol. 54, no. 4, pp. 1370–1377, 2012.

[KRY 03] KRYSTEK R., ZUKOWSKA J., "Falling accident risk in Poland", *The 3-rd Safety and Reliability International Conference*, Poland, 2003.

[LAS 01] LASSARRE S., "Analysis of progress in road safety in ten Europeans countries", *Accident Analysis and Prevention*, vol. 6, pp. 743–751, 2001.

[OEC 70] OECD, *Proceedings of the Symposium on the Use of Statistical Methods in the Analysis of Road Accidents*, Crowthorne, April 1969, 1970.

[ONI 12] ONISR, La sécurité routière en France: bilan de l'année 2011, Ministère de l'Interieur, La Défense (in French), 2012.

[SAF 07] "SAFETYNET – BUILDING THE EUROPEAN ROAD SAFETY OBSERVATORY", available at: http://www.erso.eu, 2007.

[THO 92] THORENSEN T., FRY T., HEIMAN L. *et al.*, Linking economic activity, road safety countermeasures and other factors with victorian road toll, Report No. 29, Monash University Accident Research Centre, Clyton, Victoria, 1992.

[WOR 13] WORLD BANK, *Country Report on Poland Road Safety Management Capacity Review*, Global Road Safety Facility, World Bank, Warsaw, 2013.

[WHO 13] WHO, *Global Status Report on Road Safety, Supporting a Decade of Action*, World Health Organisation, Geneva, available at: http://www.who.int/violence_injury_prevention/road_safety_status/2013/en/index.html, 2013.

[YAN 05] YANNIS G., PAPADIMITRIOU E., LEJEUNE PH. *et al.*, State of the art report on risk and exposure data, SafetyNet Deliverable 2.1, available at: http://www.erso.eu, 2005.

[ZUK 04] ZUKOWSKA J., "Traffic fatalities modelling using time-series", *International Workshop on Rehabilitation of existing urban building stock*, Gdansk University of Technology, Poland, June 17–19, 2004.

[ZUK 12] ZUKOWSKA J., "Road safety analysis in Poland using time-series modelling techniques", *Journal of Polish Safety and Reliability Association, Summer Safety and Reliability Seminars*, vol. 3, nos. 1–2, 2012.

5

Risk of Road Traffic Injuries for Pedestrians, Cyclists, Car Occupants and Powered Two-Wheel Users, based on a Road Trauma Registry and Travel Surveys, Rhône, France

We estimate the risk of traffic injuries as the ratio between amount of casualties and mobility. Crash data is sourced from a road trauma registry, including outpatients and inpatients, in the Rhône county (1.6 million inhabitants, including Lyon). These data are far more complete than police data, in particular for cyclist crashes. Mobility comes from travel surveys. Compared with car occupants and with regards to time spent traveling, the all-injury rate was times lower for pedestrians, eight times higher for cyclists, and 42 times higher for powered two-wheel (PTW) riders. For serious injuries, compared with car occupants, the risk is 2, 16 and 120 times higher for pedestrians, cyclists and PTW users, respectively. For car occupants, pedestrians and cyclists, the 18–25 year age group had the highest all-injury rate. For cyclists, the injury rates seemed higher in non-dense areas than in dense areas. Between 1996–1997 and 2005–2006 and with regards to the time spent traveling, the all-injury and serious-injury rates seem to have decreased for car occupants and cyclists, and even more so for cyclists.

5.1. Introduction

In some of France's major cities (Paris, Lyon, Lille etc.) there has been an increase in cycling, mostly as a means of transport [PAP 10]. Cycling is encouraged

Chapter written by Stéphanie BLAIZOT, Francis PAPON, Mohamed MOULOUD HADDAK and Emmanuelle AMOROS.

in the framework of sustainable development and for better health through physical activity. We therefore need to know more about cyclist road risk, compared with other road user types.

In France and in most countries, crash data are provided by the police. However, police crash data suffer from under-reporting and selection bias. This under-reporting varies mostly with injury severity, the presence/absence of a crash opponent and the road user type [HAU 88, ELV 99]. Crashes involving cyclists are particularly less reported [LAN 03, AMO 06]. In the Rhône County (population of 1.6 million, with Lyon being the main city) a medical registry provides a second source of data, far more complete than police data [AMO 06]. Indeed, over the period 2004–2007, police crash data recorded about 155 injured cyclists and the medical registry recorded 1,230 injured cyclists per year. This medical source is used here.

The objective of the present study is to estimate injury rates in the Rhône County, using medical crash data and travel surveys. Rates are estimated for car occupants, pedestrians, cyclists and powered two-wheeler (PTW) riders, separately for three injury categories (all-injury, hospitalization and serious-injury) and by gender, age groups and location (dense vs. non-dense areas). The exposure measures are number of trips, distance traveled or time spent traveling with a given means of transport. These are obtained from a regional travel survey (RTS). This survey is however restricted to the winter period of the years 2005 and 2006 and to typical weekdays. To correct for this, seasonality ratios were introduced, which were estimated from the 2007–2008 national transport survey. Trends of the injury rates are also studied, between the 2005–2006 time point and the 1994–1995 time point (previous RTS).

5.2. Material and methods

5.2.1. *Hospital-based crash data: the Rhône road trauma registry*

A road trauma registry has been in operation since 1996 and has been certified by the relevant French authority (*Comité National des Registres*). It covers all casualties from road crashes in the Rhône County who benefit from medical care in health facilities. All health care facilities (from public and private hospitals) in the County and its surrounding area which receive crash victims participate: about 280 health units, ranging from pre-hospital emergency care, emergency departments, intensive care units, surgery units, etc. to rehabilitation departments. Injury assessment is based on all diagnoses established in the different health services a casualty may have visited. Diagnoses are coded with the Abbreviated Injury Scale (AIS) which includes a severity score: 1=minor, 2=moderate, 3=serious, 4=severe, 5=critical and 6=beyond treatment [AAA 90]. In this study, the MAIS (Maximum AIS) is used (maximum severity code in case of multiple injuries). Three injury categories are studied: all-injury (whatever the severity), hospitalization and serious injury (MAIS 3+).

5.2.2. *The regional travel survey*

A regional travel survey (RTS) provides data on travel practices of the resident population of the region. Mobility is measured through all the trips made the day before the interview, for all means of transport used. A trip is defined as going from one place to another, where an activity is performed (round trips are split into two one-way trips). For each trip, the following data are collected: means of transport used, origin and destination locations of the trip and times of departure and arrival. Distance is *a posteriori* estimated from an origin–destination matrix.

The most recent RTS in the Rhône County was carried out between November 2005 and April 2006, on weekdays, outside school and public holidays. Three exposure measures were estimated: number of trips, distance traveled and time spent traveling. In the results, we favor number of hours spent traveling. The first reason is that time–budget is less elastic than distance–budget: time spent traveling of a day is quite similar between road users, whatever the type; whereas distance traveled on a day varies more by type of road user. We hence believe that time spent traveling makes comparisons more relevant. The second reason is that it allows us to compare with injury risk of other activities (e.g. sport and DIY).

5.2.3. *Seasonality correction*

Given that the RTS was carried out on weekdays (i.e. excluding weekends), outside school and public holidays, in the winter period, we sought to correct for this in order to have year-round estimates. We introduced seasonality ratios; these were estimated from the national household travel survey (NTS) performed in France in 2007–2008. We split this survey into two periods: (1) the "RTS period", i.e. November–April on weekdays, outside school and public holidays; and (2) the "outside RTS period", i.e. the rest of the year. We then divided the amount of each mobility measure associated with the "outside RTS" period by the total amount of the exposure measure associated with the "RTS" period to create seasonality ratios. These ratios were then applied to the mobility estimated from the RTS to obtain two full years of data between January 2005 and December 2006.

5.2.4. *Location: dense and non-dense areas*

The Rhône County being quite urbanized, a "dense/non-dense areas" variable was created. This distinction enables us to obtain results that do not depend on this characteristic. Moreover, this variable may be used as a proxy to identify different groups of cyclists according to their type of bicycle use: most sport or leisure cycling take place in non-dense areas, and most cycling as a means of transport takes place in dense areas [AMO 11].

We defined municipalities as being "dense areas" if their ZAUER category (as defined by INSEE – the French National Institute of Statistics and Economic Studies) was "urban cluster", with a further population density of over 500 persons per km^2 or a population of over 5,000, or if their ZAUER category was "rural employment cluster", with (more than 5,000 jobs). All other locations were considered to be "non-dense areas".

For the crash data, the crash location was classified into "dense" or "non-dense" areas. For the mobility data, trips were classified into "dense" or "non-dense" areas based on the origin and destination locations; if these two differed, the characteristic of the destination location was considered.

5.2.5. *Injury rates*

For each type of user (car occupants, pedestrians, cyclists and PTW riders), all-injury (whatever the severity), hospitalization and serious-injury (MAIS 3+) rates were estimated. These rates were estimated by dividing the number of injuries by the exposure measurement and scaled per one million trips, kilometers or hours. Rates by gender, age group and location ("dense areas" and "non-dense areas") were estimated separately for each type of user (except for PTW riders because of insufficient mobility data). Only residents of the Rhône County were included in the study. Crashes and trips were also restricted to the Rhône County.

The variability in the estimation of mobility and seasonality ratios could not be taken into account, because of complex survey design. This means that the tests may lead to conclude a difference too often.

5.2.6. *Trends of injury rates*

To study the trends of these injury rates, the last two RTSs were used. However, the previous RTS, performed in 1994–1995, was restricted to Greater Lyon (Lyon and its suburbs, mostly urban). This survey was also carried out in the winter period and on weekdays. To correct for this, seasonality ratios from the previous national travel survey (1993–1994) were used.

The trends for pedestrians could not be studied because the 1994–1995 regional survey did not investigate all types of walking (e.g. walking for transfer between two means of transport). The Rhone road trauma registry started in 1996; injury rates are estimated on 1996–1997, using crash data from the years 1996–1997 and mobility data from the years 1994–1995, making the (reasonable) assumption that mobility data did not change much between these 2 years.

5.3. Results and interpretation

5.3.1. *Seasonality ratios*

For car occupants, the seasonality ratios were around 0.7–0.8; for pedestrians, they were around 0.7–0.9. It means that exposure in the "outside RTS" period was slightly lower than in the "RTS" period (i.e. November–April on weekdays, outside school and public holidays). This might be explained by fewer commuting trips in summer and week-ends.

For cyclists, the seasonality ratios were between 1.4 and 1.6. These ratios, higher than 1, can be explained by higher bicycle use when weather conditions are good; the use is thus higher during summer than in winter. Bicycle use seemed similar on weekdays and on weekends. For PTW riders, the seasonality ratios were between 1.2 and 1.5. These ratios, higher than 1, can also be explained by the fact that PTW use is dependent on good weather conditions.

5.3.2. *Injury rates*

5.3.2.1. *Injury rates by type of road user*

The all-injury rates are given in Table 5.1. Compared with car occupants and with regards to time spent traveling, pedestrians had a lower rate by half (univariate analysis, p-value <0.01), cyclists had an 8 times higher rate ($p < 0.01$) and PTW riders had about a 42 times higher rate ($p < 0.01$). In the multivariate analyses adjusted for type of road user, gender, age group and location, cyclists had a 14 times higher injury ratio (CI = [9.5;20.6]) and pedestrians had a lower rate by half (CI = [0.4;0.7]), compared with car occupants, all other things being equal (Table 5.2). In these analyses, the adjustment increases the ratio between car occupants and cyclists due to the adjustment for gender, as women, a minority in cyclists, appeared to have a higher risk (Table 5.3).

	Car occupants		Pedestrians		Cyclists		PTW riders
	N = 6,286		*N = 1,513*		*N = 2,228*		*N = 3,776*
	Rate	*Ratio*	Rate	*Ratio*	Rate	*Ratio*	Rate
	Per one million hours						
All	**12.83**		**5.06**		**99.41**		**541.08**
Gender							
Male (ref.)	11.38	*1.0*	5.57	*1.0*	98.70	*1.0*	–
Female	14.50	*1.3*	4.59	*0.8*	96.28	*1.0*	–
Age group							
05–14	5.55	*0.5*	8.06	*2.0*	110.03	*1.6*	–
14–18	10.53	*1.0*	4.15	*1.0*	107.08	*1.6*	–
18–25	48.39	*4.5*	7.38	*1.8*	142.29	*2.1*	–
25–65 (ref.)	10.74	*1.0*	4.03	*1.0*	68.95	*1.0*	–
65+	6.40	*0.6*	5.04	*1.3*	91.46	*1.3*	–
Location							
Dense areas (ref.)	12.67	*1.0*	5.00	*1.0*	94.22	*1.0*	–
Non-dense areas	13.62	*1.1*	6.24	*1.2*	186.04	*2.0*	–
	Per one million trips						
All	**3.72**		**0.58**		**29.73**		**152.28**
	Per one million kilometers						
All	**0.48**		**1.29**		**11.30**		**14.88**

PTW, powered two-wheeler; ref, reference group; - means insufficient data to estimate rates.

Table 5.1. *All-injury rates, according to mobility measure and ratios by type of road user, gender, age and location; medical registry and road travel survey corrected for seasonality, Rhône County, 2005–2006*

The hospitalization rates are given in Table 5.4. Compared with car occupants and with regards to time spent traveling, cyclists had a 12 times higher rate ($p < 0.01$) and PTW riders had about an 80 times higher rate ($p < 0.01$); rates for pedestrians and car occupants were similar ($p = 0.20$).

	Injury ratio	**95% CI**
Type of road user		
Car occupants (ref.)	1	–
Pedestrians	0.50	[0.35; 0.73]
Cyclists	13.98	[9.50; 20.56]
Gender		
Male (ref.)	1	–
Female	1.23	[0.90; 1.68]
Age group (years)		
05–14	1.27	[0.79; 2.05]
14–18	1.13	[0.69; 1.85]
18–25	3.91	[2.44; 6.28]
25–65 (ref.)	1	–
65+	0.86	[0.53; 1.39]
Location		
Dense areas (ref.)	1	–
Non-dense areas	1.16	[0.85; 1.58]

Table 5.2. *Multivariate analysis for the number of all-injury adjusted for type of road users, gender, age groups and location, with regards to time spent traveling; medical registry and Regional Travel Survey corrected for seasonality, Rhône County, 2005–2006*

	Car occupants		**Pedestrians**		**Cyclists**	
	IR	*95% CI*	*IR*	*95% CI*	*IR*	*95% CI*
Gender						
Male (ref.)	1	–	1	–	1	–

Female	1.02	[0.84; 1.23]	0.84	[0.59; 1.18]	2.43	[1.25; 4.75]
Age group (years)						
05–14	0.49	[0.36; 0.66]	1.83	[1.10; 3.06]	1.42	[0.53; 3.80]
14–18	1.00	[0.73; 1.35]	1.04	[0.61; 1.76]	1.61	[0.54; 4.78]
18–25	4.78	[3.66; 6.26]	2.92	[1.68; 5.08]	3.00	[1.13; 7.93]
25–65 (ref.)	1	–	1	–	1	–
65+	0.65	[0.49; 0.87]	1.19	[0.71; 1.99]	0.95	[0.35; 2.60]
Location						
Dense areas (ref.)	1	–	1	–	1	–
Non-dense areas	1.08	[0.89; 1.31]	1.19	[0.84; 1.70]	1.64	[0.81; 3.32]

IR, injury ratio.

Table 5.3. *Multivariate analyses for the number of all-injury by type of road user, adjusted for gender, age groups and location, with regards to time spent on traveling; medical registry and Regional Travel Survey corrected for seasonality, Rhône County, 2005–2006*

	Car occupants		**Pedestrians**		**Cyclists**		**PTW riders**
	N=611		*N=359*		*N=332*		*N=680*
	Rate	*Ratio*	Rate	*Ratio*	Rate	*Ratio*	Rate
	Per one million hours						
All	**1.25**		**1.20**		**14.81**		**97.44**
Gender							
Male (ref.)	1.35	*1.0*	1.38	*1.0*	15.69	*1.0*	–
Female	1.13	*0.8*	1.04	*0.8*	11.26	*0.7*	–
Age group							
05–14	0.57	*0.6*	1.77	*2.0*	13.12	*1.1*	–
14–18	1.60	*1.8*	0.69	*0.8*	14.04	*1.2*	–
18–25	4.69	*5.3*	1.00	*1.1*	12.25	*1.0*	–
25–65 (ref.)	0.89	*1.0*	0.88	*1.0*	12.15	*1.0*	–
65+	1.40	*1.6*	2.05	*2.3*	30.83	*2.5*	–
Location							
Dense areas (ref.)	1.04	*1.0*	1.16	*1.0*	13.40	*1.0*	–
Non-dense areas	2.05	*2.0*	1.83	*1.6*	32.73	*2.4*	–
	Per one million trips						
All	**0.36**		**0.14**		**4.43**		**28.50**
	Per one million kilometers						
All	**0.05**		**0.31**		**1.68**		**2.68**

PTW, powered two-wheeler; insufficient data to estimate rates.

Table 5.4. *Hospitalization rates and ratios by type of road user, gender, age and location; medical registry and Regional Travel Survey corrected for seasonality, Rhône County, 2005–2006*

The serious-injury rates are given in Table 5.5. Compared with car occupants and with regards to time spent traveling, cyclists had a 16 times higher rate ($p < 0.01$) and PTW riders had a rate about 120 times higher ($p < 0.01$); rates for pedestrians and car occupants were similar ($p = 0.7$).

Relative to car occupants, cyclists are more likely to be injured, hospitalized or seriously injured per one million trips, kilometers or hours. This is consistent with other studies in France [GAB 05, MER 06, LIC 11], the United States [BEC 07], New Zealand [TIN 10] and Australia [GAR 10].

	Car occupants		Pedestrians		Cyclists		PTW riders
	N = *202*		*N* =*194*		*N* =*144*		*N* =*346*
	Rate	*Ratio*	Rate	*Ratio*	Rate	*Ratio*	Rate
Per one million hours							
All	**0.41**		**0.65**		**6.42**		**49.58**
Gender							
Male (ref.)	0.51	*1.0*	0.78	*1.0*	6.74	*1.0*	–
Female	0.30	*0.6*	0.53	*0.7*	5.09	*0.8*	–
Age group							
05–14	0.21	*0.7*	1.01	*2.2*	5.98	*1.1*	–
14–18	0.43	*1.5*	0.36	*0.8*	6.42	*1.2*	–
18–25	1.67	*5.8*	0.49	*1.1*	3.60	*0.7*	–
25–65 (ref.)	0.29	*1.0*	0.46	*1.0*	5.25	*1.0*	–
65+	0.43	*1.5*	1.17	*2.5*	16.44	*3.1*	–
Location							
Dense areas (ref.)	0.36	*1.0*	0.63	*1.0*	5.72	*1.0*	–
Non-dense areas	0.62	*1.7*	0.92	*1.5*	14.92	*2.6*	–
	Per one million trips						
All	**0.12**		**0.07**		**1.92**		**14.50**
	Per one million kilometers						
All	**0.02**		**0.17**		**0.73**		**1.36**

PTW, powered two-wheeler; ref, reference group; insufficient data to estimate rates.

Table 5.5. *Serious-injury (MAIS 3+) rates according to different mobility measures and ratios by type of road user, gender, age, and location; medical registry and road travel survey corrected for seasonality, Rhône County, 2005–2006*

With regards to each exposure measure, PTW riders had the highest all-injury, hospitalization, and serious-injury rates. In the present study, the amount of excess risk is quantified and is very high. This result is consistent with other studies in France [GAB 05, MER 06, BER 11, LIC 11], the United States [BEC 07] and New Zealand [TIN 10] using police or hospital-based crash data. This high excess of risk is giving much cause for concern.

In the present study, the risk for pedestrians compared with car occupants is not clear; it is very dependent on the choice of exposure and, to a smaller extent, on the injury category. Unlike motorcyclists and cyclists, pedestrians share the road with car occupants to a lesser extent; they are at little risk when they use sidewalks separated from the traffic by parked cars and/or trees. It would be interesting to estimate injury rates for pedestrians restricted to crossing the streets.

Ratios between injury rates for each user type and for car occupants showed differences according to which exposure measure is used. The ratios of injury rates with regards to the number of trips or hours spent were similar, for a given type of user. In contrast, with regards to the kilometer criterion, ratios of injury rates were much higher for cyclists and pedestrians, and lower among users of PTW. This discrepancy corresponds to the fact that, for the same duration, car occupants and PTW riders, having a higher average speed, travel a longer distance than pedestrians and cyclists. The time criterion is less elastic, since the average time spent on traveling a day is quite similar between types of road user. The time budget is hence the most relevant measure for comparisons between individuals, and between means of transport.

5.3.2.2. *Injury rates by gender*

Injury rates between men and women are provided in Tables 5.1, 5.4 and 5.5. Using univariate analyses, rates were similar between genders, except for cyclists, for the all-injury rate, where women have a higher risk. In the multivariate analyses, women did not have a higher injury ratio than men, all other things being equal (Table 5.3). By the type of road user, for pedestrians and car occupants separately, women did not have higher injury ratios than men (Table 5.3). For cyclists, women had an injury ratio twice as high as that of men (CI = [1.3;4.8]). A possible explanation could be that women have on average less experience in cycling than men. However, we can also wonder whether there could be an underestimation of female cycling mobility in the RTS, since this higher risk for women is not found in the literature. Indeed, in American study [BEC 07] using the police-reported crash data selected for the years 1999–2003 and the mobility data from the 2001 NTS, male cyclists had injury rates 1.7 times higher than female cyclists. An older

American study showed similar male and female rates (288 vs. 331 injuries per million bicycle trips), using crash data from the Consumer Product Safety Commission's National Electronic Injury Surveillance System in 1990 and mobility data from the 1990 Nationwide Personal Transportation Survey [LI 96].

5.3.2.3. *Injury rates by age group*

Injury rates by age group are provided in Tables 5.1, 5.4 and 5.5. Using the univariate analyses, some age groups already appear at higher risk. Using the multivariate analyses (Table 5.2), the 18–25 years age group had a four times higher all-injury ratio (IR = 3.9; CI = [2.4;6.3]) compared with the 25–65 years age group, with regards to time spent traveling. The higher risk for the 18–25 years age group was also found separately within the group of car occupants (IR = 4.8; CI = [3.7;6.3]), pedestrians (IR = 2.9; CI = [1.7;5.1]) and cyclists (IR = 3; CI = [1.1;7.9]; Table 5.3). This excess injury risk in the 18–25 years age group may correspond to stronger risk-taking behavior and, for car occupants, to a learning process (in France, driving license can be obtained from age 18). The 5–14 years age group had a lower injury ratio than car occupants by half (CI = [0.4;0.7]) but had twice as high a ratio as pedestrians (CI = [1.1;3.1]). They are protected in the car compartment, but very often injured as pedestrians (they are in a learning process of the risk). The 65+ years age group had a lower all-injury rate as car occupants (IR = 0.7; CI = [0.5;0.9]) compared with the 25–65 years age group. These results are consistent with an American and a French study [MER 06, BEC 07]. On the contrary, the 65+ years age group seemed to have higher hospitalization and serious-injury rates for car occupants, pedestrians and cyclists, compared with the 25–65 years age group and with regards to time spent traveling (significant only for pedestrians, for the serious injury rate (p = 0.03), and not significant for others, probably because of low statistical power).

5.3.2.4. *Injury rates by location*

Cyclists, with regards to time spent traveling, seemed to have a higher injury rate in non-dense areas than in dense areas using univariate analyses (between two and three times higher, non-significant; Tables 5.1, 5.4 and 5.5) or multivariate ones (IR = 1.6, CI = [0.8;3.3]; Table 5.3). Although not significant, but at the same time, lacking statistical power, a possibly higher risk in non-dense areas could be due to the higher average speed of cyclists in non-dense than in dense areas (14 vs. 8 km/h respectively, from the RTS) as well as the higher speed of motor vehicles, potential crash opponents, in non-dense areas compared with dense areas. This would mitigate the common belief that "cycling in urban areas is dangerous", usually implying that it is more secure in rural areas. A higher risk in non-dense areas also seems to be the case for hospitalization (IR = 2) for car occupants (p = 0.052). This can also be explained by higher speeds in non-dense areas.

5.3.3. *Trends*

Trends are provided in Figures 5.1 and 5.2. For car occupants, the all-injury rate seems to have slightly decreased between 1996–1997 and 2005–2006 using univariate analyses (multiplied by 0.9, $p = 0.8$; Figure 5.1) or using multivariate ones (IR= 0.89, CI = [0.73; 1.10]). For them, the serious-injury rate also seems to have decreased slightly between 1996–1997 and 2005–2006 using univariate analyses (multiplied by 0.7, $p = 0.7$).

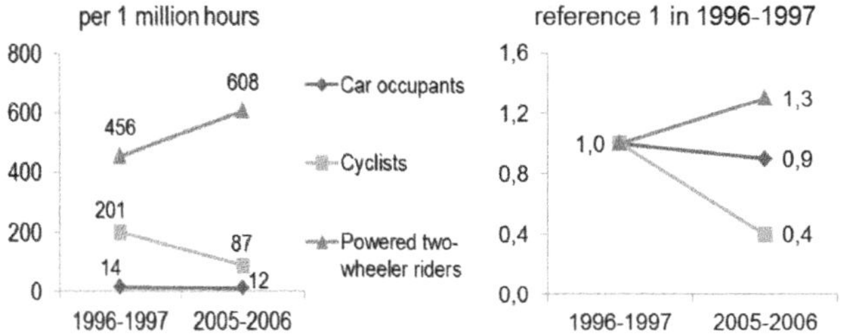

Figure 5.1. *Trends of the all-injury rates, per one million hours, and relative to 1996–1997 (reference 1), medical registry and RTSs corrected for seasonality, Greater Lyon*

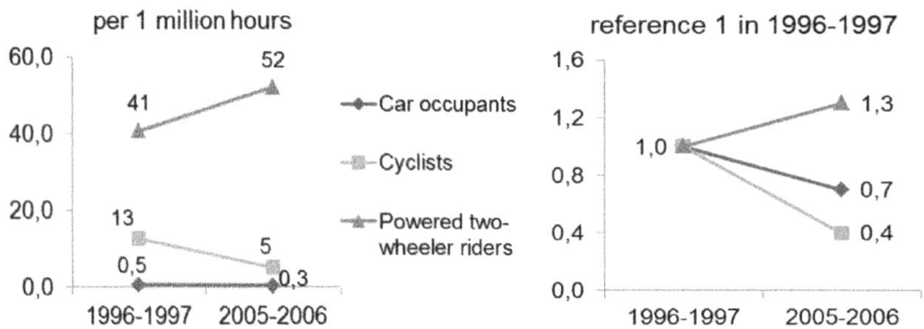

Figure 5.2. *Trends of the serious-injury rates, per one million hours, and relative to 1996–1997 (reference 1), medical registry and RTSs corrected for seasonality, Greater Lyon*

These possible decreases correspond to a similar mobility between the two periods and a slight decrease in the number of injured car occupants. This slight decrease can be explained by the fact that the period includes the years 2003 and 2004 when road safety had been greatly improved in France. There had been a massive installation of automatic speed cameras that led to a decrease in car speeds [ONI 05], and hence an improvement in road safety. The small frequencies at the regional level lead to a low statistical power, and hence to non-significant trends.

For cyclists, the all-injury rate seems to have decreased between the two periods using univariate analyses (multiplied by 0.4, p = 0.0.7; Figure 5.1) or using multivariate ones (multiplied by 0.64, CI = [0.30; 1.36]). For them, the serious-injury rate also seems to decrease between the two periods (multiplied by 0.4, p = 0.13; Figure 5.2). These possible decreases correspond to an increase in the mobility of cyclists in 2005–2006 compared with 1996–1997 (it has doubled) and a stagnation in the number of those injured. The decrease in rates seem more marked for cyclists than for car occupants. This tends to indicate the "safety in numbers" effect for cyclists: the more cyclists there are, the more they are collectively and individually seen, the more car occupants pay attention to them and thus the risk of cyclists being involved in a crash decreases [JAC 03].

For PTW riders, the all-injury and serious-injury rates may have slightly increased (both multiplied by 1.3, non-significant, using univariate analyses) between 1996–1997 and 2005–2006 in Greater Lyon. This corresponds to a similar mobility between the two periods and possibly a slightly higher number of those injured ($p = 0.82$) in 2005–2006 compared with 1996–1997. Their average speed has not decreased between the two periods [ONI 05]. These results reinforce the concern for PTW riders.

5.4. Discussion and conclusions

The Rhône County is quite urbanized with a central city (Lyon) leading to a higher risk of crashes but of a lower severity. Nevertheless, this county is an average area in terms of mobility compared with France as a whole. Injury rates may differ between regions; but orders of magnitude of ratios between rates by road user types, age categories, males–females or between dense and non-dense areas can be generalized.

5.4.1. *Study limitations*

There were missing values for some residence locations of the casualties and for some crash locations. These missing values were imputed using simple imputation from proportions estimated on observations without missing-values. The rates for PTW riders by gender, age group and location separately were not estimated: we are concerned about the reliability of the estimates because of the small number of PTW trips in the RTS. The mobility data provide information only about residents of the area studied. It would be interesting, however, to know the travel practices of the non-residents.

We were unfortunately not able to estimate confidence intervals of injury rates because of the multiplicity of the estimation steps: estimates from a mobility survey (with a complex design), estimates of seasonality ratios from another survey (with complex design), estimates (counts) of those injured from the registry and ratios between estimates.

5.4.2. *Study strengths*

This study is among the first in France to study the risk of injury using crash and mobility data [GAB 05, MER 06, BER 11] and comparing road type users. This is one of the first studies that compares these risks based on crash data from the medical registry of the Rhône County [LIC 11]. The Rhône road trauma registry is far more complete than the police crash data. The medical registry enables us to estimate the rates for various injury categories: all-injury, hospitalization and serious-injury (MAIS 3+).

The regional and national travel surveys allow us to estimate mobility based on three criteria: trips, distance traveled and time spent. The national surveys enable us to make a correction of the RTS that was restricted to the winter season (November to April), on weekdays, outside weekends and public and school holidays. This correction for seasonality appears relevant; indeed, the ratio was around 1.5 for cyclists.

5.4.3. *Conclusions*

In the present study, the amount of excess risk for PTW riders is quantified and is very high. This high excess of risk is very concerning. Moreover, the potential countermeasures will not be sufficient to fill the huge gap between the risks of PTW riders and car occupants; therefore, the question that arises is: should the use of PTW be deterred?

Although the risk is higher for cyclists than for car occupants, it should not be concluded that cycling has to be deterred. We must take into account the health benefits of this physical activity; these positive effects have been shown to balance and even outweigh the negative effects of road crashes [AND 00, DEH 10, DEN 11, ROJ 11, PRA 12]. These studies were however limited to fatalities. A French study, conducted on both those injured and those killed and using a correction for under-reporting of police crash data, showed a better cost–benefit balance for cycling compared with car use, for trips up to at least 10 km [PAP 02, PAP 02b, PAP 02c]. A French study aiming at estimating fatal rates at the national level has been carried out using fatal crash data from the police, and mobility data from the 2007–2008 national household travel survey (NTS) [BOU 15]. Exposure-based injury rates can be a tool for monitoring and evaluating the effectiveness of policies and programs, and for comparisons between countries.

5.5. Acknowledgments

We wish to thank all the people who have participated in the data collection and data entry in the framework of the Association for the Registry of road traffic casualties in the Rhône (ARVAC: president E. Javouhey) or in the framework of INRETS-UMRESTTE (B. Laumon, scientific consultant for the registry, A. Ndiaye, medical coordinator, and B. Gadegbeku, data coordinator). This work was funded by the *Délégation à la Sécurité et à la Circulation Routières (DSCR).*

5.6. Bibliography

[AAA 90] AAAM, Association for the Advancement of Automotive Medicine, The abbreviated injury scale (1990 revision), Des Plaines, Illinois, 1990.

[AMO 06] AMOROS E., MARTIN J.L., LAUMON B., "Under-reporting of road crash casualties in France", *Accident Analysis and Prevention*, vol. 38, no. 4, pp. 627–635, 2006.

[AMO 11] AMOROS E., CHIRON M., THELOT B. *et al.* "The injury epidemiology of cyclists based on a road trauma registry", *BMC Public Health*, vol. 11, pp. 653, 2011.

[AND 00] ANDERSEN L., SCHNOHR P., SCHROLL M. *et al.*, "All-cause mortality associated with physical activity during leisure time, work, sports, and cycling to work", *Archives of Internal Medicine*, vol. 160, no. 11, pp. 1621–1628, 2000.

[BEC 07] BECK L.F., DELLINGER A.M., O'NEIL M.E., "Motor vehicle crash injury rates by mode of travel, United States: using exposure-based methods to quantify differences", *American Journal of Epidemiology*, vol. 166, no. 2, pp. 212–218, 2007.

[BER 11] BERNAGAUD A.-S., HUREZ C., "Comparaison des risques d'accident entre les différents modes", *Action Territoriale, Sécurité des déplacements*, vol. 29, pp. 1–2, 2011.

[BOU 15] BOUAOUN L., AMOROS E., HADDAK M., "Fatal road traffic crashes: comparisons by road user types and measures of exposure", *Accident Analysis and Prevention*, vol. 75, pp. 217–225, 2015.

[DEH 10] DE HARTOG J.J., BOOGAARD H., NIJLAND H. *et al.*, "Do the health benefits of cycling outweigh the risks?", *Environ Health Perspect*, vol. 118, no. 8, pp. 1109–1116, 2010.

[DEN 11] DE NAZELLE A., NIEUWENHUIJSEN M.J., ANTÓ J.M. *et al.*, "Improving health through policies that promote active travel: a review of evidence to support integrated health impact assessment", *Environment International*, vol. 37, no. 4, pp. 766–777, 2011.

[ELV 99] ELVIK R., MYSEN A.B., "Incomplete accident reporting; meta-analysis of studies made in 13 countries", *Transportation Research Record*, vol. 1665, pp. 133–140, 1999.

[GAB 05] GABET P., "Méthode de calcul d'une exposition au risque d'accident en milieu urbain: application de la méthode sur la communauté urbaine de Lille", CERTU-CETE Nord-Picardie, 2005.

[GAR 10] GARRARD J., GREAVES S., ELLISON A., "Cycling injuries in Australia: road safety's blind spot?", *Journal of the Australian College of Road Safety*, vol. 21, no. 3, pp. 37–43, 2010.

[HAU 88] HAUER, E., HAKKERT A., "Extent and some implications of incomplete accident reporting", *Transportation Research Record*, vol. 1185, pp. 1–10, 1988.

[JAC 03] JACOBSEN P., "Safety in numbers: more walkers and bicyclists, safer walking and bicycling", *Injury Prevention*, vol. 9, no. 3, pp. 205–209, 2003.

[LAN 03] LANGLEY J., DOW N., STEPHENSON S. *et al.*, "Missing cyclists", *Injury Prevention*, vol. 9, no. 4, pp. 376–379, 2003.

[LI, 96] LI G., BAKER S.P., "Exploring the male–female discrepancy in death rates from bicycling injury: the decomposition method", *Accident; Analysis and Prevention*, vol. 28, no. 4, pp. 537–540, 1996.

[LIC 11] LICAJ I., HADDAK M., POCHET P. *et al.*, "Contextual deprivation, daily travel and road traffic injuries among the young in the Rhone Departement (France)", *Accident; Analysis and Prevention*, vol. 43, no. 5, pp. 1617–1623, 2011.

[MER 06] MERCAT N., "Méthode de calcul d'une exposition au risque d'accident en milieu urbain: application de la méthode sur la région urbaine de Grenoble", CERTU-Altermodal, 2006.

[ONI 05] ONISR, *La sécurité routière en France, bilan de l'année 2004*, La Documentation française, Paris, 2005.

[PAP 02a] PAPON F., 1ère partie: "le temps et l'argent", *Transports*, vol. 412, pp. 84–94, 2002.

[PAP 02b] PAPON, F., 2ème partie: "la santé et la sécurité", *Transports*, vol. 413, pp. 187–197, 2002.

[PAP 02c] PAPON F., 3ème partie : "la collectivité", *Transports*, vol. 414, pp. 232–242, 2002.

[PAP 10] PAPON F., DE SOLÈRE R., "Les modes actifs: marche et vélo de retour en ville", *La revue, Commissariat général au développement durable – Service de l'observation et des statistiques*, 2010.

[PRA 12] PRAZNOCZY C., Les bénéfices et les risques de la pratique du vélo – Evaluation en Ïle-de-France, Observatoire Régional de la Santé Ile-de-France, vol. 163, p. 163, 2012.

[ROJ 11] ROJAS-RUEDA D., DE NAZELLE A., TAINIO M. *et al.*, "The health risks and benefits of cycling in urban environments compared with car use: health impact assessment study", *British Medical Journal*, vol. 343, p. 8, 2011.

[TIN 10] TIN TIN, S., WOODWARD A., AMERATUNGA S., "Injuries to pedal cyclists on New Zealand roads, 1988–2007", *BMC Public Health*, vol. 10, p. 655, 2010.

6

Development of Safety Performance Functions for Two-Lane Rural First-Class Main Roads in Hungary

This paper describes a modeling effort to define accident prediction models for first-class main roads outside built-up areas in Hungary using variables that are available and believed to exert an influence on safety performance. The first part of the paper describes data collection and the segmentation technique. Six years of accident data are gathered for segments only; intersections with a 200 m radius are taken out in order to avoid having intersection-related crashes in the data. A total of 1,357 homogenous sections are formulated based on Annual Average Daily Traffic (AADT), road width, posted speed, horizontal curve and shoulder width.

Models are proposed using the Generalized Linear Modeling (GLM) approach assuming a negative binomial error structure. It is concluded that AADT, roadway width, horizontal curve and segment length significantly influence accident frequency. Estimated model parameters are compared with the international research context, and it is concluded that the results are in accordance with previous research findings.

6.1. Introduction

According to the Hungarian regulations road infrastructure safety management applies not only to the trans-European road network as laid down in the Directive 2008/96/EC of the European Parliament and of the Council [EUR 08], but all national main roads from 1 January 2014 onwards, as well as to all the roads with more than 10,000 Passenger Car Unit (PCU)/day from 1 January 2015. In Hungary crash rates are being used for such analyses, assuming a linear relationship between exposure and crash frequencies, which is easy to use for practitioners, but however,

Chapter written by Attila BORSOS, John N. IVAN and Gyula OROSZ.

is not founded on scientifically proven ground. Safety Performance Functions (SPFs) as accident prediction models provide professionals with the data needed to conduct road safety impact assessments, and network safety rankings, accounting for a more realistic relationship between traffic volume and accident occurrence.

To this end, this paper describes a modeling effort to define accident prediction models using variables that are available and believed to exert an influence on safety performance, including traffic volume. The first part of the paper gives a general overview on design characteristics and accident statistics of two-lane rural first-class main roads. Then it describes data collection as well as how the network is split up into homogenous sections. Then models are proposed using the Generalized Linear Modeling (GLM) approach assuming a negative binomial error structure. Five variables are used to predict accident frequencies, AADT, posted speed, curve, roadway width and shoulder width. Traffic volume is used as an exposure to accidents and the length of the homogenous sections as an offset variable in all models.

6.2. Literature review

SPFs have been widely used in the traffic safety field for analyzing how the safety performance of road facilities is related to various road characteristics. Roughly 25 years ago a few researchers (e.g. [HAU 88, MIA 93]) started to demonstrate problems associated with conventional regression techniques used for accident prediction. Miaou and Lum [MIA 93] for instance used four regression models, and demonstrated that the conventional linear regression models are not appropriate to make probabilistic statements about vehicle accidents. They also showed that if the vehicle accident data are found to be significantly overdispersed relative to its mean, then using the Poisson regression models may overstate or understate the likelihood of vehicle accidents on the road. To overcome these problems later on the negative binomial modeling method became a widely accepted modeling technique and was recommended by many researchers (e.g. [ABD 00, SAW 01]).

There have been several researchers dealing with the safety performance of two-lane rural roads [CAF 10, QIN 03, MAY 03, PAR 03, LAM 00, VOG 98]. Cafiso *et al.* [CAF 10] fitted 19 models and considered 14 variables, including geometric and operational variables (curvature change rate, paved width, tangent ratio, curve ratio, average operating speed), design consistency variables (e.g. variables considering speed differentials) and context-related variables (roadside hazard, driveway density). Three models were selected as recommended ones, where the first was a basic model with AADT only and the other two having the best fits used at least one variable pertaining to the four main groups of variables listed above. Abdel-Aty and Radwan [ABD 00] divided a 227 km long road into 566 segments with homogeneous characteristics in terms of traffic flow and geometry (degree of horizontal curvature, shoulder and median widths, rural/urban classification, lane width and number of lanes), all variables found to be strongly related to the accident occurrence. Mayora and Rubio [MAY 03] investigated the relevant variables for crash rate prediction and concluded that access density, average sight distance, average speed limit and the proportion of no-passing zones have the highest correlation. Access density is the variable that most influences the rate of head-on and lateral collisions, while in road run-off incidents and single vehicle crashes sight distance is decisive.

As far as the segmentation technique is concerned, there are a few methodologies and no one is preferred. Cafiso *et al.* [CAF 13] conducted a comprehensive study on the application of five different methods and the influence of segmentation on the performance of SPFs in terms of goodness of fit and the variables that could be modeled. They concluded that the best results were obtained for the segmentation based on having two curves and two tangents in each segment and the segmentation with fixed length. They also highlighted that a segmentation technique using constant values of all variables, therefore resulting in very short segments, led to the poorest model.

The Highway Safety Manual [AME 10] recommends the use of homogeneous segments with respect to a number of parameters, such as the AADT, number of lanes, lane width, shoulder width, shoulder type, curvature, driveway density, roadside hazard rating, median width and clear zone width. According to the manual there is no prescribed minimum segment length, however a minimum of 0.10 miles (0.16 km) is suggested. Fitzpatrick *et al.* [FIT 06] pointed out that in practice, this type of segmentation is not always easy to achieve as not all the variables are available. Koorey [KOO 09] arrived at similar conclusions, namely that variable-length road segments seem intuitively more useful than fixed-length segments, because of the mixed attributes contained in the latter. He added that this advantage is less when shorter lengths are used, and fixed-length segments are computationally easier to create from constant-interval raw data.

6.3. General overview of first-class main roads

6.3.1. *Design characteristics*

The Hungarian road network can be divided into two major categories: the national (31,833 km) and the municipal road network (approximately 170,000 km). The national road network can be further divided into two major subcategories: main perfumed consisting of motorways, express roads, first-class and second-class roads; and minor (or secondary) roads. In this paper our safety performance analysis has been performed for the rural sections of the first-class main roads, and therefore the following criteria and figures apply to this road category.

The length of the first-class rural main roads in Hungary is 1,659 km, which is 29% of the total rural main road network and 8.6% of the total rural road network. Due to the continuous expansion of the limited access road network, almost 80% of the first-class main roads have a parallel motorway or expressway, so the traffic volume in this road category is moderate (the AADT on these roads is roughly 7,500 vehicles/day on average).

A few selected criteria according to design standards are listed below:

– design speed is 90 km/h (in some cases 80 km/h depending on the terrain and land development circumstances);

– minimum curve radius must be at least 250 m (80 km/h) or 340 m (90 km/h);

– maximum gradient is 6%;

– minimum cross-slope is 2.5% and maximum superelevation is 7%;

– roadway width is 7.5 (7.0) m, lane width is 3.50 (3.25) m, shoulder width is 2.50 (2.0) m. (Values in parentheses are used under special circumstances).

The general posted speed limit is 90 km/h, 60–70 km/h adjacent to junctions and in dangerous curves. There are a few sections with higher speed limits of 100 and 110 km/h. These sections are exceptions where either the alignment of an already existing road was eligible for a higher posted speed limit or newly built sections where a more generous design was followed.

6.3.1. *Accident statistics*

In Hungary, according to recent statistics, approximately two-thirds of the accidents involving fatalities and personal injuries occur inside built-up areas, whereas 60% of fatal accidents occur outside built-up areas, 20% of which happen on first-class main roads. There has been a major decrease in the number of accidents and victims in the recent years, and in the road category investigated this decrease is above the average.

The number of fatalities was 1,232 on the entire road network in 2007 and plummeted to 641 by 2012, which is more than a 47% decrease. In the case of rural first-class main roads these results are slightly better; in the same time period the number of fatalities dropped by almost 56%. As far as the accident types are concerned, run-off road accidents represent 27%, rear-end collisions 20%, head-on collisions 19% and side impact collisions 13% of the total number of accidents.

6.4. Data collection and segmentation

As a first step of the data survey, the following data were gathered from the National Road Databank for the entire length (1,659 km):

– roadway width (pavement width);

– radius of horizontal curve;

– shoulder width;

– number of lanes;

– traffic volume;

– posted speed limit.

This database is organized in such a way that if there is a change in any of the parameters there is a new record (a new segment begins). As a result, these segment lengths can be very short, in some cases just a few meters; therefore the segmentation technique itself was an important issue to deal with.

In our analysis the following segmentation technique was used:

– crashes within a 200 m radius of intersections were taken out in order to avoid having intersection – related crashes in the data. Not only major intersections, but all minor intersections and connecting streets with a paved surface were filtered out. Intersections with connecting agricultural roads were not excluded from the sample;

– segments with more than two lanes (four-lane sections, sections with climbing lanes) were taken out;

– although using a fixed segment length might be the most practical application, it disregards the opportunity that the segment length itself can be a predictor of accidents. Homogenous sections were formulated according to AADT, road width, shoulder width and horizontal curves and posted speed limit;

– as is indicated elsewhere in the literature [KOO 09, CAF 13], too short segments can be biased due to the inaccurate identification of accident locations, and therefore segments shorter than 200 m were eliminated from the sample.

As a result of the segment delineation, roughly 55% of the entire length (919 km of 1,659 km) was used for further analysis. Although vertical alignment data were also collected, since the country has a relatively flat topography and because there were just a few sections with slopes, this characteristic was omitted. Another variable that seemed to be an important predictor according to other researchers' work was access point density (e.g. [CAF 10]). In our case access points (driveways) were hardly present as our focus was on rural sections, where access points are very limited.

Six years of accident data (2007–2012) including fatal (fatality within 30 days as a result of the accident), serious (injury healing beyond 8 days) and light (recovery within 8 days) injury accidents were gathered. Property damage accidents were not taken into consideration.

6.5. Modeling

6.5.1. *Regression technique and goodness of fit*

The GLM approach assuming a negative binomial error structure was used, and modeling was done in the R statistical software [RCO 13]. Several statistical measures were used to assess the goodness of fit of the models.

The Pearson χ^2 can be calculated by means of the following equation [6.1]:

$$Pearson\ \chi^2 = \sum_{i=1}^{n} \frac{\left[y_i - \hat{E}(y_i)\right]^2}{Var(y_i)} \quad [6.1]$$

The Pearson χ^2 of the model must be less than a critical value of the χ^2 distribution value, that is based on the model's degrees of freedom (DF) and a level of significance of α.

The Scaled Deviance (SD) is defined as the likelihood ratio test statistic measuring twice the difference between the log likelihoods of the studied model and the full or saturated model. The full model has as many parameters as there are observations so that the model fits the data perfectly. Therefore, the full model, which possesses the maximum log likelihood achievable under the given data, provides a baseline for assessing the goodness-of-fit of an intermediate model with p parameters [MCC 89].

The Akaike's Information Criterion (AIC) can be used for model selection and is applicable to compare models. The AIC value is calculated according to equation [6.2]:

$$AIC = -2\log L + 2p, \quad [6.2]$$

where $\log L$ is the maximum log-likelihood of the fitted model and p is the number of parameters in the model. The smaller the value of AIC, the better the model data fit. Given a dataset, several competing models may be ranked according to their AIC, with the one having the lowest AIC being the best.

Cumulative residual analysis was also used to evaluate the model form. The residual is equal to the difference between the observed and estimated values of the dependent variable. Cumulative residuals (CURE) were plotted versus AADT for each homogeneous section. The closer the curve stays to the x-axis, the more appropriate the model form is, and a curve that stays within two standard deviations is considered to be satisfactory [HAU 97].

Simple models of accident frequencies were estimated using five predictor variables: AADT, posted speed, curve, roadway width and shoulder width. Traffic volume is used as an exposure to accidents and is therefore present in all models. The prediction equation estimated is given in equation [6.3]:

$$E(Y) = e^{\alpha_0} \cdot L \cdot AADT^{\alpha_1} \cdot e^{\sum_{j=1}^{m} \beta_j x_j}, \qquad [6.3]$$

where E(Y) is the expected accident count/year; L is the segment length (in kilometers); AADT is the Annual Average Daily Traffic (AADT) (vehicles/day); x_j is the any of m-additional variables; a_0, a_1 and β_j are the coefficients to be estimated. AADT is scaled to an annual value (multiplied by 365 and divided by 10^7) and included with an exponent, as it is known to have a nonlinear relationship with crash incidence [IVA 04]. Here the segment length is included without an exponent to normalize for the variability from one segment to another. However, later the segment length is also used as a predictor. Equation [6.3] is transformed logarithmically to (equation [6.4]):

$$\ln E(Y) = \alpha_0 + \ln L + \alpha_1 \ln AADT + \sum_{j=1}^{m} \beta_j x_j \qquad [6.4]$$

6.5.2. *Modeling results*

As a result of the segmentation 1,357 homogenous sections were formed. Tables 6.1 and 6.2 show the variables used. AADT and length are used as continuous variables, whereas variables describing road geometry are clustered into categories for two reasons. Firstly, in many cases these characteristics change quite frequently: there are minor variations in the width of the cross-section elements within a short distance. From a road safety point of view, it can be doubted that for instance a few centimeter change in the width of the roadway would exert a significant influence on accident frequency. Secondly, had all these variations been taken into account it would have led to even shorter homogenous sections, causing problems that already have been specified in the literature review section.

Variable	Min.	Max.	Mean	Standard dev.
AADT	645	25,438	6,729	3,165.53
Length (km)	0.200	6.245	0.677	0.619

Table 6.1. *Continuous variables and their descriptive statistics*

We defined the categories for horizontal curve to have roughly even frequencies and as curves with small radii were underrepresented in the final sample, curve categories in the lower range were not further divided. For roadway width, shoulder width and posted speed limit we defined category boundaries based on the deviation from a reference level for design standards.

Variable	Categories	Frequency
Horizontal curve	Tangent	799
	<750 m	84
	750 m ≤ x < 1,500 m	151
	1,500 m ≤ x < 3,000 m	129
	≥3,000 m	194
Roadway width	<7 m	64
	7 m ≤ x ≤ 7.5 m	923
	>7.5 m	370
Shoulder width	<2 m	692
	2 m ≤ x ≤ 2.5 m	642
	>2.5 m	23
Posted speed limit	<90 km/h	49
	90 km/h	1,267
	>90 km/h	41

Table 6.2. *Categorical variables and their frequencies*

The modeling was conducted in two steps. First, each variable was entered into the model alone to see which ones significantly affect accident frequency. Then a full model was built using all the significant predictors. From the one-variable models it turned out that shoulder width and posted speed limit are not significant. Table 6.3 shows the model results for the base model (AADT only) and the two one-variable models with horizontal curve and roadway width. Not surprisingly, AADT is a highly significant predictor of accidents in all models. It has an exponent less than 1, which is in accordance with other research suggesting that the relationship between accident frequency and AADT is nonlinear, i.e. shows a decreasing slope with increasing AADT. Model 1 has the highest AIC value, likely because other predictor variables are not used.

Roadway width is a significant predictor in both categories (Model 2), suggesting that narrower pavement width (compared with the design standards) increases the expected number of accidents, whereas a wider roadway slightly decreases it. For a horizontal curve the lower curve categories have significant coefficients, and the expected signs show that accidents are more likely to occur in these curves, to a greater extent in sharp ones (with a radius of less than 750 m) than in curves between 750 and 1,500 m. It also seems that higher curve categories (over 1,500 m) are not significantly different from the tangent sections (Model 3), and therefore we decided to combine them and run a new model (Model 4).

These two significant variables (roadway width and horizontal curve) were also used in a full model (Model 5) as shown in Table 6.4.

It was also analyzed whether length as a predictor (in addition to the length as an offset variable) has an exponent significantly different from 1. To do so, the 799 tangent sections were used and length was added as a predictor to the base model. From Table 6.5 one can conclude that the tangent length itself is a significant predictor and it has a coefficient of 0.10006, which is statistically significant. This means that for each one-unit increase in length, the expected log count of accidents will increase by 0.10006. This parameter is positive, which is in line with our expectations, i.e. the longer a tangent segment is, the accident frequency will slightly increase, most likely due to speeding and overtaking maneuvers.

CURE were plotted versus AADT. Figure 6.1 shows an example for the base model. Although there are some fluctuations in the CURE, the model can be considered satisfactory.

Model	Categories	Coefficient	Standard error	*P* value	*k*	AIC	SD (DF)
1. AADT	Constant	0.14816	0.08703	*0.0887*	0.217	3,984.8	1,458 (1,355)
	LnAADT	0.76252	0.05929	*<0.0001*			
2. AADT + Roadway width	Constant	0.23719	0.09592	*0.0134*	0.211	3,979.6	1,456 (1,353)
	LnAADT	0.81115	0.06156	*<0.0001*			
	<7 m	0.28672	0.12839	*0.0255*			
	7 m ≤ x ≤ 7.5 m	Reference					
	>7.5m	−0.11543	0.06215	*0.0633*			
3. AADT + Curve	Constant	0.13870	0.08840	0.1167	0.213	3,983.8	1,454 (1351)
	LnAADT	0.76938	0.05914	*<0.0001*			
	Tangent	Reference					
	<750 m	0.30441	0.12974	*0.0190*			
	750 m ≤ x < 1,500 m	0.17915	0.09832	*0.0684*			
	1,500 m ≤ x < 3,000 m	−0.03309	0.09936	0.7391			
	≥3,000 m	−0.04004	0.08465	0.6362			
4. AADT + Curve	Constant	0.12972	0.08694	0.1357	0.213	3,980.1	1,454 (1,353)
	LnAADT	0.76898	0.05913	*<0.0001*			
	Tangent	Reference					
	<750 m	0.31281	0.12885	*0.0152*			
	750 m ≤ x < 1,500 m	0.18754	0.09714	*0.0535*			

Table 6.3. *Model parameters for the base model and one-variable models*

Model	Categories	Coefficient	Standard error	*P* value	*k*	AIC	SD (DF)
5. AADT + Roadway width + Curve	Constant	0.21916	0.09572	*0.0220*	0.208	3,975.4	1,452 (1,351)
	LnAADT	0.81636	0.06136	*<0.0001*			
	<7 m	0.26784	0.12824	*0.0367*			
	7 m ≤ x ≤ 7.5 m	Reference					
	>7.5 m	−0.11735	0.06192	*0.0580*			
	Tangent	Reference					
	<750 m	0.30278	0.12896	*0.0189*			
	750 m ≤ x < 1,500 m	0.18441	0.09699	*0.0572*			

Table 6.4. *Model parameters for the full model*

Model	**Categories**	**Coefficient**	**Standard error**	***P* value**	***k***	**AIC**	**SD (DF)**
6. AADT + Length	Constant	−0.02950	0.10417	0.777003	0.141	2,509.8	1,221 (1,119)
	LnAADT	0.73713	0.06912	*<0.0001*			
	Length	0.10006	0.03007	*0.000875*			

Table 6.5. *Model parameters for tangent sections*

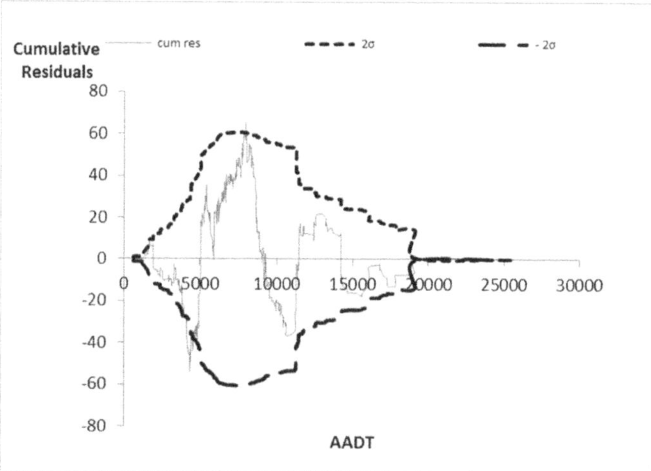

Figure 6.1. *CURE plot of the base model*

It was also tested whether or not adding a variable is overall significant. To determine if curve and roadway width are overall statistically significant, we can compare a model with and without them. Table 6.6 shows that the two degree-of-freedom chi-square tests indicate that both curve and roadway width are statistically significant predictors of the number of accidents.

Models	**Resid. DF**	**2xlog-likelihood**	**DF**	**LR stat.**	*p*
AADT	1,355	−3,978.837			
AADT + Curve	1,351	−3,969.845	4	8.992304	*0.061*
AADT	1,355	−3,978.837			
AADT + Roadway width	1,353	−3,969.637	2	9.20047	*0.010*

Table 6.6. *Likelihood ratio tests*

6.6. Discussion and conclusions

This study focused on two-lane rural first-class main roads in Hungary, with an objective to develop accident prediction models. The road network was segmented with a rigorous method, and variables that were available and believed to potentially influence road safety were collected. A GLM approach assuming a negative binomial error structure was used. Expected injury accident frequency was predicted by exposure (AADT), and length was used as an offset in all models. It was analyzed whether the variables roadway width, horizontal curve, shoulder width, and posted speed limit significantly affect the accident frequency. It was also analyzed whether length as a predictor has an exponent significantly different from 1.

The base model only includes the AADT as an exposure and can be easily used by practitioners (auditors, road administration etc.) to conduct analyses such as network safety ranking. The coefficient of AADT (0.75~0.8) is consistent with the international research results: Cafiso *et al.* [CAF 10] for instance found in their best performing models that this value is 0.62~0.75. The full model with two more variables (horizontal curve and roadway width) is a more complex one, but can still be used in practice as the predictors used are relatively easy to collect. Also in the full model the variables showed reasonable coefficients and signs. Roadway widths narrower than prescribed by the design standards tend to increase ($e^{0.26784}$ = 1.31), whereas wider roadways slightly decrease the number of accidents ($e^{-0.11735}$ = 0.89). Curves below 1500 m increase accident frequency, with a smaller value for radii between 750 m and 1500 m ($e^{0.18441}$ = 1.20) and with a higher value in the lower range below 750 m ($e^{0.30278}$ = 1.35). Our conclusion, that accident frequency tends to rise in a nonlinear fashion as the segment length increases, also complies with the international research results.

Here it should be noted that according to the international experience, there are many more variables that can be used to calibrate sophisticated models, which have their own advantages and disadvantages. An accident prediction model with more variables can help us to better understand how these factors influence and describe road safety. However, it was previously pointed out by Sawalha and Sayed [SAW 01] that many researchers find it tempting to include too many variables in a model. Such a model is not stable and will perform poorly when applied to a new sample. In this research our intention was to come up with simpler models that might help to improve road safety management. It is considered to be a starting point for similar and even more detailed studies for other road categories.

6.7. Acknowledgments

This research was supported by the European Union and the State of Hungary, co-financed by the European Social Fund in the framework of TÁMOP-4.2.4.A/2-11/1-2012-0001 "National Excellence Program".

6.8. Bibliography

[ABD 00] ABDEL-ATY M.A., RADWAN A.E., "Modeling traffic accident occurrence and involvement", *Accident Analysis and Prevention* vol. 32, pp. 633–642, 2000.

[AME 10] AMERICAN ASSOCIATION OF STATE HIGHWAY AND TRANSPORTATION OFFICIALS (AASHTO), *Highway Safety Manual*, 1st ed., vol. 2, 2010.

[CAF 10] CAFISO S., DI GRAZIANO A., DI SILVESTRO G. *et al.*, "Development of comprehensive accident models for two-lane rural highways using exposure, geometry, consistency and context variables", *Accident Analysis and Prevention*, vol. 42, pp. 1072–1079, 2010.

[CAF 13] CAFISO S., D'AGOSTINO C., PERSAUD B., "Investigating the influence of segmentation in estimating safety performance functions for roadway sections", *Proceedings of the 92nd Transportation Research Board Annual Meeting*, Washington DC, USA, 2013.

[EUR 08] EUROPEAN COMMISSION, "Directive 2008/96/EC of the European Parliament and of the Council of 19 November 2008 on road infrastructure safety management", *Official Journal of the European Union*, vol. 319, p. 59, November 29, 2008.

[FIT 06] FITZPATRICK K., SCHNEIDER IV W., CARVELL J., "Using the rural two-lane highway draft prototype chapter", *85th Transportation Research Board Annual Meeting*, Washington DC, USA, 2006.

[HAU 88] HAUER E., NG J.C.N., LOVELL J., "Estimation of safety at signalized intersections", *Transportation Research Record*, vol. 1185, pp. 48–61, 1988.

[HAU 97] HAUER E., BAMFO J., "Two tools for finding what function links the dependent variable to the explanatory variables", *Proceedings of 10th ICTCT workshop*, Lund, Sweden, 1997.

[IVA 04] IVAN J.N., "New approach for including traffic volumes in crash rate analysis and forecasting", *Transportation Research Record*, vol. 1897, pp. 134–141, 2004.

[KOO 09] KOOREY G.F., "Road data aggregation and sectioning considerations for crash analysis", *88th Transportation Research Board Annual Meeting*, Washington DC, USA, 2009.

[LAM 00] LAMM R., BECK A., ZUMKELLER K., "Analysis and evaluation of interrelationships between traffic safety and highway geometric design on two-lane rural roads", *Proceedings of the 2nd International Symposium on Highway Geometric Design*, Mainz, Germany, pp. 557–570, 2000.

[MAY 03] MAYORA J.M.P., RUBIO R.L., "Relevant variables for crash rate prediction in spain's two lane, rural roads", *Proceedings of the 82nd Transportation Research Board Annual Meeting*, Washington DC, USA, 2003.

[MCC 89] MCCULLAGH P., NELDER J.A., *Generalized Linear Models*, Chapman and Hall, New York, 1989.

[MIA 93] MIAOU S., LUM H., "Modeling vehicle accidents and highway geometric design relationship", *Accident Analysis and Prevention*, vol. 25, pp. 689–709, 1993.

[PAR 03] PARDILLO J.M., LLAMAS R., "Relevant variables for crash rate prediction in Spain's two lane rural roads", *Proceedings of the 82nd Transportation Research Board Annual Meeting*, Washington DC, USA, 2003.

[QIN 03] QIN X., IVAN J.N., RAVISHANKER N., "Selecting exposure measures in crash rate prediction for two-lane highway segments", *Accident Analysis and Prevention*, vol. 938, pp. 1–9, 2003.

[RCO 13] R CORE TEAM, *R: A language and environment for statistical computing*, R Foundation for Statistical Computing, Vienna, Austria, available at: http://www.R-project.org/, 2013.

[SAW 01] SAWALHA Z., SAYED T., "Evaluating safety of urban arterial roadways", *Journal of Transportation Engineering*, vol. 127, pp. 151–158, 2001.

[VOG 98] VOGT A., BARED J., "Accident models for two-lane rural segments and intersections", *Transportation Research Record*, vol. 1635, pp. 18–29, 1998.

PART 3

Vulnerable Road Users' Safety

7

Mobility and Safety of Powered Two-Wheelers in OECD Countries

The objective of this chapter is the analysis of powered two-wheeler (PTW) mobility and safety in the OECD countries, and the presentation of measures for the improvement of PTW mobility and safety. This research work was carried out by a working group of the OECD International Transport Forum, composed of experts from several countries. PTW use, mobility and safety figures are examined in terms of their development over time and their distribution per road user and vehicle characteristics. Moreover, contributory factors of PTW road accidents and injury risks are discussed, as well as PTW accident patterns, on the basis of an exhaustive review of international literature. On the basis of the results, a number of measures are discussed, including licensing, training and education, enforcement, traffic and speed management, vehicles and ITS, infrastructure, protective devices and conspicuity. Most importantly, improving PTW safety is compatible and should be integrated with the development of a safe system approach, and requires a toolbox of measures, which includes the safer behavior of all road users, safer infrastructure and vehicles with enhanced safety features.

7.1. Introduction

Significant growth in motorcycling has occurred during recent decades in most parts of the world [HAW 12], resulting in the powered two-wheelers (PTW) gradually becoming a true mobility tool, attracting an increasingly vast and varied population. Some riders use PTWs as their primary form of transportation, others for recreation. For many it is the only affordable or practical means of individual motorized mobility. The use of PTWs continues to grow globally each year in

Chapter written by Pierre VAN ELSLANDE, Veronique FEYPELL-DE LA BEAUMELLE, James HOLGATE, Kris REDANT, Hélène DE SOLÈRE, Dimitris MARGARITIS, George YANNIS, Eleonora PAPADIMITRIOU, Saskia DE CRAEN, Lars INGE HASLIE, Juan MUGUIRO and Per-Olov GRUMMAS GRANSTRÖM.

conjunction with multiple economic and social factors contributing to their expansion worldwide, such as increased traffic congestion and inner-city parking problems, increases in gasoline prices, the development of leisure, changes in lifestyle, etc. [SHI 12].

The result is that in spite of a remarkable improvement in traffic safety for all road users (including motorcyclists) in OECD countries, motorcyclists have seen their exposure to road risk increase to the point that in some countries the number of motorcyclists who died in road crashes actually increased over the past two or three decades [SHI 12], while the mortality of other road users declined significantly.

PTW riders are at far more risk than car drivers per kilometer ridden in terms of fatalities and severe injuries entailing long-term disability. The share of PTW fatalities is often much higher than their share in the vehicle fleet, especially in low- and middle-income countries.

Regardless of the countries concerned, however, PTW users are confronted with an excessive risk on the road, which has been qualified as “unfair” by Elvik [ELV 09], insofar as for the same number of kilometers driven they have a much higher risk of being killed or severely injured than car occupants. They are clearly overrepresented among road traffic casualty figures, even when they are not overrepresented in crash occurrences.

In this context, the objectives of this chapter can be outlined as follows:

– to review and summarize the most recent knowledge dealing with motorcycling safety;

– to review and summarize the current understanding of motorcycle crash configurations and mechanisms;

– to provide recommendations on measures that can be implemented in the short and medium term to improve the safety of motorcyclists;

– to progress toward a safe system approach for PTWs.

This research was carried out by a working group of the OECD International Transport Forum, composed of experts from several countries. The chapter starts by presenting the basic trends and figures of PTW mobility, use and safety. Then, a comprehensive review of PTW crash contributory factors is presented (user-, vehicle- or infrastructure-related). Subsequently, the integration of PTW in the Safe Systems approach is discussed, with particular focus on the challenges and particularities involved. Finally, the most promising individual measures for the improvement of PTW mobility and safety are outlined.

7.2. Mobility and safety figures of PTWs

7.2.1. *PTW mobility and use*

It is estimated that there are 313 million PTWs in the world, with a relatively uneven distribution across regions: 77% are found in Asia, 5% in Latin America, 1% in Africa, 1% in the Middle East and 16% in North America and Europe [ROG 08]. This disparity is also characterized by the uses made of this mode. Primarily recreational in North America and Australia, the two-wheeler does have a much more mixed function, for example, in Europe it is increasingly used to escape the problems of urban traffic congestion; in other regions of the world it may have a mainly utilitarian use.

In most OECD countries, over the past decade (2001–2010), the motorcycle fleet has grown much more rapidly than the passenger car fleet (see Table 7.1).

	Passenger cars	Mopeds	Motorcycles
Australia	25%	–	88%
Czech Republic	29%	10%	35%
France	11%	–22%	48%
Great Britain	13%	–27%	28%
Greece	52%	–14%	76%
Japan	11%	–20%	14%
Spain	22%	27%	82%
Sweden	8%	84%	91%
United States (excl. SUVs)	5%	–	67%

Table 7.1. *Evolution (%) in the PTW and passenger car fleets for a selection of OECD countries 2001–2010 (source: IRTAD)*

PTWs are becoming an important component of the transport system and represent in some cities up to 30% of the motor vehicle fleet [ACE 13]. They present both assets for mobility, and also challenges in terms of traffic management and safety. However, only a few countries have in place a national transport strategy for PTWs (e.g. [DFT 05, VIC 08]), though several measures may be taken at the local level.

7.2.2. *Safety development over time*

While there has been significant progress in most OECD countries in improving road safety and reducing road mortalities, PTW riders have not benefited at the same pace as car occupants from safety improvements over the last decade (see Figure 7.1).

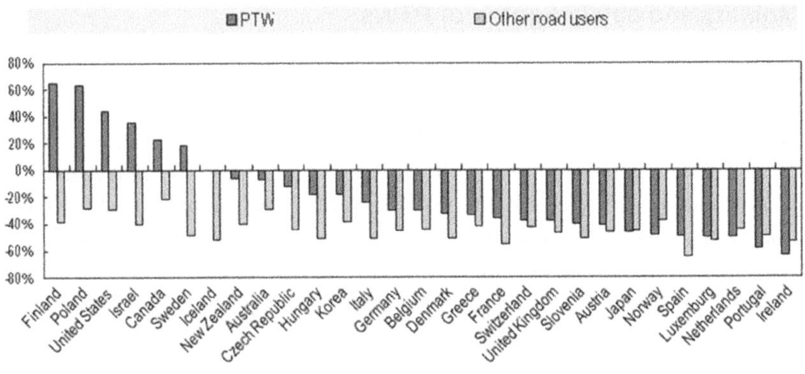

Figure 7.1. *Evolution in fatalities among PTW and other road users, OECD countries, 2001–2011 (source: IRTAD, 2001–2010 data for Canada)*

In fact, the situation of motorcyclists has deteriorated in the past decade if one compares with the significant progress with other road users. PTW trends are not following the overall fatalities decrease in the OECD countries; this may be explained by the high increase in PTW fleet and respective mobility, as PTWs have been steadily gaining in popularity in recent years, with large gains seen in PTW ownership and ridership. PTW is the only mode of transport for which the number of fatalities has increased, significantly in some countries, over the last decade, which stresses the importance of taking immediate appropriate counter measures.

7.2.3. *Crash characteristics and scenarios*

When related to the number of kilometers travelled, a motorcyclist is, depending on the country, between 9 and 30 times more likely to be killed in a traffic crash than a car driver (see Table 7.2).

On average, in OECD countries, PTW riders and passengers count for 16.5% of all road fatalities, while only representing 8% of the fleet. PTWs are clearly overrepresented in road traffic casualties. The situation in developing countries is drastically worse. In total, 90% of global deaths occur in low- and middle-income countries.

In many countries, the average age of motorcyclists killed has increased. In Europe, the number of motorcyclists killed in the 40–60 years age group has doubled in 10 years [YAN 10, NHT 12]. The fatality rates for moped riders aged 15–19 and motorcycle riders aged 20–30 are notably high. It appears that being young, male and lacking experience is associated with increased PTW fatality risk [YAN 05, LAR 05, REE 95].

	Car occupant	Motorcyclists	Mopeds (when distinction is made in statistics)	Relative risk of motorcyclists vs. car occupants
Australia	5.2	71.8		14
Austria (2010)	4.7	59.7	56.1	13 for motorcyclists
				12 for mopeds
Belgium (2010)	5.9	76.9		13
Canada (2010)	4.9	62.9		13
Czech Republic (2010)	10.5	252.6		24
Denmark	4.2	49.5		12
France	4.9	72.4	64.7	15 for motorcyclists
				13 for mopeds
Germany	3.3	59.5	14.6	18 for motorcyclists
				4 for moped
Ireland	2.5	60.8		24
Israel (2010)	5.1	45.7	26.8	9 for motorcyclists
				5 for mopeds
Netherlands (2004–08)	3.0	64	63	21
Slovenia	4.3	112.5		26
Sweden (2010)	2.2	43.9		20
Switzerland	2.3	39.2	29.6	17 for motorcyclists
				12 for mopeds
United States	5.0	150.6		30

Table 7.2. *Deaths per billion veh-km in 2011 for motorcyclists and car occupants (source: IRTAD)*

A large majority of PTW crashes are single vehicle crashes occurring on rural roads; 25% of all PTW crashes in Italy, 38% in Greece and 44% in Finland and Sweden [2BE 10]. Almost one-third of all PTW fatalities occur at junctions – a proportion notably higher compared to other road users (ACEM 2006, [YAN 10]). This stresses the importance of taking specific counter measures for junction safety improvement for PTWs.

7.3. Contributory factors of PTW crashes

The level of risk for PTW is influenced by many factors. A first general factor is the intrinsic difficulty of driving a PTW, due to the necessity to control the balance, its lower friction capacity and its greater sensitivity to environmental perturbations (wind, gravel, any change in road surface, etc.) which may destabilize the vehicle. PTW riders also have a higher risk of injury due to their greater vulnerability, resulting from a lack of protection compared to passenger cars.

Elvik [ELV 04a] has defined some basic factors which influence the level of risk of road accidents. Among these risk factors some are considered to affect PTWs more specifically, such as low friction (as mentioned above), but also lack of visibility, road-user rationality, road-user vulnerability and system forgiveness.

Moreover, by its very nature, driving a PTW may induce a specific behavior pattern on the road which is different from the drivers of four-wheeled vehicles. Such behavior is not necessarily "deviant" according to the law, but may surprise other road users. Even "normal" behavior, i.e. common to PTW riders, may be atypical for other vehicle operators (e.g. overtaking within a small space, overtaking on the right, filtering, positioning on one side of the road, intense acceleration, etc.), disturbing their normally efficient information-seeking routines. Atypical behavior also refers to "deviant" behavior, including speeding, slaloms, stunts, wheeling, etc.

Generally speaking, driver- and rider-related behavior factors are much more prevalent in PTW crashes, compared to vehicle and road infrastructure/environment factors. Speeding is a bigger problem for PTW crashes, compared to other modes. On average, motorcyclists ride at higher speeds than cars and PTW crashes usually occur at higher speeds compared with cars [HOR 05]. Similarly, consumption of alcohol is associated with an increased risk of fatal crashes among PTWs [KAS 05]. Moreover, for the same BAC, the severity of the crashes is higher for the PTW than for the other road users [MCL 93, SOD 95]. Operating a PTW requires more co-ordination and balance than operating a car, which explains that impaired riding (e.g. by alcohol or drugs) or inappropriate behavior (e.g. speeding) is even more problematic for PTW riders.

Vehicle technical failures are only a minor proportion of PTW road crash contributory factors. On the other hand, for PTWs, road design and road environment factors have a significant influence on the crash severity (e.g. roadside obstacles and barriers, speed reduction installations) even more than on crash occurrence [ACE 06]. However, PTWs are specifically sensitive to infrastructure design (e.g. alignment, curves, intersections, etc.) and maintenance (holes, gravels, vegetation, etc.). Due to this sensitivity, defects on the layout are likely to create more difficulties on PTW riders than on operators of other motorized vehicles.

A large number of crashes involve a problem of perception/appraisal by the other vehicle operator. This overrepresentation of inappropriate perception in PTW crashes suggests a specific problem of detectability (conspicuity) for these road users [PRE 95, YUA 00]. The problem of perception is complex and cannot be reduced to the simple fact that PTWs are physically less visible than other vehicles. There are many causes behind the poor detectability of PTWs and these are often connected to each other and with the general parameters of the driving context. Indeed, this problem can be explained both by the visual characteristics of PTWs, by the sensory capabilities of the human perceptual system, by the atypical behavior of PTWs and by the expectations that road users develop.

It is also noted that a more frequent combination of road crash contributory factors is found in PTW crashes, compared to other road users' crashes, which results in the multiplication of the relative risk. For example, it has been found that those without a valid license have a higher probability of not wearing a helmet, driving above the speed limit, driving under the influence of alcohol and without daytime running lights and so on ([PEE 96, REE 96].

Although human behavior and characteristics are the most frequently represented contributing factors in crashes, this does not mean that the solution to improve safety conditions for PTWs must only focus on behaviors. A safe system approach is required; it can be more efficient to change behavior by acting on a range of levels, including the infrastructure, the vehicle and the system as a whole. However, the integration of PTWs in a safe systems approach may be challenging in several ways; these are discussed in the next section.

7.4. Toward an integrated road safety strategy for PTW

7.4.1. *The safe systems approach*

The safe system is variously described in a number of jurisdictions but has a single core principle: a recognition that road users will make mistakes, or inappropriate decisions, and that the system, while also minimizing errors, should accommodate these errors so that no individual road user is exposed to crash forces likely to result in death or serious injury. The system then manages their safety by providing them with vehicles, road and roadside infrastructure and travel speeds that combine to ensure that any crashes that do eventuate result in crash forces that are below the level of human tolerance to physical harm.

Another characteristic of safe system approaches is consideration of the interactions between the different elements of the system and between the effects of different interventions. Some aspects of this are well recognized, for example, the influence of road design on chosen travel speeds. The challenge is to optimize the protection by combining the components of the road traffic system.

According to the recommended approach of the OECD's Towards Zero – Ambitious Road Safety Targets and the Safe System Approach [OEC 08], a strategic planning process on the basis of the Safe Systems approach principles can be outlined as follows:

– situational analysis: conduct sufficient data collection and analysis to understand crash risks and current performance;

– define strategic objectives: adopt a highly ambitious vision for road safety, set interim targets to move systematically toward the vision, develop a Safe System approach, essential for achieving ambitious targets;

– determine strategies and actions: exploit proven interventions for early gains, invest in road safety;

– establish supporting arrangements: strengthen the road safety management system, accelerate knowledge transfer, foster commitment at the highest levels of government.

7.4.2. *PTWs in the safe system*

Growing PTW traffic makes it imperative to adopt safety interventions targeting this mode of transport, while integrating it into a safe system approach. According to the above, improving the safety of PTWs should be a shared responsibility. All relevant stakeholders need to be actively involved in the process of drawing up and implementing a shared road safety strategy which includes safer behavior of all road users, safer infrastructure and vehicles with enhanced safety features.

A strategic approach should consider the most effective combination or measures according to the specific needs of individual jurisdictions. Several OECD countries do have strategic plans for PTW safety, although not always involving specific strategic objectives, quantitative targets, specific evidence-based measures to achieve the targets etc. (e.g. [DFT 05, DGT 07, RIJ 11, TRA 10, VIC 08])

The inclusion of motorcyclists into the Safe System yields two challenges. The first is the technical problem of providing protection from physical harm at the speeds at which collisions with other vehicles or fixed objects are likely. While this could be solved by ensuring travel speeds by, and in the vicinity of, motorcyclists are much lower, this then amplifies the second challenge. This is to ensure that any measures taken to improve motorcycle safety are supported both by the broader community and by motorcyclists in particular.

This leads to consideration of whether the conventional Safe System approach should be modified by recognizing that, in the short to medium term, motorcycling will remain an inherently risky activity and that measures should be taken to reduce risk. This may result in, for example, strategies that focus more on avoiding crashes, rather than mitigating their effects.

A toolbox of measures is required to improve the safety of PTW riders within the traffic system. These measures must take into account the specific challenges of PTW traffic, and also consider the variety of PTW users, insofar as some segments may be addressed with particular measures.

7.5. Measures for PTW safety improvement

The Safe System approach assumes that road users will enter the system competent and will take measures to ensure that they remain compliant and alert. Licensing, training, education, enforcement and communication campaigns are essential tools for improving riding safety. It is also important, in a complementarily way, to enhance the road environment and the vehicles toward this safety objective.

7.5.1. *Licensing, training and education*

Access to PTWs should be gradual, with a licensing system aiming at managing novice rider risks while riders are gaining experience and maturity [REE 95].

Novice riders of every kind of PTW should be trained. Training should not only focus on basic maneuvering skills and mastering traffic situations, but also address attitudes toward safety, putting a special emphasis on hazard perception and defensive riding [KAR 10].

Other road users should also be made aware of the specific risks associated with PTW vulnerability and crash patterns. Communication campaigns addressing required behavior change should be targeted at key groups of drivers and riders.

7.5.2. *Enforcement and communication*

Enforcement of traffic rules is an indispensable ally of other safety measures [ELV 04b] to promote homogeneous and safe traffic behavior between all road users. PTW operators, as other operators of motorized vehicles, must comply with traffic rules and typical enforcement activities, to control speeding, drinking and driving, non-respect of traffic rules, etc.

Traffic rules apply equally to operators of two- and four-wheeled vehicles and should be equally enforced. As for other motorized vehicle users, enforcement is needed to improve compliance with key safety rules like speed, drinking and driving, helmet use, proper licenses and a vehicle that meets safety standards. High-visibility enforcement accompanied by other measures, such as communication and publicity has proven to have a strong deterrent effect. Speed enforcement is key to reducing the speed and associated crash risk. Automated speed enforcement has proven its effectiveness for cars, but further adjustments are needed to make it as effective for PTWs.

The combination of enforcement-communication campaigns (on speeding, riding without a helmet and other risky behavior) has proven its effectiveness in many countries [HEN 08]. The success of enforcement and communication strategy depends on the involvement of motorcyclists themselves and motorcyclists associations have an important role to play in the diffusion of communication messages, informing riders about the rules and making enforcement acceptable.

7.5.3. *Infrastructure and traffic management*

Road and traffic management have traditionally been designed for four-wheeled vehicles. In some cases, these are not properly adapted for PTWs. Much could be done to facilitate the mobility and safety of PTWs, without compromising the mobility of other motorized vehicles.

Infrastructure should be improved with the development of self-explaining roads to guide drivers and riders to adopt appropriate speeds, traffic calming measures and PTW friendly equipment (forgiving roads), targeted at areas of highest PTW risk [ACE 06].

Engineers, road designers, road safety auditors and inspectors should be trained to consider PTWs in the design, maintenance and operation of roads, and be provided with the necessary risk assessment tools to make the right decisions. Local authorities' staff should be trained and informed on the infrastructure requirements for PTWs to compensate for the safety problems to which they are specifically subject.

Traffic management measures can have a dual purpose: facilitating PTW traffic and increasing safety. Further research is needed on the safety impact of measures such as advanced stop lines and traffic filtering. When implementing any new measure in favor of PTW mobility, caution must be paid that no new risk is induced for themselves or for any other road users.

7.5.4. *Vehicles, ITS and protective devices*

There are a number of developments within the motorcycle industry to improve the passive and active safety of motorcycles.

The anti-lock braking system (ABS) is a well proven technology which can significantly improve the safety of PTWs in certain situations [RIZ 09]. While it is currently offered as an option on new high-end bikes of major PTW manufacturers, with a slow penetration rate in most OECD countries, it can certainly benefit all powered two-wheelers and should become a standard. Cost is however an issue, and industry and government should work together to facilitate a quicker penetration of these technologies, which anyway will become mandatory in some regions in the coming years (expected in EU for the years 2016–2017). Other advanced braking systems may also help in reducing injury risk, but the priority today is to keep ensuring the penetration of ABS in the fleet.

There has been little advancement of intelligent transportation systems research dedicated to motorcycle safety. Motorcycle ADAS (Advanced Driver Assistance Systems) could improve the safety of the rider as well. There is however a number of obstacles that will likely lead to a slower uptake compared to passenger cars, including the challenges posed by the Human Machine Interface requirements, costs and the required support from the motorcyclists community. In spite of these obstacles, ITS has a role to play to increase motorcycle safety in the future. e-Call, blind spot detection, curve and collision warning systems are suitable applications for the motorcycle – once sufficiently developed for them.

While research into the benefits of protective clothing is unequivocal [DER 11, ACE 06], there are rider willingness-to-pay issues with mandatory requirements for protective clothing. Further research and development into clothing and equipment with lower weight and improved ventilation should be encouraged.

The helmet is the most important source of protection against severe injuries for both motorcyclists and moped riders [LIU 07]. The use of helmets and other protective equipment with adequate safety standards should be promoted and regulated where required.

7.6. Key messages and recommendations

The OECD/ITF group on PTW mobility and safety proposed a number of key messages and recommendations, which are summarized in Table 7.3, and analyzed below.

Key messages and recommendations	
1)	The powered two-wheeler population increases and plays a significant role in mobility
2)	PTW riders are at far more risk than car drivers
3)	PTW crashes are mainly due to perception and control failures
4)	A safe system approach is required to improve the safety of PTWs
5)	A toolbox of measures is required to improve the safety of PTW riders
6)	Promoting appropriate behaviors of road users is a prerequisite
7)	Self-explaining and forgiving roads contribute to lower crash risk
8)	Protective equipment and vehicles with enhanced safety features save lives
9)	It is essential to extend the knowledge on PTW mobility and crash mechanisms

Table 7.3. *Key messages and recommendations on PTW mobility and safety*

Motorcycling has become an integrated part of the traffic system offering certain benefits over other modes of transport; consequently, they need to be properly integrated into mobility plans and safety strategies. As the economic costs associated with PTW crashes are significant, investing in PTW safety can bring important societal and economic benefits.

However, PTW safety figures have not followed the impressive improvement trends of the last decade that other users' safety figures demonstrate. Moreover, per kilometer driven, PTW riders have a much higher risk of being killed than car occupants, between 9 and 30 times higher. PTW riders are also more likely to be very seriously injured in a road crash with long term disabilities than other motorized road users. They are also more vulnerable to impairment by e.g. alcohol.

Young, inexperienced and male riders are over represented in crashes. The most frequent PTW fatal crashes are single-vehicle crashes, partly due to intrinsic difficulties of riding a PTW (e.g. necessity to keep the balance) and to the higher sensitivity of riders to external perturbations (e.g. wind or poor pavement condition). The other most frequent crash type occurs at intersections with other traffic, involving, for a large number of crashes, a problem of perception and appraisal by the driver and/or the rider.

A safe system approach is proposed for improving PTW mobility and safety in OECD countries; it can be more efficient to change crash and injury outcomes by implementing a range of interventions, including road users, infrastructure, the vehicle and the system as a whole, i.e. their interactions. The implementation of this approach involves dealing with a number of challenges in balancing the different objectives of the safe system design with the PTW particularities and needs. In each case, a selection of specific measures for implementing the strategy will be required.

Priorities in individual measures may include:

– promoting appropriate behaviors of road users (licensing, training and education, enforcement of traffic rules, communication campaigns addressing required behavior);

– self-explanatory and forgiving roads (roads to guide drivers and riders to adopt appropriate speeds, traffic calming measures and PTW friendly equipment, training of engineers and local authorities etc.);

– protective equipment and vehicles with enhanced safety features (use of helmets and other protective equipment with adequate safety standards, enhanced safety features in vehicles, notably with the general introduction of advanced braking systems, etc.).

Additional research is needed to better understand current challenges related to PTW mobility and safety problems. This involves a need to develop and apply relevant methods, tools and indicators to measure PTWs in traffic flows and analyze their mobility and behavior (exposure data). More in-depth investigations will allow a better understanding of fatal and serious injury crash patterns and causes. Conspicuity and other perception problems deserve further study in order to identify key contributing factors and effective countermeasures.

7.7. Bibliography

[2BE 10] 2-BE-SAFE, Rider/Driver behaviours and road safety for PTW. Deliverable D1 of the 2-BE-SAFE project. European Commission, Brussels, Available online at: http://www.2besafe.eu/sites/default/files/deliverables/2BES_D1_RiderDriverBehavioursAndRoadSafetyForPTW.pdf, 2010.

[ACE 09] ACEM, Motorcycle Accident In-Depth Study: In-depth investigations of accidents involving powered two wheelers, MAIDS project Final report 2.0 – Association des Constructeurs Européens de Motocycles (The Motorcycle Industry in Europe), Brussels, 2009.

[ACE 13] ACEM, *The Motorcycle Industry in Europe: Statistical Overview*, version 18/4/2013. The Motorcycle Industry in Europe, Brussels, 2013.

[DFT 05] DFT, *The Government's Motorcycling Strategy*, Department for Transport, London, 2005.

[DGT 07] DGT, *Strategic Plan for the Road Safety of Motorcycles and Mopeds*, Direccion General del Trafico, Ministerio del Interior, Madrid, 2007.

[ELV 04a] ELVIK R., "To what extent can theory account for the findings of road safety?", *Accident Analysis and Prevention,* vol. 36, pp. 841–849, 2004.

[ELV 04b] ELVIK R., VAA T., *The Handbook of Road Safety Measures.* Amsterdam: Elsevier Science, 2004.

[ELV 09] ELVIK R., "Benefits and fairness: are the high risks faced by motorcyclists fair?", in VAN ELSLANDE P. (ed.), *Les deux-roues motorisés: nouvelles connaissances et besoins de recherche*, Bron: Les collections de l'INRETS, 2009.

[NHT 12] NHTSA, *Traffic Safety Facts, 2010 Data, Motorcycles*, DOT HS 811 639, 2012.

[HAW 12] HAWORTH N., "Powered two wheelers in a changing world: challenges and opportunities", *Accident Analysis and Prevention*, vol. 44, no. 1, pp. 12–18, 2012.

[HEN 08] HENKENS N.C., HIJKOOP S., *Monitoring Bromfietshelmen,* In opdracht van Bureau Verkeershandhaving Openbaar Ministerie, Grontmij Verkeer en Infrastructuur, De Bilt, 2008.

[HOR 05] HORSWILL M.S., HELMAN S., ARDILES P. *et al.*, "Motorcycle accident risk could be inflated by a time to arrival illusion", *Optometry and Vision Science*, vol. 82, no. 8, pp. 740–746, 2005.

[KAR 10] KARDAMANIDIS K., MARTINIUK A., IVERS R. *et al.*, Motorcycle rider training for the prevention of road traffic crashes: review, Cochrane Injuries Group, 2010.

[KAS 05] KASANTIKUL V. *et al.*, "The role of alcohol in Thailand motorcycle crashes", *Accident Analysis & Prevention*, vol. 37, no. 2, pp. 357–366, 2005.

[LAR 05] LARDELLI-CLARET P., JIMENEZ-MOLEON J.J., LUNA-DEL-CASTILLO J.D. *et al.*, "Driver dependent factors and the risk of causing a collision for two wheeled motor vehicles", *Injury Prevention*, vol. 11, no. 4, pp. 225–231, 2005.

[LIU 07] LIU B.C., IVERS R., NORTON R. *et al.*, "Helmets for preventing injury in motorcycle riders", in *Cochrane Database of Systematic Reviews*, no. 4, 2007.

[MCL 93] MCLELLAN *et al.*, "The role of alcohol and other drugs in seriously injured traffic crash victims", *International Conference on Alcohol, Drugs and Traffic Safety-T92, Proceedings of the 12th Conference*, 1993.

[OEC 08] OECD/ITF, *Towards Zero – Ambitious Road Safety Targets and the Safe System Approach* , OECD/ITF, Paris, 2008.

[PEE 96] PEEK-ASA C., KRAUS J.F., "Alcohol use, driver, and crash characteristics among injured motorcycle drivers", *Journal of Trauma,* vol. 41, no. 6, pp. 989–93, 1996.

[PRE 95] PREUSSER D.F., WILLIAMS A.F., ULMER R.G., "Analysis of fatal motorcycle crashes: crash typing", *Accident Analysis and Prevention*, vol. 27, no. 6, pp. 845–851, 1995.

[REE 95] REEDER A.I, CHALMERS D.J, LANGLEY J.D., "Young on-road motorcyclists in New Zealand: age of licensure, unlicensed riding, and motorcycle borrowing", *Injury Prevention,* vol. 1, pp. 103–108, 1995.

[RIJ 11] RIJKSOVERHEID, Action plan for improving road safety for motorcyclists – Strategic approach, Ministry of Infrastructure and Environment, The Hague, 2011.

[RIZ 13] RIZZI M., STRANDROTH J., KULLGREN A. *et al.*, "Effectiveness of Anti-Lock Brakes (ABS) on motorcycles in reducing crashes: a multinational study", *2013 ESV Conference*, Seoul, 27–30 May 2013.

[ROG 08] ROGERS N., Trends in motorcycle fleets worldwide, Presentation to Joint OECD/ITF Workshop on Motorcycling safety, available at: http://www.internationaltransportforum.org/jtrc/safety/Lillehammer2008, 2008.

[DER 11] DE ROME L., IVERS R., FITZHARRIS M. *et al.*, "Motorcycle protective clothing: Protection from injury or just the weather?", *Accident Analysis & Prevention* vol. 43, no. 6, pp. 1893–1900, 2011.

[SHI 12] SHINAR D., "Safety and mobility of vulnerable road users: pedestrians, bicyclists, and motorcyclists", *Accident Analysis and Prevention*, vol. 44, no. 1, pp. 1–2, 2012.

[SOD 95] SODERTROM C.A., DISCHINGER P., KERNS T. *et al.*, "Marijuana and other drug use among automobile and motorcycle drivers treated at a trauma center", *Accident Analysis and Prevention*, vol. 27, no. 1, pp. 131–135, 1995.

[TRA 10] TRAFIKVERKET, Improved Safety for Motorcycle and Moped Riders. Joint strategy for the period 2010–2020, Swedish Transport Administration, 2010:043, Borlange, 2010.

[VIC 08] VICTORIAN GOVERNMENT, Victoria's Road Safety and Transport Strategic Action Plan for Powered Two Wheelers 2009–2013, Victorian Government, Melbourne, 2008.

[YAN 05] YANNIS G., GOLIAS J., PAPADIMITRIOU E., "Driver age and vehicle engine size effects on fault and severity in young motorcyclists accidents", *Accident Analysis and Prevention*, vol. 37, no. 2, pp. 327–333, 2005.

[YAN 10] YANNIS G. *et al.*, Basic Fact Sheet "Mopeds and Motorcycles", Deliverable D3.9 of the EC FP7 project DaCoTA, European Commission, Brussels, 2010.

[YUA 00] YUAN W., "The effectiveness of the 'Ride Bright' legislation for motorcycles in Singapore", *Accident Analysis and Prevention* vol. 32, no. 4, pp. 559–63, 2000.

8

Comparison of Car Drivers' and Motorcyclists' Drink Driving in 19 Countries: Results from the SARTRE 4 Survey

Previous studies showed that motorcyclists drink and drive as often as car drivers but with lower BAC. Moreover, when they know that they may drink heavily, motorcyclists typically leave their motorcycle and use their car. They may be trying to compensate for the greater vulnerability of their habitual transport mode by more cautious behavior. In total, 12,507 car drivers and 4,483 powered two-wheeler drivers from 19 countries were asked about their attitudes and behaviors related to driving while impaired. The results showed large differences between countries in alcohol consumption and reported drink driving. More specifically, the drivers from Northern Europe reported far less drink driving than Southern drivers. In most countries motorcyclists drink and drive almost as often as car drivers do. The comparison of motorcyclists and car drivers showed similar patterns of drink-driving processes.

8.1. Introduction

Road fatalities in Europe are often associated with driving under the influence of alcohol (DUI, [SAF 09]) in spite of considerable efforts from member states to cope with this phenomenon [INT 08]. Indeed, since Borkenstein (1964, cited by International Transport Forum [INT 08]) established a direct link between Blood Alcohol Content (BAC) and road deaths, several studies have confirmed the high risk of drinking and driving (see for example Laumon *et al.* [LAU 05]). It has been estimated that 25% of road fatalities in Europe are linked to alcohol-related road

---

Chapter written by Julien CESTAC, Cécile BARBIER, Gian-Marco SARDI, Richard FREEMAN, Sami KRAÏEM and Jean-Pascal ASSAILLY.

traffic accidents, with great variations among countries – from 5% in Bulgaria to 30% in France, Slovenia and Ireland [INT 08]. According to a European study conducted within the Driving Under the Influence of Drugs, Alcohol and Medicines (DRUID) project in 13 countries [HOU 06], an average of 3.48% of European drivers are driving under the influence of alcohol. In such a context, it is crucial to upgrade our knowledge about the determinants of impaired driving. It could indeed contribute to improving the efficacy of future road safety actions and thus reduce the number of alcohol/legal drug-related fatalities on European roads.

It is well known that alcohol consumption in a country is related to the overall accident mortality rate [SKO 01]. Fifty years ago, alcohol consumption was very heterogeneous in Europe: in 1961, alcohol consumption ranged from the equivalent of 2 l of pure alcohol per year per inhabitant in Finland to 17.7 l in France. But over those 50 years, European countries saw their annual alcohol consumption converging toward 10 l of pure alcohol per person [WOR 11]. However, the increased homogeneity in overall consumption masks different patterns of consumption in the countries by various groups of road users such as differing preferences for wine, beer and spirits as well as the incidence of binging on alcohol. These inter-country differences need to be understood to facilitate effective policies for different types of road users.

Considering driving while impaired, some studies have showed that car drivers and motorcyclists do not behave the same way. Indeed, motorcyclists have been found to drink and drive as often as car drivers but with lower BAC [WAT 92]. Moreover, when drinking heavily, motorcyclists typically leave their motorcycle and use their car so as to feel safer: a car is perceived as easier to drive, does not fall over and offers more protection [SYN 01]. Motorcyclists may thus be trying to compensate for the greater vulnerability of their transport mode by more cautious behavior (theory of risk homeostasis, Wilde [WIL 82], Trimpop [TRI 94]).

8.2. Method

8.2.1. *The SARTRE surveys*

The Social Attitudes to Road Traffic Risk in Europe (SARTRE) project started in 1991. It consists of a European wide survey about knowledge of road traffic laws and road traffic risks, attitudes regarding road safety issues, reported road traffic behaviors, transport habits and needs in several European countries. Various topics related to road safety are the focus of the project such as alcohol, drugs or phone use while driving, speeding, use of advanced driver assistance systems and the transport infrastructure and environment. After the first edition of SARTRE, a follow-up was performed in 1996 (SARTRE2) and 2002 (SARTRE 3). For this fourth edition of the survey [CES 12], the target groups were car drivers, drivers of "powered two-wheelers", pedestrians, cyclists and public transport users. Data was collected between September 15th 2010 and November 30th 2010 among 19 countries: Austria, Belgium, Cyprus, Czech Republic, Estonia, Finland, France, Germany, Greece, Hungary, Ireland, Israel, Italy, Netherlands, Poland, Serbia, Slovenia, Spain, and Sweden.

In this chapter, we focus only on the car drivers and powered two-wheeler behaviors and attitudes related to drink driving.

8.2.2. *Procedure*

Each interviewee was surveyed by a face-to-face interview only in one single role, i.e. as car driver, or as powered two-wheeler user. The car drivers' sample (CD) was composed of car driving license holders that reported having driven during last 12 months. The powered two-wheelers' sample (PTW) was composed of PTW >50 cc license holders that reported having ridden a PTW > 50 cc during last 12 months.

8.2.3. *Sample*

The total sample for the 19 countries included 12,507 car drivers and 4,483 powered two-wheelers. The target population was sampled by: sex, age and occupation (non-active, independent and salaried). The proportions of those sampling variables were representative of their distribution in the population for each subsample. Although each interviewee was only surveyed in one role, subgroups were not artificially distinct. This means that the CD sample included a certain proportion of interviewees who also ride a PTW and also that the proportion of male and female respondents is different among car drivers and powered two-wheelers (see Table 8.1). National samples had to be geographically stratified according to at least NUTS1 regions. Agglomeration sizes were to be included to respect a balance between rural and urban areas.

8.2.4. *Measures*

An English reference version of the questionnaire has been established by the partners and translated into the respective languages. The translation itself was verified by back translation.

The respondents answered questions about their:

– reported drink-driving frequency "even after a small amount" (from 1 "never" to 6 "always");

– friends' behavior: "Most of your friends would drink and drive a car" (from 1 "strongly agree" to 4 "strongly disagree");

– estimation of the number of alcohol units allowed in order to remain under the legal BAC;

– attitudes toward drink-driving law: "Penalties for drink-driving offences should be more severe" (from 1 "strongly agree" to 5 "strongly disagree");

– attitudes toward alcohol interlock (from 1 "strongly agree" to 4 "strongly disagree");

– sociodemographic characteristics: age, gender, driving experience and mileage;

8.3. Results

If the proportion of women among CD is near parity for most countries, their proportion among PTWs is very low (see Table 8.1). It seems that women are far less interested than men by this kind of transport mode. This is of high importance when comparing the two groups because we know that men's and women's relations with alcohol and drink driving are very different. Gender distribution among countries varies widely and, again, when comparing PTWs between countries we have to keep in mind the differences in gender distributions.

We asked respondents to estimate the number of units that they were allowed to drink before driving in order to comply with legal BAC. We then compared their answers to the actual number of units allowed in their country (two units for countries with a 0.5-g/l legal BAC, 1 unit for countries with a 0.2-g/l legal BAC and 0 unit for countries with a 0-g/l legal BAC) and found that some respondents overestimated this number. There was no overall difference between PTW and CD with 18 and 15% of overestimating answers respectively. However, there was again a great variability among countries and in some of them, differences between PTW and CD (see Table 8.1).

The between-country variability was also found for perceived friends' drink-driving behavior (see Table 8.1). Indeed, Swedish drivers estimated rather low drink driving among their friends (4% for PTW and 14% for CD) whereas Italian drivers' estimations were very high (90% for PTW and 81% for CD). For these questions,

PTW friends are most often perceived as drink driving less frequently than CD friends, with the exception of Cyprus and Italy where it is the contrary.

Regarding attitudes toward alcohol interlocks and reinforcing drink-driving penalties, the differences between countries are far lower, with an overall relatively high level of acceptability. CD seemed to accept these measures even more than PTW though (see Table 8.1).

	Women		Overestimation of alcohol units		Drink-driving Friends, other than "Not at all"		Alcolock in cars (very + fairly)		Reinforcing DD penalty (very + fairly)	
	PTW	CD	PTW	CD	PTW	CD	PTW	CD	PTW	CD
Austria	14	48	26	28	41	59	58	69	72	73
Belgium	10	44	16	20	58	74	72	72	70	75
Cyprus	14	45	50	30	80	69	66	80	65	88
Czech Rep	17	43	29	26	48	56	68	76	84	88
Estonia	8	64	6	4	59	71	61	76	84	93
Finland	11	50	2	3	9	25	64	81	88	91
France	23	49	6	6	39	67	78	84	68	79
Germany	12	43	12	12	35	54	72	81	79	81
Greece	13	43	17	19	71	70	82	85	74	79
Hungary	4	36	8	9	12	13	80	83	93	92
Ireland	7	50	7	7	27	34	86	89	83	85
Israel	15	44	2	5	39	38	75	80	88	91
Italy	30	44	39	13	90	81	47	79	52	74
Netherlands	26	52	5	6	41	79	75	78	85	92
Poland	7	34	16	12	50	58	80	78	85	85
Serbia	4	27	31	–	70	84	66	78	78	84
Slovenia	6	44	4	4	47	63	80	87	59	77
Spain	19	43	43	35	54	71	82	82	76	80
Sweden	19	54	8	11	4	14	80	89	86	86
TOTAL	*13*	*45*	*18*	*15*	*47*	*58*	*73*	*80*	*78*	*84*

Table 8.1. *Frequencies by country, percentages*

On average, one in ten Europeans admitted to driving under the influence of alcohol at least once in the previous week. This rate has decreased since the previous survey (SARTRE 3: 20%) but the ranking of countries remained unchanged: the declared drink-drive behavior was more frequent in southern than in eastern and northern countries (see Figure 8.1). Both CD and PTW claimed to rarely drink and drive, but drink drivers are consistently less frequent among motorcyclists (with Cyprus a notable exception).

At the regional scale, motorcyclists claimed to drive under the influence of alcohol less often than car drivers in five countries ($p<0.001$): Austria, Finland, France, Germany and The Netherlands. Among those countries some have high rates of mean reported drink driving (Austria and France) whereas others have low rates (Finland, Germany and The Netherlands). Cyprus constitutes an exception: the effect is reversed in this country, with motorcyclists who report drinking and driving more often than car drivers do ($p<0.001$). In the other countries, the differences were not significant.

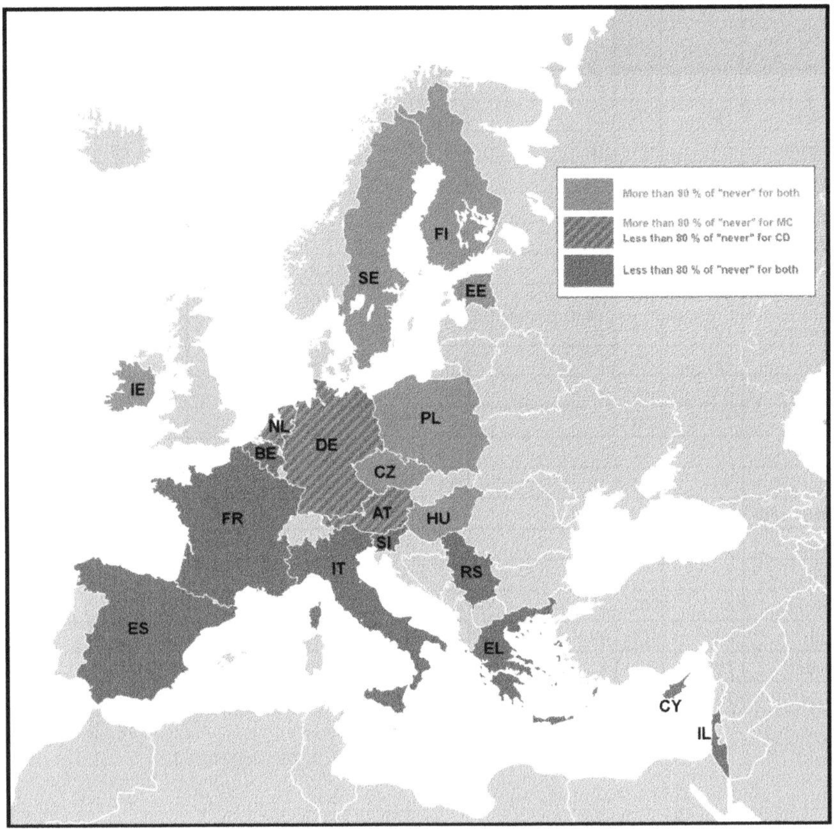

Figure 8.1. *"Over the last month, how often have you driven a car/motorcycle after having drunk even a small amount of alcohol?" percentage responding "never". For a color version of the figure, see www.iste.co.uk/jacob/safety.zip*

We conducted a linear regression analysis (see Table 8.2) in order to identify the factors explaining drink driving. It appeared that the three main predictors of driving under influence for both of them are:

– friends' drink-drive behavior;

– the level of legal BAC;

– the accuracy of drivers' estimation of the maximum number of allowed units.

	β CD	β PTW
Accuracy of legal units estimation	0.19***	0.22***
Gender	−0.11***	-0.07**
Attitude toward alcohol interlocks	0.07***	0.00 ns
Attitude toward drink-driving penalty	0.11***	0.12***
Driven (km)	0.03**	0.08***
Friends drink driving	−0.26***	−0.26***
Age	−0.09***	−0.08***
Driving experience	0.08***	0.03ns
Legal BAC	0.21***	0.19***
R^2	*0.25*	*0.23*

NOTE.– Gender is coded as a dummy variable with 1: male and 2: female. Accuracy of legal units estimation is coded as: 1, underestimation; 2, correct estimation; 3, overestimation; ***$p<0.001$, **$p<0.01$, ns, not significant.

Table 8.2. *Linear regression on drink-drive behavior ("even a small amount")*

The influence of friends' behavior on drivers' behavior is the main effect. The social influence of peer groups is well known in road safety issues [FOR 09] and our results confirmed it with the same impact of friends' behavior for both car drivers and motorcyclists.

The effect on behavior of the accuracy of estimation of maximum alcohol units allowed is interesting because this is linked to one of the main problems with alcohol legislation: the limit of BAC while driving is fixed to some acceptable level by authorities but drivers have no means to measure what their actual level of BAC is. This may lead some drivers to overestimate the number of glasses or bottles of a particular alcoholic beverage they are allowed to drink before taking the wheel [ASS 95]. Moreover, we should note that there were no apparent differences between car drivers and motorcyclists on this question.

The national level of BAC tolerance while driving also had a positive impact on driving under the influence of alcohol. This finding not only emphasizes the usefulness of legislating to regulate this kind of behavior but also confirms that this behavior is still, to some extent, voluntary. In countries with a higher level of tolerated BAC, people tend to make use of this "right" to drink small amounts of alcohol before driving.

The question of causality is often raised when dealing with such analyses. Here, it seems more logical to interpret the relationship between drink-drive behavior and attitude toward drink-drive penalty as the first causing the second. That is to say: the more we drink and drive, the less we are in favor of penalty enforcement. This interpretation leads to the conclusion that people are aware of their behavior, they know what they are doing but they are not willing to change it.

Analyses on a country level reproduced the general pattern with only slight variations, e.g. Poland where these effects are very small. This appears to be due to few people admitting drinking and driving in Poland, so there is little variance to explain.

8.4. Discussion

Drink driving in European countries is a rare behavior. Indeed, if we consider the percentage of people found to be above the legal limit by police testing, it ranges from 1 to 6% for all SARTRE-4 countries except for Poland in which the percentage is 9.5% [EUR 08]. Nevertheless, we know that those roadside breath tests are not random and performed in selected at-risk places and times. According to a random testing study by Houving and his colleagues [HOU 06] from the DRUID project there is a great variability in alcohol use by drivers in Europe (from 0.15% in Hungary to 8.59% in Italy). SARTRE 4 results allow for an estimation of individual drink-driving frequency based on declared behavior (71.2% declare never having driven after having drunk even a small amount of alcohol during the last month and 17% "rarely") with again a great variability between countries (from 98.5% in Poland and 94.7% in Hungary to 43.7% in Italy). It is important to note here that there is a correlation of $r = .90$ ($p<0.001$) between DRUID results based on random breath tests and SARTRE4 results based on declared behavior. Nevertheless, despite this behavior being relatively uncommon, it is responsible for approximately 25% of European road deaths [SAF 09].

In this chapter, we mainly explored whether car drivers (CD) and powered two-wheelers (PTW) were thinking or claiming to behave differently regarding driving under the influence of alcohol. Our results showed that in most countries drivers of PTWs drink and drive almost as often as CD do. Where there are differences (in Austria, Cyprus, Finland, France, Germany and The Netherlands) usually it is the PTWs who declare less often to be drinking and driving, with the exception of Cyprus in which it is the opposite. CD and PTW differed on their attitudes regarding alcohol interlocks and reinforcing the drink-driving penalty. Even if respondents were mainly favorable to it, PTW drivers were less favorable than CD. It seems that PTW drivers are keen on self-regulation because they prefer less stringent laws but they are slightly more careful than car drivers (see also, Syner & Vegega [SYN 01]).

The drink-drive best predictors are identical for car CD and PTW. PTW drivers are influenced by their friends' behavior just as CD are. Moreover, in both groups, those who overestimate the allowed amount of alcohol before driving are those who drink and drive the more frequently. Finally, both CD and PTW drivers drive under the influence of alcohol more often in countries with the highest legal BAC threshold. These results lead us to suggest undifferentiated measures for CD and PTW for drink-driving enforcement. Nevertheless, it suggests potential useful measures concerning peer group influence on drink driving (communication campaigns), concerning accuracy of BAC estimation (communication campaigns and development/spreading of alcohol metering devices) and concerning the legal BAC threshold.

A legal BAC of 0.2 g/l seems to be the best option regardless of the category of road user. Indeed, people living in countries with 0.2 g/l threshold do not drink and drive more often than those in countries with a 0 g/l threshold, but people in countries with a 0.5 g/l threshold do drink and drive more frequently than the other two levels. Nevertheless, the often raised question of a lower limit for PTW [COL 93, SUN 98] could be considered again in the light of the SARTRE 4 results. Indeed, if PTW tend to self-regulate their behaviors, they still drink and drive more frequently in 0.5 g/l countries than in lower BAC countries. The consequences of drink driving are much more severe for them than for CD because of the "effects on balance, motor coordination, and judgment and [the] more-basic skills [that] are needed to operate [their] inherently unstable vehicle" [LIN 09].

8.5. Acknowledgments

This research was funded by the European Commission (TREN/09/SUB/E3/229/SI2.544555/SARTRE4). This chapter reflects the authors' views. The European Commission is not liable for any use that may be made of the information contained therein.

8.6. Bibliography

[ASS 95] ASSAILLY J.-P., "Les jeunes, l'alcool et la conduite: un risque pris, non perçu ou… accepté? [Young people, alcohol and driving: a chosen, not perceived or ... accepted risk?]", *Revue Transports Sécurité*, vol. 49, pp. 43–51, 1995.

[CES 12] CESTAC J., DELHOMME P. (ed.), "European road users' risk perception and mobility", *The SARTRE 4 survey*, Lyon, France: Public Imprim, 2012.

[COL 93] COLBURN N., MEYER R.D., WRIGLEY M. *et al.*, "Should motorcycles be operated within the legal alcohol limits for automobiles?", *The Journal of Trauma*, vol. 35, no. 2, pp. 183–186, 1993.

[EUR 08] EUROPEAN TRANSPORT SAFETY C., "Tackling the three main killers on the roads: a priority for the forthcoming EU Road Safety Action programme". PIN Flash 16. Brussels: ETSC, Available at: http://www.etsc.eu/documents/05.05%20-%20PIN%20Flash%2016.pdf, 2008.

[FOR 09] FORWARD S., "The theory of planned behaviour: the role of descriptive norms and past behaviour in the prediction of drivers' intentions to violate", *Transportation Research Part F: Traffic Psychology and Behaviour*, vol. 12, no. 3, pp. 198–207, 2009.

[HOU 06] HOUVING S., HAGENZIEKER M., MATHIJSSEN *et al.*, "Prevalence of alcohol and other psychoactive substances in drivers in general traffic", Part I: General Results (Project N°TREN-05-FP6TR-S07.61320-518404-DRUID, Deliverable D2.2.3.). Available at: http://www.druid-project.eu/cln_031/nn_107534/Druid/EN/deliverales-list/deliverables-list-node.html?__nnn=true, 2006.

[INT 08] INTERNATIONAL TRANSPORT FORUM, *Towards Zero: Ambitious Road Safety Targets and the Safe System Approach*, OECD Publications, Paris, 2008.

[LAU 05] LAUMON B., GADEGBEKU B., MARTIN J.-L., "Stupéfiants et accidents mortels (projet SAM). [Drugs and fatal crashes (SAM project)]", Convention OFDT/CEESAR , Rapport final – Analyse épidémiologique, 2005.

[LIN 09] LIN M.-R., KRAUS J.F., "Methodological issues in motorcycle injury epidemiology", *Accident Analysis and Prevention*, vol. 40, pp. 1653–1660, 2009.

[SAF 09] SAFETY NET, Alcohol. Project co-financed by the European Commission, available at: http://ec.europa.eu/transport/road_safety/specialist/knowledge/alcohol/index.htm, 2009.

[SKO 01] SKOG O., "Alcohol consumption and overall accident mortality in 14 European countries", *Addiction*, vol. 96, pp. S35–S47, 2001.

[SUN 98] SUN S.W., KAHN D.M., SWAN K.G., "Lowering the legal blood alcohol level for motorcyclists", *Accident Analysis and Prevention*, vol. 30, no. 1, pp. 133–136, 1998.

[SYN 01] SYNER J., VEGEGA M., "Impaired motorcycle riding: what motorcyclists think about alcohol and motorcycling", *American Public Health Association*, Boston, November 2000, 2001.

[TRI 94] TRIMPOP R.M., "Risk compensation and the interaction of personality and situational variables", in TRIMPOP R.M., WILDE G.J.S. (eds), *Challenges to Accident Prevention*, Styx Publications, Groningen, pp. 135–146, 1994.

[WAT 92] WATSON W.A., GARRIOTT J.C., "Alcohol and motorcycle riders: a comparison of motorcycle and car/truck DWIs", *Veterinary and Human Toxicology*, vol. 34, no. 3, pp. 213–5, 1992.

[WIL 82] WILDE G.J.S., "The theory of risk homeostasis: implications for safety and health", *Risk Analysis*, vol. 2, no. 4, pp. 209–225, 1982.

[WOR 11] WORLD HEALTH O, Available at: http://apps.who.int/ghodata. Global Information System on Alcohol and Health. Recorded adult per capita consumption, from 1961, Total, 2011.

9

Trajectories of Multiple People in Crowds Using Laser Range Scanner

In everyday life, we see an amazing choreography of the movements in a crowd of pedestrians. Pedestrians cross and avoid vehicles but do not appear to voluntarily cooperate with each other. In this paper, we track a crowd of pedestrians in a crowded area to understand their social behavior in such areas. The observations are gathered from the data base provided by a network of laser rangefinders with a sampling frequency of 50 Hz. We introduce a stable feature extraction method based on accumulated distribution of successive laser frames. To isolate pedestrians, we propose a non-parametric method to exploit the Parzen windowing technique. We apply the new method of Rao-Blackwellized Monte Carlo data association for tracking a highly variable number of pedestrians. The algorithm is quantitatively evaluated through a social behavior experiment taking place in the lobby of a school. During this experience, nearly 300 students were tracked.

9.1. Introduction

During the last decade, human behavior has been studied in various fields such as architectural design, prevention and traffic engineering. Pedestrian management requires the ability to predict the evolution of behavior in different situations, especially in terms of traffic flow analysis. Most prediction models lack experimental data to be validated. It is important in this context to provide a system for detection and robust tracking. The trajectory, walking speeds and direction changes of pedestrians are the data for such an analysis. Indeed many monitoring systems have been developed and various monitoring methods have been proposed. Bu and Chan [BU 05] identified a wide range of different sensors used for the

---

Chapter written by Ladji ADIAVIAKOYE, Patrick PLAINCHAULT, Marc BOURCERIE and Jean-Michel AUBERLET.

detection and tracking of pedestrians. The vast majority of work on the detection focuses on the analysis of motion video image, since it has the ability to capture the shape and texture of a target.

However, the high sensitivity to weather conditions, limited opening angle of the cameras and the inability to obtain accurate information on depth directly are major obstacles to the reliability and robustness of the system based on video camera. Indeed, many cameras are needed to cover a large area and also reduce occlusions due to objects or other people on the move. However, Brückner *et al.* [BRÜ 11] explained that it is difficult to digitally merge data from multiple cameras, which requires precise calibration and high computational cost and complicated calibration.

On the basis of that fact, we focus on an active sensor such as a laser scanner. This sensor is an instrument that has received increasing attention to the problems of monitoring. It measures the distances of nearby objects by emitting laser beams eye safety in controlled directions. To analyze this information on a distance a variety of technical processes are possible, (flight time) and depend on the magnitude of the distances measured. Several authors, Prassler *et al.* [PRA 99], Schulz *et al.* [SCH 01], Mon *et al.* [MON 02], often use this type of sensor in many practical applications of robotics.

In this chapter, we propose an algorithm for detecting and tracking a crowd of pedestrians in a relatively large and crowded area, e.g. the lobby of a school district. An algorithm for image registration allows us to automatically merge data from sensors. Then, a non-parametric method exploiting the Parzen windowing technique is performed to detect targets. We trace the most complex trajectory of pedestrian tracking based on the use of the extended Kalman filter and Rao-Blackwellized Monte Carlo data association.

9.2. Approach

This section first presents briefly the sensor system for tracking pedestrians. Emphasis is placed on the original operating data from our sensors. A description of the tracking algorithm starting with the choices made for the detection of pedestrians is given. Rao-Blackwellized Monte Carlo data association can track and identify objects in time with its various uncertainties.

9.2.1. *Measurement system*

In our tracking system, we used a network of laser rangefinder scanners of the SICK LMS 111 corporate type. The application context that motivates us in this chapter is the technique of pulse telemetry also called "time of flight measurement", which corresponds to a return of the light between the laser source and the photoreceptor through the target awesomeness that is widely used. The time of flight technique is one of the most appropriate for distance measurement in a wide range of relative values. On the other hand, this technique offers the advantage of being able to use surfaces of any kind, scattering or non-cooperative. Thus, a fraction of the optical energy received rockets from the surface of the target in the direction of the incident laser beam.

When scanning, a profile of 541 points on the distance corresponding to 270 degrees in terms of scanning is prepared. Distance data are collected in the form (ρi, θi), θi = 0°, 0.5°...270°, where ρi represents the measured distance and θi the corresponding angle. By combining the position and direction of the laser scanner, the telemetry data can easily be converted into Cartesian coordinates. The manufacturer suggests a maximum range of 20 m with an accuracy of ± 4 cm. To obtain an angular resolution of 0.5°, the acquisition rate is set at 50 Hz.

In our experiments, four laser scanners scan the scene horizontally at a height of between 16 and 18 cm. This choice is justified by the fact that occlusions are higher at ankle height. We use several laser rangefinders placed at different points of view in order to reduce occlusions of the target and cover a larger area than one sensor can cover. Overlapping lasers add another vision angle to observe inaccessible targets in the field of vision of another sensor. Another advantage is that multiple lasers offer the potential for higher detection and therefore can be used in an area with a high density of pedestrians.

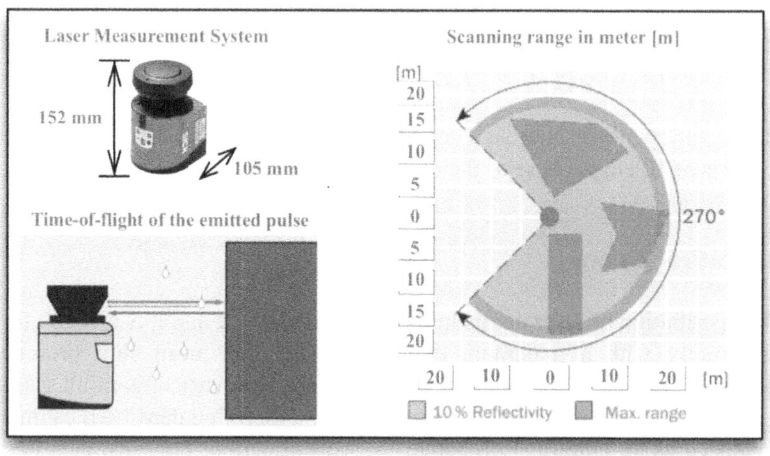

Figure 9.1. *Laser measurement systems. For a color version of the figure, see www.iste.co.uk/jacob/safety.zip*

9.2.2. *Data fusion*

This part is roughly divided into three parts: background subtraction, clustering and image registration. It exposes our method of mapping telemetry profiles to create a global perspective.

9.2.3. *Background subtraction*

Telemetry data contain both moving objects (e.g. the legs of pedestrians) and static objects that we call bases (e.g. walls poles). The first step of background subtraction is designed to remove stationary (or nearly stationary) objects, and acts on the data of each individual rangefinder. Thus, it can run independent of the registration steps outlined above. We generate a background image $I_0(x\ y,\ t)$ by a scan of the scene without the target.

Then we put five small diameter rolls on the stage so that they are seen by each sensor (Figure 9.2(b)). These images contain the substance and the object will be named $I(x, y, t)$. Thus, we can run the formula described below independently.

Estimate the background for a time t.

Subtract the estimated from the plot of $I(x, y, t)$ background.

Apply a threshold "Th", the absolute difference for the mask foreground.

The formula for extracting the background is written as:

$$\left|I_0(x,y,t)\text{-}I(x,x,t)\right|>Th \qquad [9.1]$$

The accuracy of this approach depends heavily on the threshold. In our case it is of the order of 10 cm because of the error in the sensor.

9.2.4. *Control points*

The distribution of laser points in the image is discrete and it is difficult to locate points directly with a maximum density around. The mean shift presented by Comaniciu and Meer [COM 99] is a powerful non-parametric algorithm that allows us to find the local maximum of a function of the underlying density. It estimates the feature space as an empirical probability density function. For each data point, the mean shift defines a window and calculates the average of data within the window. Then it moves the center of the window to the average in the direction of the gradient and repeats until convergence. After each iteration, it can be considered that the window moves to a region of the densest dataset.

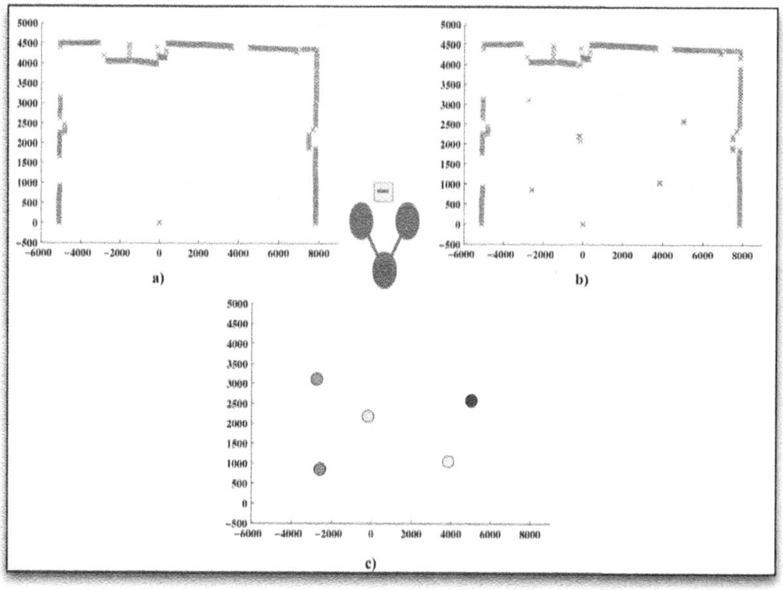

Figure 9.2. *a) Background image; b) image of the scene; c) background subtraction. For a color version of the figure, see www.iste.co.uk/jacob/safety.zip*

Kernel density estimator for a given set of d-dimensional points is:

$$\hat{f}(x) = \frac{1}{n.h^d}\sum_{i=1}^{n} K\left(\frac{x-x_i}{h}\right) \qquad [9.2]$$

$$\nabla\hat{f}(x) = \frac{1}{n.h^d}\sum_{i=1}^{n} K'\left(\frac{x-x_i}{k}\right) \qquad [9.3]$$

Setting it to 0 we get,

$$\sum_{i=1}^{n} K'\left(\frac{x-x_i}{h}\right)\vec{x} = \sum_{k=0}^{n} K'\left(\frac{x-x_i}{h}\right)\vec{x_i} \qquad [9.4]$$

Finally, we get:

$$\vec{x} = \frac{\sum_{k=0}^{n} K'\left(\frac{x-x_i}{h}\right)\vec{x_i}}{\sum_{i=1}^{n} K'\left(\frac{x-x_i}{h}\right)} \qquad [9.5]$$

$(x_1 \dots x_n)$ is a set of samples of size d (d = 2 in this case), $K(\bullet)$ is the kernel (window function) and h is the parameter window width. The estimator $f(x)$ converges to the true density if $K(\bullet)$ and h are properly chosen. We perform the algorithm exactly as it was described by Comaniciu and Meer [COM 99], using the Gaussian distribution because of its good features.

9.2.5. *Image registration*

It is to match two or more images or in order to combine their respective information. This is to find the transformation T, such that the transformed coordinates (x', y') of a point P of coordinates (x, y) of the reference image better correspond to the same point of the target image. Three main criteria are used for this method: the attributes, the similarity criterion and the transformation model.

For attributes, we rely on geometric approach, i.e. the extraction of coordinates of the five cylinders detected before. Subsequently, these points are matched to determine the transformation between the two images.

The criterion of similarity between images depends on the attributes considered. The goal is to match the information based on the attributes. The operation is to minimize the distances between the corresponding attributes. For this we calculate the centroid of the five coordinated and we do a sort indicator according to Euclidean distance from the centroid.

The transformation model we use is the similarity to align the input image to the base image.

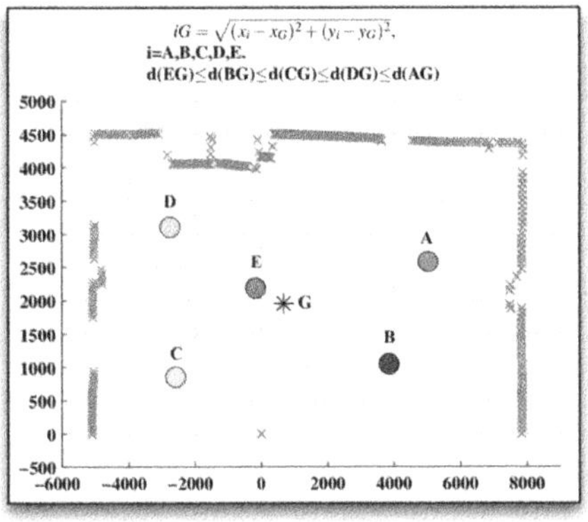

Figure 9.3. *Classification of attributes according to their Euclidean distances from the centroid. For a color version of the figure, see www.iste.co.uk/jacob/safety.zip*

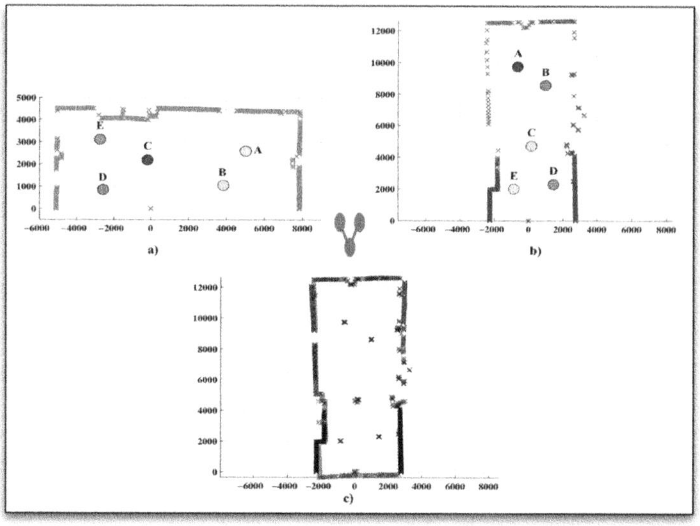

Figure 9.4. *Laser fusion of different frames. For a color version of the figure, see www.iste.co.uk/jacob/safety.zip*

9.3. Detection

Taking advantage of the high data sampling rate, we cumulate successive frames to have only subtle changes. Depending on the human model, when a normal person walks forward, one of the typical aspects is that at any time, a foot swings pivoting on the other as shown in Figure 9.5. Two feet exchange their duty by posing and moving changes in a rhythmic pattern.

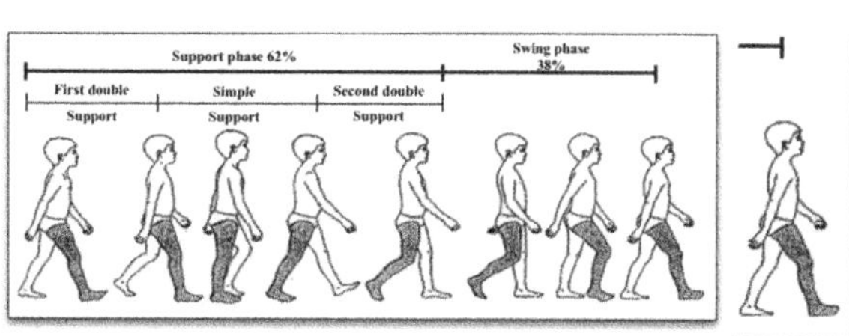

Figure 9.5.

The stance phase during which the foot is in contact with the ground represents 62% of the duration of the complete cycle. Swing phase during which the foot leaves the ground represented 38% of the cycle. If we locate precisely the points of support during this phase, they can provide us with very direct and stable trajectories to infer people's walking indices.

Given the prominent points without background, we use the mean shift clustering method for local maximum points identified on the feet. Indeed, two groups of points should be obtained for one person, corresponding to two feet because there may be several laser impacts on one foot due to the high angular resolution of the sensor. The mean shift clustering allows us to combine these impacts in one central point that will be treated as a foot. Once both feet are detected, we assume that the local maxima are 30 cm from each other and form one person.

We evaluated our detection algorithm through a sequence of 2,500 frames. The detection performance is evaluated by comparing the true count and the estimated count of persons in the scene. As a result of difficulty and hard workload in obtaining the ground-truth count of multiple people, we evaluated our detection algorithm with sampling results at an interval of 50 frames. The count of correctly detected persons is compared with the true count at every 50 frames. The errors exist for two reasons: occlusion and noise. Most of the detection failures from occlusions can be recovered with time accumulation computation. Through the sequence of 1,000 frames we evaluated the detection rate. Only four people are forgotten in the entire sequences.

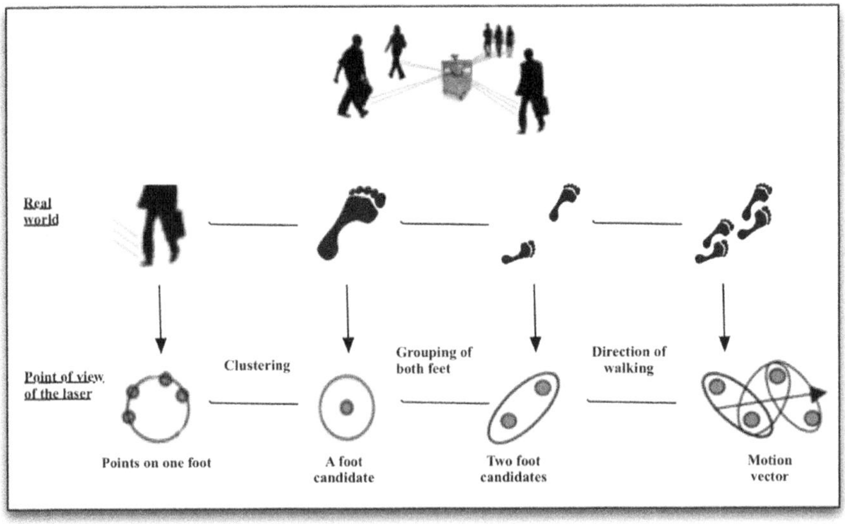

Figure 9.6. *Laser-based people detection*

9.4. Multiple tracking

To estimate the trajectories, we take into account the optimal distribution of Gaussian noise approximately using a tracking algorithm called Rao-Blackwellized Monte Carlo Data Association.

Our goal is to estimate the posterior distributions of the states of all targets at each instant k. The state vector used summarizes all the information observed in the scene, that is to say the number of pedestrians observed and their characteristics

$$X_k = (x_k\ y_k\ \dot{x}_k \dot{y}_k w_k)^T \qquad [9.6]$$

where $(x_k\ y_k)$ is the position of the pedestrian and $(\dot{x}_k \dot{y}_k)$ velocity in Cartesian coordinates in 2D.

Regarding the dynamic movements of a pedestrian, we assume we have no *a priori* information on its unpredictable trajectory (change of pace, direction reversal, crash). In order to predict all the best possible path changes, we use a model of evolution in circular motion where the heading angle is used as a perturbation of the predicted trajectory.

$$x_{k+1} = \begin{pmatrix} 1 & 0 & \frac{sin(w_k \Delta t)}{w_k} & \frac{cos(w_k \Delta t)-1}{w_k} & \frac{\partial x_{k+1}}{\partial w_k} \\ 0 & 1 & \frac{1-cos(w_k \Delta t)}{w_k} & \frac{sin(w_k \Delta t)}{w_k} & \frac{\partial y_{k+1}}{\partial w_k} \\ 0 & 0 & cos(w_k \Delta t) & -sin(w_k \Delta t) & \frac{\partial \dot{x}_{k+1}}{\partial w_k} \\ 0 & 0 & sin(w_k \Delta t) & cos(w_k \Delta t) & \frac{\partial \dot{y}_{k+1}}{\partial w_k} \\ 0 & 0 & 0 & 0 & 1 \end{pmatrix} X_k + \begin{pmatrix} 0 \\ 0 \\ 0 \\ 0 \\ 1 \end{pmatrix} v_k \qquad [9.7]$$

$v_k \sim \mathcal{N}(0, \sigma_w^2)$ is univariate white Gaussian process noise for the turn rate parameter, where the partial derivatives to turning rate are:

$$\begin{cases} \frac{\partial x_{k+1}}{\partial w_k} = \frac{w_k \Delta t\, cos(w_k \Delta t) - sin(w_k \Delta t)}{w_k^2} \dot{x}_k - \frac{w_k \Delta t\, sin(w_k \Delta t) + cos(w_k \Delta t) - 1}{w_k^2} \dot{y}_k \\ \frac{\partial y_{k+1}}{\partial w_k} = \frac{w_k \Delta t\, sin(w_k \Delta t) + cos(w_k \Delta t) - 1}{w_k^2} \dot{x}_k - \frac{w_k \Delta t\, cos(w_k \Delta t) - sin(w_k \Delta t)}{w_k^2} \dot{y}_k \\ \frac{\partial \dot{x}_{k+1}}{\partial w_k} = -\Delta t sin(w_k \Delta t)\, \dot{x}_k - \Delta t cos(w_k \Delta t)\, \dot{y}_k \\ \frac{\partial \dot{y}_{k+1}}{\partial w_k} = -\Delta t cos(w_k \Delta t)\, \dot{x}_k - \Delta t sin(w_k \Delta t)\, \dot{y}_k \end{cases} \qquad [9.8]$$

where Δt is the stepsize of the discretization.

We assume that the measurement accuracy of our sensor can be modeled by:

$$Y_{k+1} = \begin{pmatrix} 1 & 0 & 0 & 0 & 0 \\ 0 & 1 & 0 & 0 & 0 \end{pmatrix} X_k + r_k \qquad [9.9]$$

$r_k \sim \mathcal{N}(0, \sigma_r^2)$, with $\sigma_r^2 = 0.04$ m, where r_k is white, univariate Gaussian noise with zero mean and variance σ_r^2.

To illustrate the use of RBMCDA we create data structures necessary for the particles. The essential part of the filtering procedures is re-sampling. For this, we calculate the actual number of particles. Then, the re-sampling is performed if the number is smaller than the mentioned fourth number of particles. After the actual re-sampling, we normalize the weight of the particles. In the last line, we save the current structures of particles, which conclude the loop filter.

9.5. Experimental results

Two sets of data are used to assess the accuracy of the monitoring system. Both sets of data were collected with four laser scanners placed in the lobby of a high school. Students move and interact in different ways in the lobby. The first data set includes students and their parents with a waiting line which lasts for approximately 30 min. In the second set of data, 300 people were invited to walk and run, following a prearranged script describing various social behaviors to perform for a duration of two hours. Data sets each provided their own set of challenges. However, the most difficult to resolve are:

Case 1: People walk closely. This is a case that happens very often, especially in a crowded environment. Because lasers distributed the points, the clustering based on a single image frame cannot correctly extract the target.

Case 2: Two people from opposite directions and their feet are too close to the intersection points. The data are mixed and a person can be lost in the extraction.

Case 3: Temporal occlusion with loss of data.

In the experiment, using the cumulative distribution of 10 successive frames, legs can be extracted stably and efficiently. We plotted the results of filtering RBMCDA algorithm using 100 particles. Since the particles are a mixture of Gaussian distribution estimates visualized with samples drawn from Gaussian mixtures, it can be seen that the algorithm has no problems following the target trajectory despite the fairly high noise and clutter values. The filter RBMCDA effectively tracks people who walk side by side. The accuracy of the tracker is evaluated in three ways: accuracy of detection, the average length of the track and the number of people present on stage.

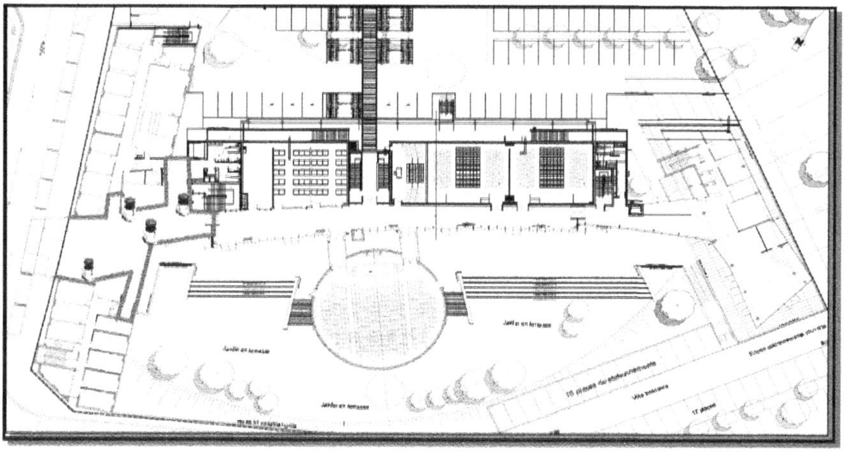

Figure 9.7. *Layout of laser scanners and their coverage at an exhibition hall*

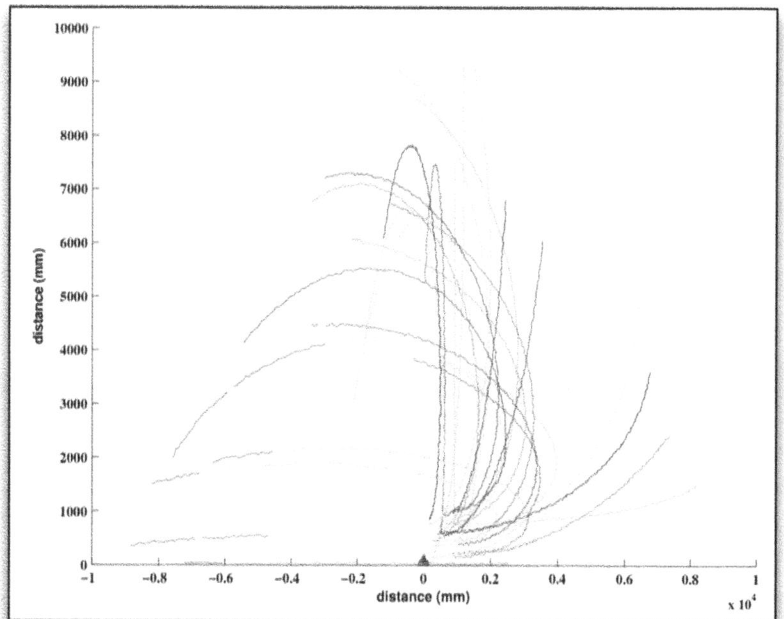

Figure 9.8. *Screen copy of reproduced trajectories. For a color version of the figure, see www.iste.co.uk/jacob/safety.zip*

9.6. Conclusions

The algorithms presented in this chapter provide a solid mechanism for the automatic creation of trajectories of a crowd of pedestrians in a broad coverage area. The process involves collecting data from a series of laser rangefinders. Compared with existing laser-based trackers, the extracted feature based on accumulated distribution of successive laser frames is very stable and deals with the measurement noise very well. When two or more tracks are so close together that they cannot be clearly differentiated in the data, they are split into new tracks, preventing a single track from inadvertently representing more than one target.

The tracker is tested in experiments with four laser rangefinders observing the entrance of a school involving 300 people who walk and run in scenarios of social behavior. Although there are a few jumps on the track, they have occurred every few minutes at most and 90% of pedestrians are detected. Compared with existing trackers, this work stands out in a number of ways. First, it is designed to track an unknown number of targets unlike many other trackers who consider each independent rangefinder. The experimental results show that the proposed method is very effective and robust.

Applications such as:

– the quantitative prediction on the dynamic behavior in various environments (subway, airport);

– spatial distribution of flows, traffic management;

– evacuation dynamics, building design or crowd behavior analysis need reliable experimental data (trajectories, speeds, directions, densities) to improve the realism of the models of existing behavior.

9.7. Bibliography

[BRÜ 11] BRÜCKNER M., BAJRAMOVIC F., DENZLER J., "Self-calibration of camera networks: active and passive methods", in CHEN C.H. (ed.), *Emerging Topics in Computer Vision and its Applications*, World Scientific Publishing, pp. 447–469, 2011.

[BU 05] BU F., CHAN C-Y., "Pedestrian Detection in Transit Bus Application: Sensing Technologies and Safety Solutions", in *IEEE Intelligent Vehicles Symposium (IV)*, Las Vegas, June 2005.

[COM 99] COMANICIU D., MEER P., "Distribution free decomposition of multivariate data", *Pattern Analysis and Applications*, vol. 2, pp. 22–30, 1999.

[MON 02] MONTEMERLO M., THUN S., WHITTAKER W., "Conditional particle filters for simultaneous mobile robot localization and people-tracking", *Proceedings of the IEEE International Conference on Robotics & Automation (ICRA)*, 2002.

[PRA 99] PRASSLER E., SCHOLZ J., FIORINI P., "Navigating a robotic wheelchair in a railway station during rush hour", *International Journal on Robotics Research (IJRR)*, vol. 18, no. 7, pp. 760–772, 1999.

[SCH 01] SCHULZ D, BURGARD W., FOX D. *et al.*, "Tracking multiple moving targets with a mobile robot", *Proceedings of the IEEE Computer Society Conference on Computer Vision and Pattern Recognition (CVPR)*, Kauwai, Hawaii, 2001.

10

Safety of Urban Cycling: A Study on Perceived and Actual Dangers

This work combines data from studies on threat perception by cyclists with data of actual traffic accidents involving cyclists. It presents survey results on the perceived dangers of urban cycling based on interviews with cyclists and polls conducted in the city of Vienna (Austria) over the last two years. These results are contrasted with an analysis and evaluation of records of injury accidents over the last 10 years. While the number of bicycle accidents per year has remained stable, the data shows increases of up to 225% for certain accident characteristics. Additionally, the study also reveals potential extensions to the accident database, which would enable more detailed analyses of certain aspects such as differences in cyclist safety on different types of bicycle routes.

10.1. State of urban cycling

Constantly growing city populations and increasing mobility requirements generate more and more urban traffic. This challenge must be addressed adequately to prevent a degradation of the city environment through mobility-related effects, such as associated noise and air pollution, traffic accidents, loss of time due to congestions or reduction of public space, which reduce quality of life for all inhabitants. Many cities therefore encourage citizens and commuters to switch private motorized vehicles for alternative means of transport such as public transport, walking and cycling. In many cities, the number of urban cyclists has been growing steadily over recent years. Even so, cycling rates are still far below cities, such as Amsterdam, which have a longstanding tradition of urban cycling.

Chapter written by Anita GRASER, Michael ALEKSA, Markus STRAUB, Peter SALEH, Stephan WITTMANN and Gernot LENZ.

In the city of Vienna, the annual bicycle-transport performance was 178 million kilometers according to the Austrian study "Transport in figures" [TRA 13]. Bicycle counts are conducted both at fixed locations using automated counting sensors as well as manually at locations of special interest. Table 10.1 shows the development of mean daily bicycle count numbers at these fixed counting locations over the last ten years from 2002 to 2012. The data show that bicycling is growing in popularity (+44% from 2002 to 2012) but the observant reader will notice a drop in 2010 which is attributed to the exceptionally bad weather conditions during the summer of 2010. Also the numbers for Opernring are lower from 2012 on because in this year a bicycle path on the opposite side of the street was opened.

	2002	2003	2004	2005	2006	2007	2008	2009	2010	2011	2012
Opernring	2,443	2,672	2,241	2,428	2,545	2,719	2,527	2,663	2,451	2,981	2,741
Lassallestraße	1,295	1,334	1,462	1,539	1,654	1,752	1,798	1,847	1,676	1,844	2,118
Argentinierstraße	823	980	884	948	1,039	1,084	1,192	1,211	1,104	1,321	1,411
Kennedybrücke		296	293	287	370	430	476	480	438		
Westbahnhof	811	977	948	1,028	1,106	1,193	1,299	1,146	1,118	1,380	1,461
Donaukanal		911	877	763	851	905	927	1,106	1,111	1,206	1,227
Langobardenstraße		328	361	311	343	328	330	323	293	403	372
Liesingbach		431	479	437	457	437	466	478	461	422	407
Weighted average index based on 2002	1	1.09	1.04	1.07	1.16	1.23	1.27	1.29	1.20	1.38	1.44

Table 10.1. *Mean bicycle count per day and year (January to December, Monday to Sunday) at Viennese counting stations, based on data published by the [GES 13a] for 2002–2010 and [RAD 13c] for 2011–2012*

Regarding accident data analysis and identification of critical scenarios, the currently ongoing EU project "Improving the safety and mobility of vulnerable road users through ITS applications" [VRU 13], gives an overview on European cycling accident data. Analysis of a range of databases has been conducted to identify scenarios for cyclists. The CARE database (European Road Accident Database) has been used as the most widely available database for EU accidents and data from national databases of Austria, Finland, Spain, Sweden and the UK have been compared to the CARE data for consistency of results. The main findings from the database are summarized as follows:

– the majority of cycling accidents in the accident analysis were found to occur at intersections;

– one of the most common scenarios involved vehicles pulling out into the path of the oncoming cyclist at an intersection. CARE data suggests that the most common scenario involves both cyclist and vehicle heading in the same direction but the vehicle then turns into the cyclist's path;

– overall, males are overrepresented in the data;

– the majority of the accidents occur in fine dry weather during daylight hours;

– the majority occur in urban areas on roads with relatively low speed limits.

The Austrian study "Radverkehr in Zahlen" [RAD 13a] confirms the conclusion published in the VRUITS report that most cycling accidents occur at intersections. According to this study, of all bicycle accidents in the year 2010, 45% happened while turning or with perpendicular traffic and 19% happened with traffic in the same or opposite direction. However, considering casualties, the proportions change with 35% while turning or with perpendicular traffic and 31% with traffic in the same or opposite direction. This suggests that accidents with traffic in the same or opposite direction tend to be more severe. In 2011, in Austria 35,129 accidents occurred in which 45,025 people were injured and 523 were killed [ROA 11]. In total, 8% of Austrian road fatalities and 13% of all road injuries concern the road user group "cyclists" given a bicycle mode share 5% (Transport in Figures [TRA 13] reports). Over the last 10 years, the number of cyclists killed in Austria has been cut almost in half, whereas the number of cycling accidents and thereby injured cyclists hardly declined at all in the same period. In 2011 – which currently is the most recent year for which accident data has been published – 5,745 cyclists were injured and 42 were killed. Considering age groups, the four groups with the highest number of accidents with injuries are children aged between 10 and 14 and three consecutive groups of adults aged between 40 and 54. Fatal accidents however are shifted toward adults older than 50. In 2011, one-third of the fatal accidents fell into the age group of 70–74.

The remainder of this chapter is structured as follows: section 10.2 presents insights into the perceived dangers of urban cycling gained through interviews and polls conducted in 2012 and 2013 for two national research projects. Next, section 10.3 introduces the Austrian accident database and presents an overview of common accident types as well as the development of accident numbers for the years 2002–2011. A comparison of perceived dangers and recorded accident numbers is covered in section 10.4. Finally, section 10.5 presents conclusions and offers an outlook as well as open questions.

10.2. Perceived safety of urban cycling

This chapter presents study results on perceived dangers of urban cycling based on interviews with cyclists and polls conducted in the city of Vienna, Austria. The interviews and polls were conducted as part of the nationally funded projects COEXIST and Com-oVer in 2012 and 2013.

In the project [COE 13], interviews were conducted at three separate locations: at the Vienna State Opera (June 18th 2012), in Lasallestraße (June 20th 2012) and at Westbahnhof (October 1st and 3rd 2012). These locations were selected based on the bicycle counts published by the city of Vienna [RAD 13b]. Cyclists who stopped at the red traffic light next to the counting section were asked to complete a questionnaire in cooperation with the interviewer. While this mode of interviewee selection could lead to biased results due to underrepresentation of cyclists who tend to ignore red lights, this is not an issue at the intersections used in this survey since the high traffic volume forces every participant to obey the traffic lights. The questionnaire covered questions regarding cyclists' behavior as well as their opinions concerning the safety of cycling in the city. Some questions were multiple-choice, others open-ended.

In total, 366 cyclists were interviewed: 164 (45%) identified themselves as female, 198 (54%) as male. Of the cyclists that were interviewed, 78% used their bicycles every day and 98% at least once a week. As expected, interviewees selected for this study tend to use the bicycle much more frequently than the general population (compare Table 10.2). When asked how safe they felt cycling in city traffic, 65% of the cyclists answered that they felt "safe" or "very safe". However, 32% answered that they felt "less safe" and 3% "unsafe". On routes they use regularly, cyclists feel safer with 84% answering that they feel "safe" or "very safe". The interviews did not show the differences in perceived safety between female and male cyclists: 64% of female cyclists and 66% of male cyclists felt "safe" or "very safe". It has to be noted though that these results might not reflect the attitude of the whole population, especially of less experienced cyclists or people who currently don't cycle. Of the 366 interviewed cyclists, 260 also stated the postal code of their home address. An evaluation of statements by geographic home area shows considerable differences in perceived safety (compare Table 10.6).

Additionally, cyclists were asked to classify different situations into "dangerous", "not dangerous" or "no opinion". Amongst the situations considered most dangerous are cycling on roads with a lot of traffic (72%), roads with high traffic speeds (68%), roads with rail infrastructure (68%) and unclear/confusing intersections (67%). Darkness and missing cycling infrastructure at intersections are considered dangerous by 45 and 41%, respectively. Cycling lanes which are painted on the road are considered only slightly more dangerous than combined pedestrian and cycling ways (37 and 35%, respectively). Doorways/garage entrances (32%), legally cycling against the one-way street direction (30%) and cycling on winding/curvy roads (22%) are only of concern to a minority of cyclists. Respondents could also provide additional situations they consider dangerous: 25 cyclists used that possibility. The most commonly mentioned situations are bad road surface conditions (three mentions), suddenly ending cycling infrastructure (three mentions), cycling lanes along parked cars (three mentions) and conflicts with pedestrians (three mentions).

In the [COM 13] survey, cyclists were asked to fill in an online questionnaire. The questionnaire was the second stage of a data collection task which included the collection of bicycle GPS tracks using a dedicated smartphone app. The questionnaire was distributed to all participants of the first data collection stage and was completed by 26 participants, 21 male and five female, who consider themselves "allrounders" (as opposed to "inexperienced or careful cyclists"). The self-categorization as "allrounders" is supported by the fact that 62% of the respondents use their bike nearly every day and 89% at least several times a week. Most of the participants were young adults that also use public transport (46% at least several times a week) but little motorized individual traffic (8% at least several times a week). The averaged subjective fitness level was above average (6.5/10).

Choice of bicycle for passenger transport	Vienna 20007 [TRA 13] (%)	COEXIST (%)	Com-oVer (%)
Daily	3	78	62
Several times a week	10	20	27
Several times a month	13	2	11
Infrequently	17		
Never	57	–	–

Table 10.2. *Population by bicycle usage frequency*

When asked if safety concerns influenced whether they choose to cycle or not, 43% of respondents answered that these concerns would keep them from cycling. Similarly, unattractive routes keep 42% from cycling. In the ranking of all reasons covered by the questionnaire, danger and attractiveness ranked 6th and 7th out of 9, the top three deterrents being icy conditions, rain and long distances.

Preferences for cycling infrastructure representing subjective safety could be stated in five classes from "strong like" (grade 1) over "indifferent" (grade 3) to "strong dislike" (grade 5). Table 10.3 shows the average grade as well as the most commonly selected classes. Separated cycle paths (grade 1.6) and residential areas with a low speed limit of 30 km/h (2.2) got the best grades and are (strongly) liked by a majority of the participants. A majority of around 50% is indifferent to legal cycling in pedestrian zones (2.8) and legal cycling against one-way streets (2.9), with no participant strongly disliking them.

Infrastructure	Average grade (95% confidence interval)	Majority of ...% for	COEXIST mentions
Separated cycle path	1.6 (± 0.35)	65% strong like	–
Residential area (30-km/h zone)	2.2 (± 0.26)	62% like	–
Main road without cycling infrastructure (little traffic)	2.7 (± 0.36)	42% indifferent	–
Main road with cycle lane	2.8 (± 0.40)	31% like	37%
		31% indifferent	
Cycling against one-way street direction	2.8 (± 0.33)	54% indifferent	30%
Pedestrian area	2.9 (± 0.35)	50% indifferent	35%
Roads with tramway (rails)	3.89 (± 0.28)	46% dislike	68%
Main road with cycle lane next to parked cars	4.1 (± 0.26)	50% dislike	(3 mentions in comments)
		31% strong dislike	
Main road without cycling infrastructure (a lot of traffic)	4.34 (± 0.35)	27% dislike	68/72%
		68% strong dislike	

Table 10.3. *Infrastructure preferences of the 26 participants of the Com-oVer questionnaire*

For main roads, several infrastructure variants were considered. Main roads without infrastructure and little traffic got similar but worse grades than residential areas (2.7). Missing infrastructure in combination with a lot of traffic got the worst grade (4.3) and a majority of participants strongly disliking it. The popularity of cycle lanes on main roads heavily depends on the proximity to parked cars and the resulting danger of "being doored". While one-third each likes or is indifferent to cycle lanes on main roads in general (2.8), when parked cars are near half of the participants dislike and another one-third strongly dislikes them (4.1). Similar to the results from COEXIST, the responses showed once more that the main subjective safety concerns are high traffic volumes and roads with rail infrastructure.

Additionally, a more differentiated view on cycle lanes emerges when explicitly asking for the preferences of cycle lanes with and without parked cars. Cycle lanes with parked cars are identified as a major safety concern in that case.

The results on perceived dangers presented in this section represent only a subset of the topics covered in the interviews and polls that were conducted in the projects Com-oVer and COEXIST. It is worth noting that these interviews and polls were designed without prior knowledge of the structure of the Austrian accident database, which is described in the following section before section 10.4 aims to compare different aspects of the perceived dangers of urban cycling to these accident records.

10.3. The Austrian accident database

For this study, we analyzed all records of injury accidents from the national Austrian accident database, where cyclists were involved. The data covers the years 2002 until 2011 since data for 2012 was not available at the time of writing. While studies [ELV 99, LAN 03] recommend the use of hospital data for safety assessments, police data for accidents was used in this study since hospital data is not available for analysis in Austria. While accident insurance providers do collect data from hospitals, current data privacy laws do not allow for a consolidation of accident data from police and hospital sources. For the city of Vienna, the resulting dataset contains a total of 6,287 accidents. The database consists of several tables which describe the accidents. The accident table contains information about the location of the accident, the weather conditions, as well as the date and type of the accident. The participants table contains information about the participants such as age, degree of injury and type of vehicle. Table 10.4 shows the trends of the yearly number of accidents. Figure 10.1 presents a comparison of the trends of accident counts and bicycle counts based on the initial values for 2002. In the analysis time frame between 2002 and 2011, the data show no correlation between trends of accident counts and bicycle counts (R^2 = 0.03). This suggests the validity of the concept of "safety in numbers", which states, that an increase in the modal share of bicycles leads to a decrease in the number of accidents per cycled kilometer.

	2002	2003	2004	2005	2006	2007	2008	2009	2010	2011
Accidents per year	635	663	601	682	660	646	616	573	552	659

Table 10.4. *Trends of bicycle accident counts in Vienna between 2002 and 2011*

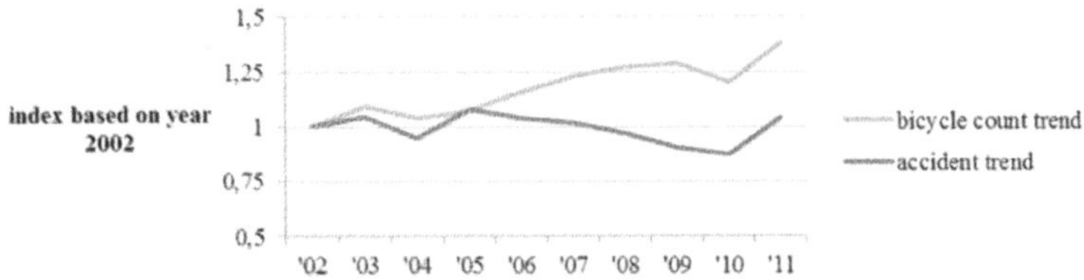

Figure 10.1. *Trends of accident counts compared to bicycle counts based on year 2002. For a color version of the figure, see www.iste.co.uk/jacob/safety.zip*

In the Austrian accident database, it is possible to encode the location of the accident in up to four different ways. For example, by using road type, number and kilometer or road and house number. However, not all four kinds of information are filled out for every accident, which makes the data difficult to analyze. The preferred way in the guidelines for the data collector is using the road number and the road kilometer, if it is possible. Some of the accident records contain geographic coordinates but for the majority of records these are missing. Starting from 2012 there will be changes in the accident data schema, amongst other changes every record collected from 2012 onward will have to contain geographic coordinates.

The accident data also contain location characteristics, e.g. intersection, cycle path, bridge, pedestrian zone, etc. An accident can be described with up to five location characteristics. The most common location characteristics in the analyzed records are: cycle path or lane (2,106 accidents), intersection with yield-sign (2,109 accidents), normal intersection (2,062 accidents), T-shaped intersection (1,594 accidents) and one-way street (1,583 accidents). Note that one accident location can share multiple of these characteristics, for example in the case of a location at a normal intersection with yield signs. Table 10.5 shows the development of accident counts for the most common location characteristics. The graphical representation in Figure 10.2 shows the differences in development of accident counts for these location characteristics. While the number of accidents at intersections is relatively stable, accidents on bicycle routes seem to be on the rise.

Location characteristic	2002	2003	2004	2005	2006	2007	2008	2009	2010	2011
Intersections	363	389	351	395	410	378	379	350	311	402
One-way street	203	224	169	146	115	132	112	156	129	197
Cycle path or lane	184	172	172	223	185	196	223	239	224	288
Street with rail infrastructure	101	98	96	94	89	78	72	95	96	127
Doorway or garage entrance	57	66	70	72	60	57	62	53	49	66
Sidewalks	55	81	63	63	44	59	56	67	61	78

Table 10.5. *Trends of accident counts for the most common location characteristics*

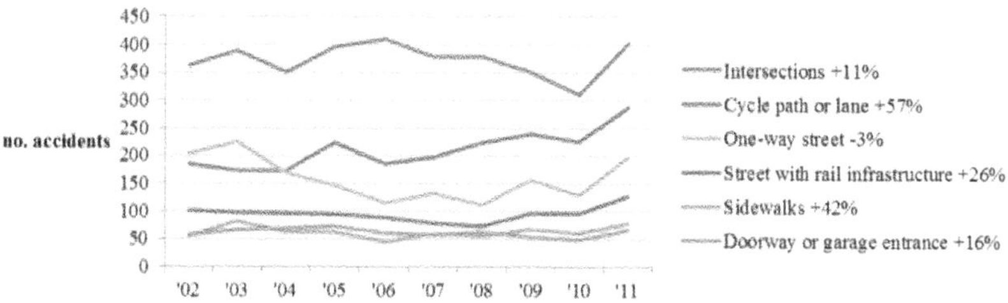

Figure 10.2. *Trends of accident counts for the most common location characteristics. For a color version of the figure, see www.iste.co.uk/jacob/safety.zip*

Besides the accident location characteristics, the records also contain information about how the accident happened, e.g. the type of the collision or ignoring the traffic rules, etc. A participant can have up to three circumstances. The most common circumstances in the analyzed records are: orthogonal collision (1,137 participants), collisions with pedestrians (534 participants), lateral collisions (482 participants), falling from the vehicle (478 participants), ignoring right-of-way (451 participants), merging into flowing traffic (436 participants) and collisions with parking vehicles or objects (359 participants).

Since 2012, the accident database has been improved and extended. In the new version, there is only one valid way go georeference and accident and geographical coordinates have to be included. The cycling and driving direction is noted for accidents in one-way streets, and different types of cycle infrastructure are distinguished, such as cycle path, cycle lane, marked bike crossing. In general, there are more attributes to describe the circumstances of the participant, more attributes to characterize the location of the accident and more means of transports (MicroCar, MicroScooter). The new attribute presumed accident cause will also be helpful.

10.4. Comparison of perceived safety and recorded accidents

This section presents comparisons of perceived safety as collected by the COEXIST and Com-oVer surveys and the accident records in the Austrian accident database for the city of Vienna, Austria. We first look into regional aspects before we go into a more detailed analysis of location characteristics which are perceived as dangerous.

10.4.1. *Regional aspects*

As noted in the previous section, the COEXIST evaluation by geographic home area shows considerable differences in perceived safety. Table 10.6 connects this perceived safety to the recorded number of accidents as well as distance risk (accidents per one million bicycle kilometers). The number of bicycle kilometers traveled by year was derived from the traffic model of the city of Vienna for 2011 [HOL 11]. The model describes the average traffic demand during work days. For weekend days, a demand reduction of 65% due to a lack of work and education-related trips was assumed. The average distance risk in the city of Vienna according this data is 2.95 accidents per one million bicycle kilometers with values ranging from 1.47 in the 12th district to 5.57 in the 17th district. In a related study, [ROB 97] reported a distance risk of 3.7 for whole Austria.

For some districts, the number of interviewees in the COEXIST survey is very low and the 23rd district was not represented at all. A statistical evaluation of the data shows no correlation between distance risk and the perceived safety stated by the districts' inhabitants. Since the interviews were conducted at three different locations, in the 1st, 2nd and 15th district, it is expected that the cyclists' responses to the safety question are not specific to their home districts.

District	Accidents (2002–2011)	bicycle km per year (in thousands)	Distance risk	Cyclists feeling safe or very safe (%)	COEXIST sample size
1	672	14,328	4.69	67	6
2	387	16,449	2.35	75	40
3	370	14,361	2.58	79	19
4	240	8,102	2.96	50	6
5	189	7,133	2.65	77	13
6	267	6,687	3.99	60	15
7	221	8,366	2.64	31	13
8	140	5,050	2.77	78	9
9	302	12,979	2.33	56	9
10	371	16,937	2.19	20	5
11	181	7,696	2.35	50	2
12	178	12,119	1.47	71	17
13	107	5,323	2.01	33	3
14	157	4,999	3.14	70	10
15	227	8,541	2.66	43	14
16	254	8,947	2.84	61	23
17	175	3,143	5.57	55	11
18	107	1,918	3.67	67	9
19	170	5,156	3.30	100	3
20	194	10,461	1.85	50	10
21	537	11,802	4.55	63	8
22	639	14,166	4.51	60	15
23	202	7,571	2.67	–	–
Total	6,287	213,234	2.95	65	

Table 10.6. *Perceived safety and accident counts per home district (source bicycle km: traffic model of the city of Vienna 2011)*

10.4.2. *Location characteristics*

Responses showed that the main safety concerns voiced by cyclists are high traffic volumes and speeds, unclear or confusing intersections, cycling lanes next to parked cars and roads with rail infrastructure. This section aims to compare this perceived risk to the number of recorded accidents.

The accident database does not directly contain information about whether an accident happened on a road with high traffic volume or speed. The available road type field only distinguishes between highways, state roads, highway ramps and one class for all other roads. Furthermore, it is not possible to tell whether an accident happened directly on the road or rather on the separate cycle lane which runs along the road since there is only one class which does not distinguish between separate cycle lanes and cycle lanes which run directly besides or even between lanes of motorized traffic.

Unclear or confusing intersections were considered dangerous by 67% of cyclists in the COEXIST survey. Additionally, 41% stated that they think missing cycling infrastructure at intersections is dangerous. The records show that 3,728 of 6,287 bicycle accidents in Vienna happened at intersections making it the most common accident location characteristic. Notably, the number of accidents at intersections has increased considerably more slowly (+11%) than the observed bicycle counts (+44%). Potential explanations include improved bicycle routes as well as a rising awareness and higher alertness toward the presence of cyclists, an effect known as "safety in numbers".

Similar to intersections, rail infrastructure on streets is considered dangerous by 68% of respondents in the COEXIST survey and 66% of respondents in the Com-oVer survey. During the course of ten years, 946 accidents occurred on roads with rail infrastructure and from 2002 to 2011 the yearly number of accidents increased by 26%. While the absolute number of accidents amounts to only one-fourth of the number of accidents at intersections, the fact that rail infrastructure is rarer and can often be avoided by choosing a different route makes it clear that the perceived danger posed by rails on the street is confirmed by the accident data.

Doorways and garage entrances are considered to be a danger by 32% of respondents in COEXIST survey. In ten years, 612 accidents indicate traffic accidents at doorways and garage entrances. As Figure 10.2 shows, the number of these accidents has remained relatively stable in the analysis period with a rise of 16% which is well below the observed rise in bicycle counts of 44%.

According to the Com-oVer poll, 30% of cyclists think legal cycling against the one-way street direction is dangerous. In ten years, one-way streets accounted for

1,583 accidents. Accident records show a decrease in accident numbers from 2002 to 2006 followed by an increase between 2006 and 2011. Overall, one-way streets accounted for 197 accidents in 2011 down from 203 accidents in 2002. At the same time, the length of bicycle routes which can be used against the one-way street direction has been steadily growing (compare Table 10.7). This seems to indicate that the danger related to legal cycling against the one-way street direction is overrated.

In general, the surveys show that cycling on bicycle infrastructure is considered safe with the exception of cycle lanes next to parked cars but while the number of accidents at intersection has remained relatively stable and even decreased slightly on one-way streets, accidents on cycle paths and lanes are up by 57% from 184 to 288 accidents per year. Only parts of this increase can be attributed to the expansion of the bicycle network which has grown in length by 34% between 2002 and 2011 (compare Table 10.7).

	2002	2003	2004	2005	2006	2007	2008	2009	2010	2011
Total length of the bicycle network	895	919	953	1,009	1,051	1,090	1,135	1,153	1,174	1,197
Length of one-way streets with legal Bicycling in the opposite direction	106	115	131	148	161	175	186	199	209	215

Table 10.7. *Trends of km of bicycle routes in Vienna (source: Radverkehr in Zahlen 2013)*

The danger of cycling next to parked cars was acknowledged by 82% of respondents in the Com-oVer survey and while this situation was not part of the COEXIST questions, it was still mentioned by three out of the 25 participants who used the comments section of the questionnaire. According to the Austrian accident database, 359 accidents (in 10 years) are collisions with parking vehicles or objects. Table 10.8 shows that collisions with parking vehicles or objects have increased by 225% since 2002. Part of this increase might be due to the continued construction of more so-called “Mehrzweckstreifen” in Vienna [RAD 13d]. “Mehrzweckstreifen” are a variety of Austrian cycle lanes, measuring between 1.2 and 1.7 m that lie between lane(s) for motorized traffic and an optional parking lane. Motorized traffic is not allowed to drive on cycle lanes in Austria, with the exception of wide vehicles, i.e. busses and trucks. This type of infrastructure is typically used in Vienna, where the remaining lane(s) for motorized traffic are too narrow for wide vehicles. When more space is available, “Radfahrstreifen” are used. These are similar to “Mehrzweckstreifen”, but may not be used by any motorized traffic.

	2002	2003	2004	2005	2006	2007	2008	2009	2010	2011
Collisions with parking vehicles or objects	24	17	9	21	18	25	30	50	87	78

Table 10.8. *Trends of collisions with parking vehicles or objects*

Traffic-reduced areas are preferred by 74% of respondents in the Com-oVer survey. The accident database contains 328 accidents in traffic-reduced areas. While the absolute numbers are quite low compared to the accident numbers of other location types, the data indicate a rise in accidents in traffic-reduced areas. This rise seems to be at least partially due to an increase in traffic-reduced area assignments in Vienna in the previous years.

	2002	2003	2004	2005	2006	2007	2008	2009	2010	2011
Accidents in traffic-reduced areas	26	27	23	23	22	31	24	42	58	52

Table 10.9. *Trends of accidents in traffic-reduced areas*

A majority of around 50% of Com-oVer respondents is indifferent to legal cycling in pedestrian zones. The accident data in Table 10.10 seem to support this attitude with no value over seven accidents per year.

	2002	2003	2004	2005	2006	2007	2008	2009	2010	2011
Collisions with pedestrians	45	54	38	49	52	56	57	52	66	65
Accidents in pedestrian zones	3	2	1	3	1	2	1	6	7	2
Accidents on sidewalks	55	81	63	63	44	59	56	67	61	78

Table 10.10. *Trends of accidents with pedestrians*

10.5. Conclusion and outlook

Bicycle count data for the city of Vienna shows the rising popularity of this mode of transport. Counts increased by 39% between 2002 and 2011. During the same period, the overall number of recorded bicycle accidents with physical injuries rose by 4%. The most notable increase in recorded accidents regarding the location characteristics (57%) can be found for bicycle routes. It has to be noted, though, that part of this increase has to be attributed to the expansion of the bicycle network which took place during the same timeframe.

Responses to interviews with cyclists and polls conducted in Vienna showed that the main safety concerns are high traffic volumes and speeds, unclear or confusing intersections, cycling lanes next to parked cars and the presence of rail infrastructure on streets used by cyclists. The threat posed by intersections is confirmed by the high number of recorded accidents.

Suggestions for a further extension of the accident database include the distinction of bicycles, pedelecs and vehicles classified as bicycles such as segways, and noting if rails were actually involved in the accident. Another limitation for in-depth analysis is that many fields describing the accident circumstances are optional and never filled in. For example, there are zero accidents with car doors in the database. This type of accident is intermixed with collisions with objects. Future studies should make use of the newly available data from 2012 and explore the impact of traffic volume and speed. Special attention should be given to accidents at junctions. Safe junction designs and properties leading to accidents should be identified and compared to international best practices in infrastructure design.

10.6. Acknowledgments

The authors would like to thank Veronika Friedl and Anna Aytan for their contribution to designing the COEXIST survey. COEXIST is funded by the Verkehrssicherheitsfonds/BMVIT (Federal Ministry for Transport, Innovation and Technology), Call "sicher.elektro.mobil" under project number 199.755. Com-oVer is funded by the BMVIT (Federal Ministry for Transport, Innovation and Technology), 3rd call "ways2go" (April 2010) under project number 828980.

10.7. Bibliography

[COE 13] COEXIST, Verkehrssicherheitsfonds/BMVIT project, information from 2, Midterm Report, AIT, Vienna, Austria, 2013.

[COM 13] COM-OVER, bmvit project, information from final report, AIT, Vienna, Austria, 2013.

[ELV 99] ELVIK R., MYSEN A.B., "Incomplete accident reporting: meta-analysis of studies made in 13 countries", *Transportation Research Record: Journal of the Transportation Research Board*, vol. 1665, no. 1, pp. 133–140, 1999.

[GES 13a] Gesamtvergleich – automatische Raddauerzählungen 2002 bis 2010 (bicycle counts 2002 to 2010), published by the city of Vienna, Austria, Available at: http://www.wien.gv.at/stadtentwicklung/projekte/verkehrsplanung/radwege/erhebungen/gesamtvergleich/index.html, 2013.

[HOL 11] HOLZAPFEL P., RIEDEL R., Traffic model of the city of Vienna, Stadtplanung Wien MA18, Vienna, 2011.

[LAN 03] LANGLEY J.D., DOW N., STEPHENSON S. *et al.*, "Missing cyclists", *Injury Prevention*, vol. 9, no. 4, pp. 376–379, 2003.

[RAD 13a] RADVERKEHR IN ZAHLEN (cycling in figures), published by Federal Ministry for Transport, Innovation and Technology (bmvit), Vienna, Austria, Available at: http://www.bmvit.gv.at/service/publikationen/verkehr/fuss_radverkehr/downloads/riz_2013.pdf, 2013.

[RAD 13b] RADVERKEHRSZÄHLUNGEN (bicycle counts), published by the city of Vienna, Austria, Available at: http://www.wien.gv.at/verkehr/radfahren/zaehlstellen/index.html, 2013.

[RAD 13c] RADVERKEHRSZÄHLUNGEN (bicycle counts), published by Nast consulting, Vienna, Austria, Available at: http://www.nast.at/verkehrsdaten/radverkehrszaehlungen/jahresauswertung, 2013.

[RAD 13d] RADWEGBAUPROGRAMM 2013 (bicycle route construction program), published by the city of Vienna, Austria, Available at: http://www.wien.gv.at/verkehr/radfahren/bauen/programm/, 2013.

[ROA 11] ROAD SAFETY IN AUSTRIA, ANNUAL REPORT 2011, published Federal Ministry for Transport, Innovation and Technology (bmvit), Vienna, Austria, Available at: http:// www. bmvit.gv.at/en/service/publications/downloads/roadsafety_report2011.pdf, 2012.

[ROB 97] ROBATSCH K., KRÄUTLER C., Forschungsarbeiten aus dem Verkehrswesen, Band 71, Radverkehrsunfallanalyse und daraus resultierende Planungsgrundlagen, BMVIT Wien, 1997.

[TRA 13] Transport in Figures, Austria – Edition 2011, published by Federal Ministry for Transport, Innovation and Technology (bmvit), Vienna, Austria, Available at: http://www.bmvit.gv.at/verkehr/gesamtverkehr/statistik/downloads/TIF_11_english_2011_20130627.pdf, 2013.

[VRU 13] VRUITS, DG-MOVE FP7 project, information at the VRUITS First Interest Group Workshop, ERTICO, Brussels, 2013.

PART 4

Road Infrastructure Safety

11

Speed Distribution and Traffic Safety Measures

The main traffic safety priority in many countries is the implementation of measures that reduce road user speed and increase speed limit compliance. This study compares changes in the speed distribution and estimated accident risk brought about by three measures: speed limit change from 110 to 100 km/h on roads without speed cameras, speed limit change from 90 to 80 km/h on roads with speed cameras and introducing new speed cameras on roads with a speed limit of 90 km/h. The accident risk is estimated from mean speed, as the impacts of a change in mean speed on the number of accidents and of injured and killed people are well known. The results indicate that installing speed cameras can reduce average speed more than reducing the speed limit by 10 km/h, at least at the camera locations. Moreover, speed cameras reduce the standard deviation of speed and the percentage of drivers exceeding the speed limit by far more than those that lower the mean speed. This implies that the relationship between the change in the speed distribution, not only the mean speed, and traffic safety merits further investigation.

11.1. Introduction and aim of the study

It is well known that speed is a crucial road safety factor. Many implemented safety measures aim to induce road users to reduce their speed and comply with speed limits. With respect to the change in the mean speed, the impacts on road safety in terms of number of accidents and the number of injured and killed people are well known. For example, this relationship is described by the Power model [ELV 04, NIL 04], which is often used to estimate the traffic safety effects of speed changes. However, it is not certain that only the mean speed is affected by particular traffic safety measures; measures such as the 85th percentile, standard deviation of speed and shape of the speed distribution can also be affected.

Chapter written by Anna VADEBY and Åsa FORSMAN.

This study investigates and compares the changes in speed distribution and estimated accident risk for three traffic safety measures, namely:

– speed limit change from 110 to 100 km/h on rural roads without speed cameras;

– speed limit change from 90 to 80 km/h on rural roads with speed cameras; and

– introducing new speed cameras on rural roads with a speed limit of 90 km/h.

The Power model is applied to aggregate data from a road site, but other models estimate how speed affects accident risk at the individual level [AAR 06]. Several studies in the 1960s [CIR 68, SOL 64] analyzed individual risks in relation to the choice of speed and found a U-shaped relationship between speed and accident risk. More recent studies suggest that the relationship increases fairly monotonically, the slope becoming steeper at higher speeds [KLO 01, MAY 98, QUI 99]. This means that the higher the driving speed, the higher the risk of being involved in an accident. However, overall risk does not increase if one drives below the average speed on the road. Different studies have come up with quite different sizes of the increased risk when driving above the average speed, and there is still considerable uncertainty as to the actual appearance of the individual risk curve.

The different models described by Aarts and Schagen [AAR 06] reflect different views of how speed affects accident risk. In somewhat simplified terms, one approach claims that the speed itself is most important, while the second approach claims that speed variation between vehicles on the same road is most important. The main argument supporting the first approach is that increased speed causes longer reaction times and stopping distances if something unexpected happens, and that increased speed leads to more serious consequences if an accident does occur. One argument supporting the second approach is that large variations in speed between adjacent vehicles give rise to conflicts, which in turn can lead to accidents. Shinar [SHI 98] describes these two approaches in great detail. One aspect highlighted in several studies is that speed variation is probably relevant only to certain types of accidents, while absolute speed affects every accident. Another difficulty when distinguishing between the effects of mean speed and speed variation is that these variables are highly correlated [AND 97, FIN 94, SHI 98], making it difficult to isolate the effects of speed from those of speed variation.

This study seeks a better understanding of the relationship between traffic safety measures and speed distribution, to obtain a better basis for planning and evaluating various road safety measures. Changes in estimated accident risk are calculated using the Power model [ELV 04, NIL 04]. However, our aim is not to estimate the absolute traffic safety effect of these measures, but to illustrate how the speed distribution and various related measures, such as the 85th percentile and standard deviation of speed, can change due to these measures.

11.2. Method

Speed limit changes and speed enforcement are two available measures to reduce the speed of the road users. Here, decreased speed limits on roads with and without speed cameras and the introduction of new speed cameras are examined.

11.2.1. *Data*

To perform the study, speed data from two sources were used: a national evaluation of new speed limits in Sweden and an evaluation of movable speed cameras. The speed of all vehicles was considered in the analyses.

11.2.1.1. *National evaluation of new speed limits*

Between 2008 and 2011, the Swedish Transport Administration reviewed the speed limits on the state rural road network in Sweden. Guidelines were established for different types of roads, and the long-term vision was that speed limits should be adapted to the safety classification of each road. A new set of speed limits (i.e. 80, 100 and 120 km/h) was introduced on rural roads to complement the previously used limits of 70, 90 and 110 km/h. Earlier studies have investigated the effects of new speed limits on the rural road network [VAD 13, VAD 10, VAD 12a]. In the present study, we reuse speed data collected in the national evaluation of new speed limits; these speed data come from both tube measurements and speed cameras.

The aim is to illustrate how the speed distribution and other measures (e.g. mean speed) can be changed on roads where the speed limit is lowered, not to estimate the total effect on the network. Therefore, on roads where the speed limit was decreased from 110 to 100 km/h, the 8 sites with the greatest speed reduction (of a total of 20 sites) were selected. The measurements are from 2 + 1 roads (i.e. continuous three-lane roads with alternating passing lanes and the two directions of travel separated by a flush divider with a median barrier) and standard two-lane rural roads. At each measurement site, the measurements were made both before (August 2008) and one year after (August, 2009) the new speed limit was instituted. To be considered an acceptable measurement, the speed at each road site had to be measured for at least three whole weekdays between noon Monday and noon Friday, and the results presented had to represent daytime conditions, i.e. from 06.00 to 20.00. The speed of passing vehicles was measured using pneumatic tubes stretched across the road. Overall, speeds were registered from approximately 47,000 vehicles in 2008 and approximately 59, 000 vehicles in 2009. For a more detailed description of the data collection, see [VAD 13].

Speed data from speed cameras (radar) come from three sites. The speed limits at these sites were 90 km/h before the speed limit change and 80 km/h after. Speed data were collected both before (August 2008) and after (August 2009) the new speed limit was instituted. In total, speeds from approximately 66,000 vehicles were registered both before and after the new limit.

11.2.1.2. *Evaluation of movable speed cameras*

In 2010, moveable speed cameras were trialed on five road stretches in Sweden. Along each stretch, nine fixed mountings were installed on which cameras could easily be mounted. The idea was to move three cameras around between the nine mountings so that road users could not be sure where the cameras were located. Warning signs were placed a few 100 m before the mountings, regardless of whether there were cameras mounted on them.

This study uses speed data from six sites, all of which had a camera in place for the after measurements. All sites were on rural roads with a speed limit of 90 km/h. Speed data were collected using pneumatic tubes stretched across the road. The speed measurements used here were made before (August/September 2010) and after (September/October 2010) the speed cameras were installed. At each site and on each measurement occasion, the speeds of passing vehicles were registered for one week. In total, the speeds of approximately 115,000 vehicles were registered per occasion (before and after). Only speeds of vehicles travelling in the direction of the camera orientation were considered.

11.2.2. *Measures*

It is unclear in the literature what measure is the most appropriate to use when studying the relationship between speed and road safety, although the mean speed and/or the standard deviation of speed are often calculated. Other dimensions, such as the 85th percentile, variance, coefficient of variation, drivers exceeding the speed limit, and the mean speed of those exceeding the speed limit, are also proposed in the literature [AAR 06]. To characterize how the speed of all vehicles varies, one can study the speed distribution, which indicates the proportions of vehicles driving at certain speeds. Besides the speed distribution, the measures listed in Table 11.1 are considered here.

Measure	Description
Mean speed	Arithmetic mean speed
Standard deviation of speed	
Coefficient of variation	Ratio between mean speed and standard deviation
Mean speed of those obeying speed limit	
Mean speed of those exceeding speed limit	
Percentage exceeding speed limit	
Percentage exceeding speed limit by 6 km/h or more	
Percentage exceeding speed limit by 30 km/h or more	
P85	Speed exceeded by 15% of drivers
P15	Speed exceeded by 85% of drivers

Table 11.1. *Description of measures*

11.2.3. *The Power model*

The changes in accident risks are calculated using the Power model [ELV 04, NIL 04], according to which the relative speed changes affect the number of accidents or injured people, y_A, as follows:

$$\frac{y_A}{y_B} = \left(\frac{v_A}{v_B}\right)^d$$

where y_B is the number of accidents or injured people before the speed change, v_B and v_A are the mean speeds before and after the speed change and the power d depends on the outcome of interest. The powers are set according to [ELV 04] and are $d = 4.5$ for the number of fatal, $d = 3.0$ for the number of severe and $d = 1.5$ for the number of minor injuries.

11.3. Results

11.3.1. *Speed distribution*

Figures 11.1, 11.2 and 11.3 show the speed distributions before and after the following three traffic safety measures:

– speed limit change from 110 to 100 km/h on rural roads without speed cameras;

– speed limit change from 90 to 80 km/h on rural roads with speed cameras;

– introducing new speed cameras on rural roads with a speed limit of 90 km/h.

The figures cover all vehicles passing all measurement sites for each safety measure. Note that these examples are intended to illustrate simply how the speed distributions change for different safety measures and should not be viewed as illustrating the general effects of the measures. Figure 11.1 shows that as the speed limit was lowered from 110 km/h (red curve) to 100 km/h (blue curve), the entire speed distribution shifted to the left, indicating a general speed decrease. The biggest change occurred in the mid-speed range, while the changes were smaller at speeds below 90 km/h. The percentage of vehicles traveling below the speed limit declined from approximately 70% to approximately 50%.

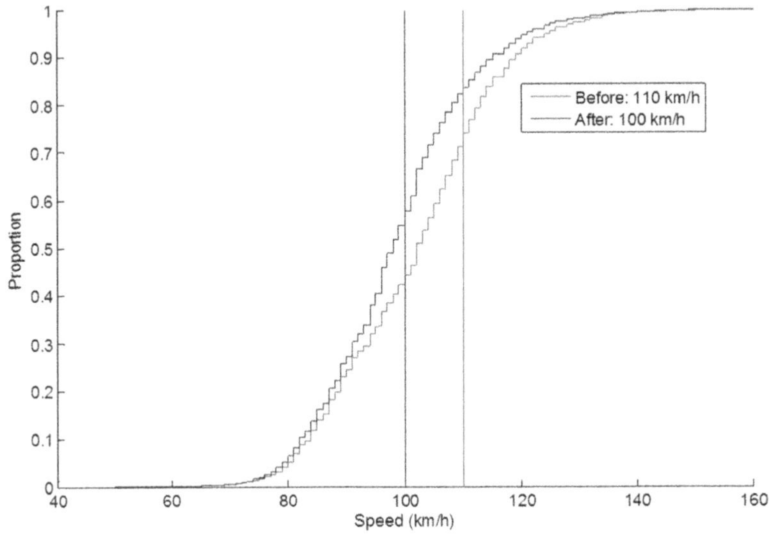

Figure 11.1. *Speed limit change from 110 to 100 km/h on rural roads without speed cameras. Speed distribution for all vehicles before and after new speed limit 100 km/h. For a color version of the figure, see www.iste.co.uk/jacob/safety.zip*

Figure 11.2 shows how the speed distribution changed when the speed limit was lowered from 90 to 80 km/h on roads with speed cameras. Compared with Figure 11.1, the speed distribution is more upright both before and after the speed limit was lowered and the speed limit compliance is much better. Approximately 95% of the road users complied with the speed limit before the change and approximately 85% afterwards. One should also note the relatively large shift to the left even at speeds under the new speed limit of 80 km/h, possibly because speedometers tend to overestimate the speed.

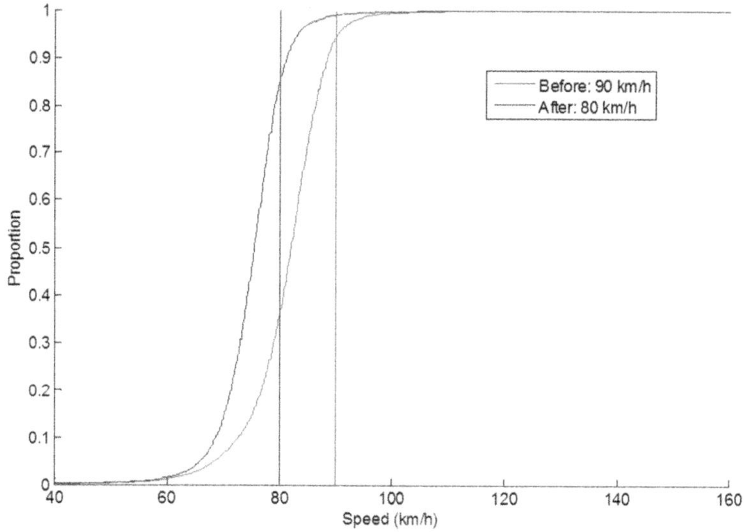

Figure 11.2. *Speed limit change from 90 to 80 km/h on rural roads with speed cameras. Speed distribution for all vehicles before and after new speed limit of 80 km/h. For a color version of the figure, see www.iste.co.uk/jacob/safety.zip*

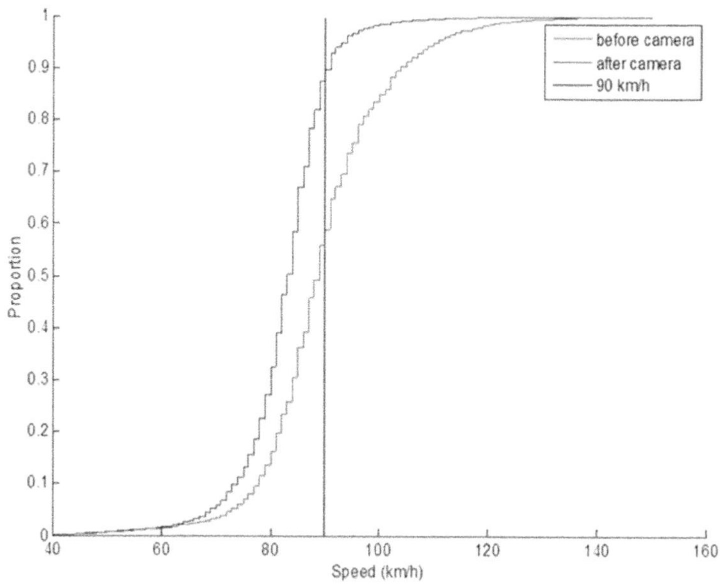

Figure 11.3. *Introducing new speed cameras on roads with a speed limit of 90 km/h. Speed distribution for all cars before and after new cameras. For a color version of the figure, see www.iste.co.uk/jacob/safety.zip*

Figure 11.3 illustrates what can happen to the speed distribution when speed cameras are introduced on a rural road with a speed limit of 90 km/h. This distribution displays a larger displacement to the left (i.e. toward lower speeds) at high speeds, meaning that those driving at the highest speeds reduced their speed the most. As in Figure 11.2, one can also see that this distribution moves to the left even for speeds that, before the change, were already within the new speed limit. The figure indicates that approximately 90% of the drivers drove within the speed limit after the introduction of speed cameras, while before it was approximately 60%.

11.3.2. *Speed measures*

In addition to the speed distributions, the levels of various measures before and after camera installation or speed limit changes as well as the absolute and relative changes in the measures are shown in Tables 11.2–11.4. Table 11.2 shows that the mean speed decreased by more than 3 km/h when the speed limit was lowered from 110 to 100 km/h, representing a decrease of approximately 3%. The change in the standard deviation of speed was slightly greater, approximately 7%; therefore, the coefficient of variation has decreased. The mean speed decreased slightly more at speeds above the current speed limit than at those under the current limit. This is also reflected in the percentiles, where the speed reduction is greater for the 85th percentile (P85) than for the 15th percentile (P15). This result was expected, because those who were already driving below the new speed limit had no incentive to reduce their speed further. The percentage of drivers exceeding the speed limit increased from 0.26 to 0.42, while the percentage of those exceeding the limit by more than 6 km/h increased from 0.14 to 0.24. Looking at the relative change, the increases in these measures are of the same magnitude. Few people drive more than 30 km/h above the speed limit; therefore, these changes should be carefully interpreted.

The results for the speed limit change from 90 to 80 km/h on roads with speed cameras are shown in Table 11.3. The speed levels were adjusted by a standard increment of 3 km/h (due to the angular displacement of the radar measuring equipment) so that the levels should be roughly comparable to those from the tube measurements used in other cases. The results indicate that the mean speed decreases from approximately 81 to 75 km/h. Note that the mean speed before the change was already close to the new speed limit of 80 km/h. The percentage of drivers exceeding the speed limit increased by 9 percentage points. Observing the change in speed levels in the tails of the distribution, namely, the 15th and 85th percentiles, note that P85 decreased by approximately 7 km/h, slightly more than the decrease in mean speed. P15 decreased slightly less than the mean speed, by approximately 5 km/h.

Measure	Before	After	Change: after–before	Relative change (%)
Mean speed (km/h)	101.7	98.6	–3.1	–3%
Standard deviation of speed (km/h)	14.1	13.1	–1.0	–7%
Coefficient of variation	0.14	0.13	–0.01	–4%
Mean speed of those obeying speed limits (km/h)	95.6	89.9	–5.7	–6%
Mean speed of those exceeding speed limits (km/h)	119.1	110.4	–8.7	–7%
Percentage exceeding speed limits	26%	42%	16% points	62%
Percentage exceeding speed limits by 6 km/h or more	14%	24%	10% points	69%
Percentage exceeding speed limits by 30 km/h or more	1%	2%	1% points	185%
P85 (km/h)	115.54	111.83	–3.71	–3%
P15 (km/h)	86.84	85.55	–1.28	–1%

Table 11.2. *Change in various measures when the speed limit was decreased from 110 to 100 km/h on rural roads without speed cameras*

Measure	Before	After	Change: after–before	Relative change (%)
Mean speed (km/h)	81.4	75.3	–6.1	–7%
Standard deviation of speed (km/h)	6.9	5.7	–1.1	–17%
Coefficient of variation	0.08	0.08	–0.01	–10%
Mean speed of those obeying speed limits (km/h)	80.6	73.9	–6.7	–8%
Mean speed of those exceeding speed limits (km/h)	94.5	83.7	–10.8	–11%
Percentage exceeding speed limits	6%	15%	9% points	166%
Percentage exceeding speed limits by 6 km/h or more	1%	2%	1% points	110%*
Percentage exceeding speed limits by 30 km/h or more	0.08%	0.14%	0.06% points	69%*
P85 (km/h)	87.3	80.0	–7.2	–8%
P15 (km/h)	75.4	70.4	–5.0	–7%

*small numbers – interpreted with care

Table 11.3. *Change in various measures when the speed limit was decreased from 90 to 80 km/h on rural roads with speed cameras*

In Table 11.4, we can see that the mean speed of all vehicles passing the sites equipped with new speed cameras declined by nearly 7 km/h or approximately 8%. In comparison, we see that P85 decreased even more by nearly 11.5 km/h or 11%. This can be compared with the national evaluation of speed cameras [SWE 09], according to which P85 fell by approximately 10%. P15 fell by nearly 4 km/h or approximately 5%. All studied speed violation measures declined sharply. The percentage of drivers exceeding the speed limit declined from over 40% before the speed cameras were installed to 10% after, a 75% decrease that agrees well with the results of the [SWE 09] national evaluation. The percentage exceeding the speed limit by 6 km/h or more decreased by 21 percentage points. As noted earlier in relation to both the national evaluation and the speed distribution shown in Figure 11.3, those who drove at the highest speeds were affected the most by the installation of speed cameras.

Measure	Before	After	Change: after–before	Relative change (%)
Mean speed (km/h)	89.5	82.6	–6.9	–8%
Standard deviation of speed (km/h)	12.0	8.3	–3.7	–31%
Coefficient of variation	0.13	0.10	–0.03	–25%
Mean speed of those obeying speed limits (km/h)	82.3	81.1	–1.2	–2%
Mean speed of those exceeding speed limits (km/h)	99.9	95.7	–4.2	–4%
Percentage exceeding speed limits	41%	10%	–31% points	–75%
Percentage exceeding speed limits by 6 km/h or more	24%	3%	–21% points	–86%
Percentage exceeding speed limits by 30 km/h or more	0.5%	0.05%	–0.5% points	–91%
P85 (km/h)	101.1	89.6	–11.5	–11%
P15 (km/h)	80.5	76.7	–3.8	–5%

Table 11.4. *Change in various measures when new speed cameras were introduced on roads with a speed limit of 90 km/h*

For all measures presented above, the same tendency concerning P85 and P15 can be observed, i.e. the relative and absolute changes in P85 are greater than the comparable changes in P15. This effect is much greater when new speed cameras are introduced (Table 11.4) than when only the speed limit is lowered (Tables 11.2 and 11.3). When lowering the speed limit on a road with speed cameras, drivers seem to adapt well to the new speed limit.

11.3.3. *Relative risks*

Table 11.5 shows the estimated change in the number of fatalities, severe and minor injuries. The changes in risk are calculated using the Power model [ELV 04], and the powers used are $d = 4.5$ for the number of fatal, $d = 3.0$ for the number of severe and $d = 1.5$ for the number of minor injuries.

Severity	Decreased speed limit from 110 to 100 km/h	Decreased speed limit from 90 to 80 km/h, with existing cameras	Speed limit of 90 km/h with new cameras
Minor injury	–4.6%	–10.9%	–11.4%
Severe injury	–8.9%	–20.7%	–21.4%
Fatality	–13.1%	–29.4%	–30.4%

Table 11.5. *Change in risk of injured of differing severities with three traffic safety measures: speed limit change from 110 to 100 km/h on roads without speed cameras, speed limit change from 90 to 80 km/h on roads with speed cameras and introduction of new speed cameras on roads with a speed limit of 90 km/h. Risk changes calculated using the Power model*

Results indicate that the largest decreases in injury risk can be achieved when new speed cameras are introduced, due to the larger decrease in mean speed (see Table 11.4; compared to Tables 11.2 and 11.3). When new speed cameras were introduced on roads with a speed limit of 90 km/h, the mean speed decreased by almost 7 km/h or 8%, and the number of fatalities decreased by an estimated 30.4% and the number of severely injured by 21.4%. On roads where the speed limit decreased from 110 to 100 km/h, the mean speed decreased by approximately 3 km/h or 3%, and the number of fatalities decreased by an estimated 13.1% and the number of severely injured by 8.9%.

11.4. Discussion

This study sought a better understanding of the relationship between traffic safety measures and speed distribution, to obtain a better basis for planning and evaluating various road safety measures. The study compares changes in the speed distribution and estimated accident risk for three different measures: speed limit change from 110 to 100 km/h on roads without speed cameras, speed limit change from 90 to 80 km/h on roads with speed cameras and introducing new speed cameras on roads with a speed limit of 90 km/h. The results indicate that introducing new speed cameras resulted in a larger reduction of average speed than decreasing the speed limit by 10 km/h, at least at the camera sites. When new speed cameras were introduced on roads with a speed limit of 90 km/h, the mean speed decreased by almost 7 km/h or 8%. On roads with existing cameras where the speed limit changed from 90 to 80 km/h, the mean speed decreased by approximately 6 km/h (7%), while on roads where the speed limit decreased from 110 km/h to 100 km/h, the mean speed decreased by only approximately 3 km/h or 3%, despite the fact that we selected the measurement sites with the largest decrease. The results of the national evaluation of the new speed limits [VAD 13] indicate that, in general, when reducing the speed limit by 10 km/h on 2 + 1 roads and rural roads with a speed limit of 110 km/h, the mean car speed decreases by 2 km/h. This chapter also demonstrates that speed cameras reduce the standard deviation of speed and the percentage of drivers exceeding the speed limit by far more than those that reduce the mean speed. The speed distribution indicates a larger displacement to the left at high speeds, meaning that those who drive at the highest speeds reduce their speed the most. These results are in line with recent findings concerning average speed enforcement. Soole *et al.* [SOO 13], who reviewed the effect of average speed enforcement, concluded that "the approach has been noted as a particularly effective countermeasure in reducing excessive speeding behavior".

The results also indicate that, due to the accompanying speed reductions, the estimated change in injury risk is considerably greater for the introduction of new speed cameras than for reducing speed limits. However, this estimated risk reduction is based on the Power model, which considers only changes in mean speed. The true difference in risk reduction is possibly even greater, as speed cameras also change the speed distribution, causing those who drive at the highest speeds to reduce their speed the most. No currently established model can take this into account. The individual risk models reviewed by Aarts and Schagen [AAR 06] were developed from different types of data, and each produces quite different results. It can also be noted that in [ELV 13], a re-parameterization of the Power model was done. It is beyond the scope of this study to compare our results to this new model. However, as only relative risk changes are studied and the models are rather similar, the assessment is that this does not affect the results and conclusions of our study to any significant extent.

One objection related to speed variation is that the measures are often generic, such as variance or percentage of drivers exceeding the speed limit. Shinar [SHI 98] notes that most studies estimate the variance across days, which means that a large variance may reflect speed variations due to differences in speed between high and low traffic volume periods. To avoid this problem, alternative measures are suggested by Lu and Chen [LU 09], and Munden [MUN 67] that reflect the speed variation between adjacent vehicles. Vadeby and Forsman [VAD 12b] compare the traditional standard deviation of speed calculated over the entire time period (several days), the standard deviation calculated on an hourly basis, i.e. average speed difference (ASD) [LU 09], and the measure suggested by Munden. The data used for the comparison are the same as used in the present study, i.e. collected on roads where the speed limit decreased from 110 to 100 km/h. Vadeby and Forsman's [VAD 12b] results indicate the difficulty of drawing any general conclusions from the comparison. However, for the roads studied, there were only minor differences between an overall standard deviation of speed and a standard deviation calculated on an hourly basis, although there may be differences in the overall standard deviation and the variation between adjacent vehicles.

In conclusion, the results of this and earlier studies indicate a need for further research, examining both speed variation measures and the relationship between speed distribution change and traffic safety.

11.5. Acknowledgments

The authors are grateful to the Swedish Transport Administration for funding the original study.

11.6. Bibliography

[AAR 06] AARTS L., VAN SCHAGEN I., "Driving speed and the risk of road crashes: A review", *Accident Analysis and Prevention*, vol. 38, pp. 215–224, 2006.

[AND 97] ANDERSSON G., NILSSON G., *Speed Limits, Speeds and Safety*, VTI, Linköping Sweden, 1997.

[CIR 68] CIRILLO J.A., "Interstate system accident research study II, interim report II", *Public Roads*, vol. 35, no. 3, pp. 71–76, 1968.

[ELV 04] ELVIK R., CHRISTENSEN P., AMUNDSEN A.H., Speed and road accidents: An evaluation of the power model, TØI report no. 740/2004, Oslo: TØI, 2004.

[ELV 13] ELVIK R., "A re-parameterisation of the Power Model of the relationship between the speed of traffic and the number of accidents and accident victims", *Accident Analysis and Prevention*, vol. 50, pp. 854–860, 2013.

[FIN 94] FINCH D.J., KOMPFNER P., LOOCKWOOD C.R. *et al.*, Speed, speed limits and accidents, Project report no. 58, Transport Research Laboratory TRL, Crowthorne, Berkshire, 1994.

[KLO 01] KLOEDEN C.N., PONTE G., MCLEAN A.J., Travelling speed and the risk of crash involvement on rural roads, Report no. CR 204, Road Accident Research Unit, Adelaide University, Adelaide, Australia, 2001.

[LU 09] LU F., CHEN X., "Analysing the speed dispersion influence on traffic safety", *2009 Internal Conference on Measuring Technology and Mechatronics Automation*, 2009.

[MAY 98] MAYCOCK G., BROCKLEBANK P.J., HALL R.D., Road layout design standards and driver behaviour, TRL Report no. 332, Transport Research Laboratory TRL, Crowthorne, Berkshire, 1998.

[MUN 67] MUNDEN J.M., The relation between a driver's speed and his accident rate, RRL Report LR 88, Road Research Laboratory, Crowthorne, Berkshire, 1967.

[NIL 04] NILSSON G., Traffic safety dimensions and the power model to describe the effect of speed on safety, Bulletin 221, Lund, Sweden: Lund Institute of Technology, Department of Technology and Society, Traffic Engineering, 2004.

[QUI 99] QUIMBY A., MAYCOCK G., PALMER C. *et al.*, The factors that influence a driver's choice of speed: A questionnaire study, TRL report no. 325, Transport Research Laboratory, Crowthorne, UK, 1999.

[SHI 98] SHINAR D., "Speed and crashes: A controversial topic and an elusive relationship", *Appendix B in Managing speed: Review of current practice for setting and enforcing speed limits*, Special report no. 254, Transportation Research Board, Washington, DC, 1998.

[SOL 64] SOLOMON D., Crashes on main rural highways related to speed, driver and vehicle, Bureau of Public Roads, U.S. Department of Commerce, United States Government Printing Office, Washington DC, 1964.

[SOO 13] SOOLE D.W., WATSON B.C., FLEITER J.J., "Effects of average speed enforcement on speed compliance and crashes: A review of the literature", *Accident Analysis and Prevention*, vol. 54, pp. 46–56, 2013.

[SWE 09] SWEDISH ROAD ADMINISTRATION, The effects of automated road safety cameras on speed and traffic safety, Publication no. 2009: 162, Borlänge, Swedish Road Administration, Sweden, 2009.

[VAD 10] VADEBY A., FORSMAN Å., Utvärdering av nya hastighetsgränssystemet: Effekter på hastigheter, etapp 1 [Evaluation of the new speed limits: the effect on vehicle speeds, Phase 1], VTI notat 14-2010, VTI, Linköping, Sweden, in Swedish, summary in English, 2010.

[VAD 12a] VADEBY A., FORSMAN Å., Utvärdering av nya hastighetsgränssystemet: Effekter på hastighet, etapp 2 [Evaluation of the new speed limits: the effect on vehicle speeds, Phase 2], VTI notat 16-2012. VTI, Linköping, Sweden, in Swedish, summary in English, 2012.

[VAD 12b] VADEBY A., FORSMAN Å., Hastighetsspridning och trafiksäkerhet [Traffic safety and speed distribution], VTI report no. 746–2012, Linköping, Sweden: VTI, in Swedish, summary in English, 2012.

[VAD 13] VADEBY A., FORSMAN Å., "Speed management in Sweden: Evaluation of a new speed limit system", *Proceedings of the 16th International Conference on Road Safety on Four Continents*, Beijing, China, 15–17 May, 2013.

12

Ex-ante Assessment of a Speed Limit Reducing Operation – A Data-driven Approach

Reducing speed limits often leads to a decrease in the injury and fatality accident counts. This depends on the type of accident, compliance in the speed limits and traffic conditions, before and after the implementation of the speed limit. In France, the speed limit is 130 and 110 km/h on interurban and urban motorways. The safety assessment of a speed limit reducing operation before its implementation is the subject of this chapter. Such an assessment was required before the decision to reduce speed limits in a part of an urban motorway network in the north of France. The safety assessment process consists of four steps: first, accident analysis for the prior period, in order to count the speed- or density-related accidents; second, analysis of the average speed and density before the introduction of the speed limit reduction; third, predicting new traffic conditions using fundamental diagrams established from other sites where the new speed limit has already been experimented; fourth, using available relationships linking the accident count to changes in speed and density, estimating the expected changes in accidents. These relationships, calibrated on another urban motorway network, located in the South of France, are given in the appendix. The results predict a decrease in the accident count.

12.1. Introduction

Driving at an inappropriate speed contributes to the occurrence of road accidents. A number of authors have used traffic conditions and accident databases to analyze the relationship between accidents and speed [GOL 04, ABD 05], accidents and volume/capacity ratio [LOR 05] and accidents and traffic states [YEO 13]. The relationship between speed limits and accidents has been analyzed from national

---

Chapter written by Maurice ARON, Régine SEIDOWSKY and Simon COHEN.

accident databases before and after the setting of new speed limits [NIL 04, ELV 09, ELV 13, PAU 13], who suggested and used the Power model, relating the after/before ratio of accident counts to the speed limit ratio.

The safety impact of a speed limit reduction (prior to its implementation) is assessed in this paper. The safety assessment process consists of four steps that are introduced in section 12.2. Such an assessment was required before making the decision to reduce the speed limits in parts of the ALLEGRO urban motorway network in the North of France – the site, traffic and accident data are presented in section 12.3. The method requires some relationships between accidents and traffic conditions: their calibration is given in section 12.4, based on data coming from another motorway set, the Marius network in the South of France – more details are in the annex. Section 12.5 gives the results for the ALLEGRO network; section 12.6 outlines the validation of this approach and the links with a future ex-post assessment which has not yet been carried out. Concluding remarks are given at the end.

12.2. Method for predicting the injury or fatality accident count

Predicting begins by identifying what things might change, formulating and calibrating a model for the changes. The hypotheses should be validated or accepted in the community. The prediction process consists of the following four steps.

12.2.1. *Accident analysis*

Golob *et al.* [GOL 04] have shown strong links between accident counts (by type of accident) and traffic conditions. The first step of our method consists of simplifying his approach by classifying the accidents into just three types and reducing their links to traffic conditions to two dimensions (speed and density):

Type 1: some accidents are mainly due to the driver suffering from fatigue or falling asleep, to alcohol or to a vehicle failure. Let us consider here that the number of such accidents is independent of any speed limit.

Type 2: other accidents are linked to a high speed (run off the road and rollovers); they occur in a low traffic flow and concern a single vehicle. Let us consider that changes in speed affect only those single vehicle accidents.

Type 3: the remaining accidents concern several vehicles (side collisions, rear-end collisions and multi-vehicle collisions) and are linked to a high relative speed, a high traffic density or a lane change. Let us consider that these accidents, concerning two or more vehicles, are affected only by the changes in traffic density.

12.2.2. *Empirical speed analysis*

Empirical speeds are analyzed in two components:

– fundamental diagrams (FDs), which link speed to traffic flow; it consists in two branches, whether the traffic is not congested (index $s = 1$) or congested ($s = 2$). FDs vary according to the speed limits;

– a deviation round the FD, because the FD is valid only on average and not at every single moment. These deviations are different at each 6-minute period. Let us consider that they do not depend on the speed limit.

12.2.3. *Traffic condition prediction*

This results from the compliance in new speed limits. Our task is to predict what will be and what should be.

12.2.3.1. *Predicting what will be*

Let $V^{FD}(q)$ be the FD giving the speed corresponding to the traffic flow q. The empirical speed is divided here into two components, the first given by the FD and the second being the deviation from the FD.

For every 6-minute period i, let q_i, v_i, and k_i be the empirical traffic flow, speed and density, respectively:

$$v_i = v^{FD,s}_{speed\_limit}(q_i) + \left(v_i - v^{FD,s}_{speed\_limit}(q_i)\right) \; s \text{ being the congestion index (1 or 2)} \quad [12.1]$$

The relationship $q_i = k_i v_i$ implies that an average 6-minute speed decrease leads to an increase in density k_i. For the non-congested branch of the FD, provided the new density (after the speed limit implementation) remains lower than the critical density, the traffic demand q_i goes through. If not, the period passes into congestion, with a queue that needs to be managed. However, in the simulation study carried out on the same subject Cohen [COH 14] has shown that the increase in congestion was negligible. In addition, capacity was virtually maintained (see the FDs in Figure 12.4), except for the slow lane, when the new speed limit is 90 km/h. When the demand is greater than the supply, the FD-congested branch is not impacted by the reduction of the speed limit (Figure 12.4). Let us therefore assume that after the implementation of the new speed limit, the traffic flow and the traffic state remain constant for a 6-minute period. The new speed is obtained by adding to the empirical one the change in FD (see Figure 12.1).

$$v^{FD,s}_{new\_speed\_limit}(q_i) - v^{FD,s}_{old\_speed\_limit}(q_i) \qquad [12.2]$$

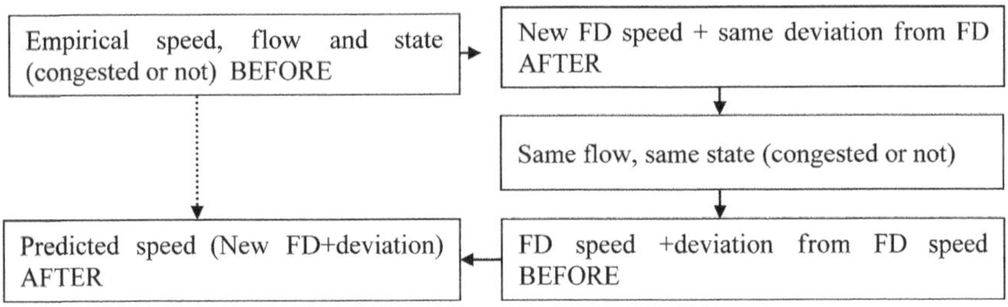

Figure 12.1. *Traffic conditions analysis and prediction*

Thus, the traffic density is deduced from the equation *traffic flow = speed density*. The variability of the speed and density is conserved, as empirical 6-minute deviations from the FD are taken up again.

12.2.3.2. *Predicting what should be*

In section 12.2.3.1, the new empirical FD takes into account actual compliance with the new speed limit where already implemented. Computing the ideal result of the speed limit reduction (with full compliance) requires another model, where the empirical speed (when higher than the new speed limit) is replaced by the limit (or by a function of the limit, in the case of partial compliance). Compliance is said to be partial either when the rate τ of compliant drivers or periods is <100% or when the speed decrease is only ρ $(V_{old\_lim}-V_{new\_lim})$ $(0 < \rho < 1)$. The empirical speed v, when greater than the new speed limit, is changed to v' or v'':

$v' = (1-\tau)\,v + \tau\,.\,V_{new\_lim},\ 0\% \leq \tau \leq 100\%$ – see results in Tables 12.3–12.6 for τ = 100 or 50% [12.3]

$v'' = Max\{V_{new\_lim},\ v - \rho\,(V_{old\_lim} - V_{new\_lim})\}$ see results in Tables 12.3–12.6 for ρ =50% [12.4]

12.2.4. *Calibration of the average speed–accident and density–accident relationships*

As accidents often result from just one person driving at an inappropriate speed, the link between the *average* speed of a set of drivers and accidents is more tenuous, which might jeopardize the approach. As the initial lane of the accident is not available in the accident database, the traffic–accident relationships have been calibrated independently by lane, as if each accident was independently related to the traffic conditions on each lane. The distributions of the 6-minute average speed and occupancy have been established for the full year.

12.2.4.1. *Relationships between single vehicle accident risk per vehicle kilometer and speed*

The speed Power model is calibrated from type 2 accidents and speed data of another motorway network, the Marius network (see the annex).

Let *nb_acc*(v) be the injury accident count at speed v, and let v_b be the "before" speed; the Power model gives:

$$nb\_acc(v)/nb\_acc(v_b) = (v/v_b)^{\alpha} \text{ or } nb\_acc(v)=\beta v^{\alpha} \text{ with } \beta = nb\_acc(v_b)/(v_b)^{\alpha} \quad [12.5]$$

The risk per vehicle kilometer appears when dividing *nb_acc*(v) by the number of vehicle kilometers:

$$Risk(v) = nb\_acc(v) / nb\_veh\_km(v) = \beta \cdot v^{\alpha} / nb\_veh\_km(v) = \beta' \cdot v^{\alpha} \quad [12.6]$$

with $$\beta' = nb\_acc(v_b) / (v_b^{\alpha}.nb\_veh\_km(v)) \quad [12.7]$$

The (theoretical) number of accidents *Nb_acc* is the sum, over the speed classes, of the risks per vehicle kilometer, multiplied by the number of vehicle kilometers of the class *nb_veh_km*(v); so β' is solution of:

$$Nb\_acc = \sum_v nb\_acc(v) = \sum_v nb\_veh\_km(v) \cdot \beta' \cdot v^{\alpha} = \beta' \cdot \sum_v nb\_veh\_km(v) \cdot v^{\alpha} \quad [12.8]$$

Nilsson [NIL 04] determined that $\alpha = 2$. R. Elvik [ELV 09] gave the following values for α: for all injury accidents [1.6], for slight injury accidents [1.1], for serious injury accidents [2.6] and for fatality accidents [4.1]. The following table gives the α value obtained, with the SAS software, by Aït-Belkacem [AIT 12], on the Marius network, another French urban motorway network. Other estimations, taking into account the impact of rain, are in the Appendix (Tables 12.7 and 12.8) for the power and the exponential models or in Aron *et al.* [ARO 15].

Exponent α and (5% confidence interval) of the power model	Lane close to the hard shoulder	Middle lane	Median Lane (fast lane)
Speed (Daylight, no rain, based on 45 single vehicle accidents)	0.2 (-1.7; 2.7)	0.7 (-1.5; 2.9)	6.0 (2.4; 9.7)
(Night, no rain, based on 17 single vehicle accidents)	4.1 (0.6; 8.8)	6.1 (-5.1; 17.3)	6.1 (2.1; 10.0)

Table 12.1. *Calibration of the exponents of the Power model – Marius network, excluding rainy periods and accidents*

As sometimes the time recorded for an accident is in reality the time of the police arrival, the 6-minute average speed which has been considered in Aït-Belkacem [AIT 12] or in Aron [ARO 15], is generally the one recorded 18 minutes before the accident, at the nearest upstream or downstream traffic station.

The value $\alpha = 2.3$, the average of the 3 α-values by lane, is used here. This relatively high value was not unexpected because accidents involving several vehicles were discarded from the calibration set: this tends to sharpen the speed–accident relationship. Note that Elvik [ELV 13], following Hauer [HAU 09], found an even better fit with an exponential model, the number of accidents being proportional to $e^{0.034V}$.

12.2.4.2. *Relationships between accident risk per vehicle kilometer and traffic density or sensor occupancy*

Traffic stations provide occupancy and not density so relationships will be defined, calibrated and used between accident and occupancy. There is a gross relationship between occupancy and density: assuming a constant vehicle length of 5 meters and a constant speed of individual vehicles on the section – the density is the occupancy multiplied by 200. When predicting the new traffic conditions on the ALLEGRO network, we remarked that while the speed limit reduction affects the proportion of high-density vehicle kilometers (a density is said "high" when greater than the threshold $K_h = 24$ vehicles per lane), it does not much affect the traffic density distribution (conditionally to be in the low/high class). This simplifies the calculation. Let us note:

$nb\_veh\_km(k)$ the number of vehicle kilometers for density class k;

$Nb\_Veh\_Km(K_h)$ (with a capital N) the cumulated number of vehicle kilometers for $k > K_h$;

$Risk(k)$ the risk per vehicle kilometer for density k;

f_k the conditional density (conditionally to $k > K_h$): $f_k = nb\_veh\_km(k)/Nb\_Veh\_Km(K_h)$.

$$Nb\_Acc(K_h) = \Sigma_{k|k>Kh}\ nb\_acc(k) = \Sigma_{k|k>Kh}\ nb\_veh\_km(k).Risk(k) = Nb\_Veh\_Km(K_h).\ \Sigma_{k|k>Kh}\ f_k.Risk(k) \quad [12.9]$$

$\Sigma_{k|k>Kh} f_k.Risk(k)$ remains constant, equal to $Nb\_acc/Nb\_Veh\_Km(K_h)$, provided theconditional density f_k does not change. Thus, the number of high-density accidents is proportional to the number of high-density vehicle kilometers and to the risk by high density, which was determined, thanks to historical data. Symmetrically, the number of low-density accidents is proportional to the number of low-density vehicle kilometers and to the risk by low density. The predicted accident count *Nb_acc* is obtained through the scheme given in Figure 12.2.

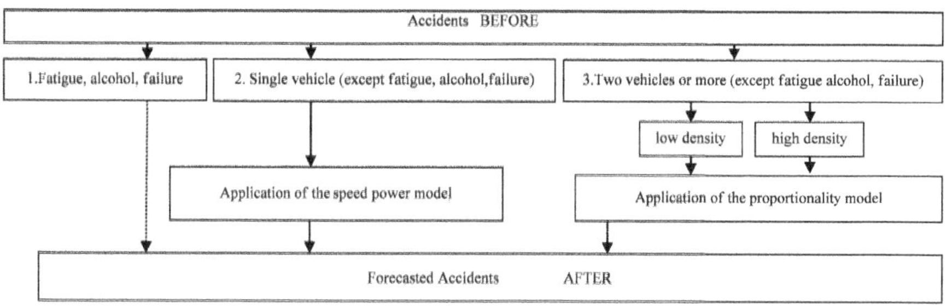

Figure 12.2. *Predicting accidents by type*

12.3. The part of the ALLEGRO motorway network concerned with speed limit reduction

The speed limit will decrease from 130 to 110 km/h in three motorway areas: A (A25 motorway), C (A23) and D (A27); it will decrease from 110 to 90 km/h in areas E (A25), F (A1), G (weaving section A22–A27–A23) and I (A22); additional sections PC1 and PC2 (A1), B (national road 41) and H (national road 356) are not considered here.

In total, 345 injury accidents, described in the French accident database BAAC (*Bulletins d'Analyse des Accidents Corporels*), occurred during the period of 2003–2010 on this 23 km long part of the ALLEGRO Network. Traffic data from 15 stations were available only during the period of 2009–2010, covering lane flow, speed and occupancy every 6 minute. However, it was not possible to link an accident to the traffic conditions before, because very few traffic stations were available for this research.

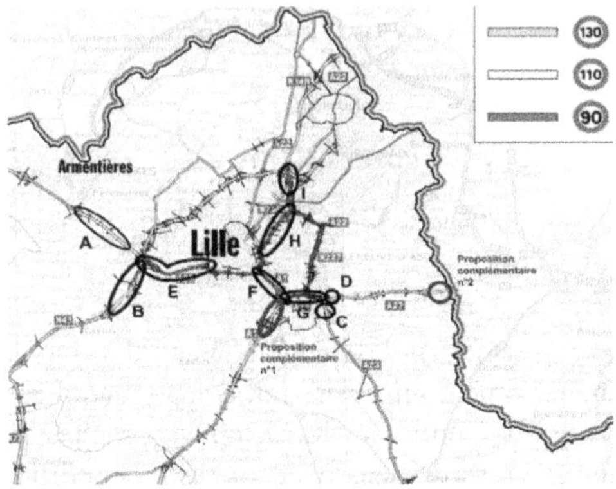

Figure 12.3. *Speed limit reduction scheme near Lille. For a color version of the figure, see www.iste.co.uk/jacob/safety.zip*

12.4. *Ex-ante* assessment results of the speed decrease in the ALLEGRO motorway network

12.4.1. *FDs on the ALLEGRO network*

One way to obtain the relationship between speed and density is to classify the traffic flow, then to divide each class into two subclasses (one congested and the other not) and to assign a subclass to each 6-minute period according to its traffic flow. The FD speed is the average of speeds by subclass. Congestion is selected when the speed is lower (or density greater) than the critical speed (or density). These critical speeds and density can be:

– derived from the calibration of an analytic curve which globally models the FD;

– estimated from the data subset corresponding to the maximum observed flow or;

– the values which correspond to the minimum of the sum of square deviations between the empirical data and the FD resulting from these critical values.

Here this method has been used, while forcing the same critical value on the three lanes (the value 35 vehicles/lane per kilometer has been obtained).

Unlike usual traffic simulation, no global form of a speed–density relationship is required. The main constraint – a decrease in speed when density increases – is not always satisfied with empirical data, and it is necessary to process monotonically the obtained average speeds by class. We have developed a heuristic which does so.

On some parts of the ALLEGRO network, the 90, 110 and 130 km/h speed limits are already used, which make it possible to build 3 FDs (see Figure 12.4).

Overestimation of the critical speed induces overestimation of the number of congested periods and the speed levels of both branches of the FD. This is because the average speed for a given flow class corresponds to the weighted average of the mean non-congested and congested speeds, the weights being the proportions of vehicles traveling during non-congested and congested periods. Overestimation of the critical speed implies that some non-congested periods, with a rather low speed, are wrongly classed in congestion; thus, the non-congested average speed increases, because the lowest speed values are not integrated in this average; the congested average speed increases too, because of the addition of these wrongly classified speeds, which are rather high for the congested case. Of course, these two overestimations do not lead to any overestimation of the "average of the two averages", which does not vary, because of the increase in the weight of the lowest average.

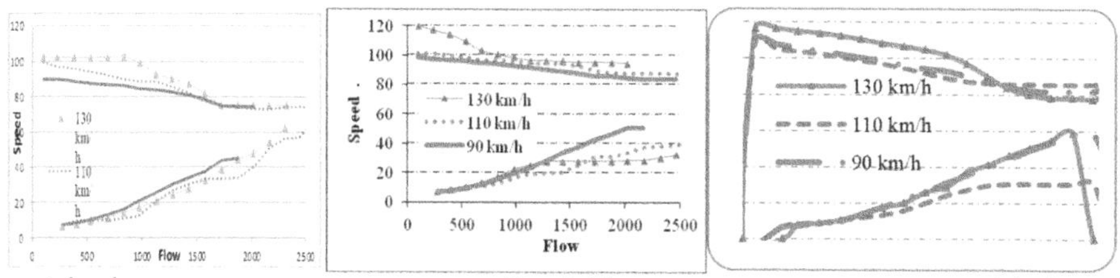

a) slow lane b) central lane c) fast lane

Figure 12.4. *FD in the ALLEGRO motorway network according to the speed limit for daylight, no rain periods. For a color version of the figure, see www.iste.co.uk/jacob/safety.zip*

In the middle and median lanes, capacity is the same whatever the speed limit; in the slow lane, capacity is the same at 110 and 130 km/h, but decreases at 90 km/h. As the sections used for determining the FDs were different, the FD modification cannot be completely attributed to the change in speed limits, and part of it is probably due to local geometric or traffic characteristics which are not independent of the speed limit decisions.

12.4.2. *Splitting the accidents into three types and predictions*

On the Marius network where detailed traffic data exist, a density can be assigned by accident. This was not the case for the ALLEGRO Network, where traffic data are more rare; however, that available traffic data are assumed to be representative of the whole (unknown) set of data; 70% of type 3 accidents concerning at least two vehicles have been assigned in a low-density period and 30% in high-density periods.

Accidents, areas A to I	Observed before	Predicted by FDs	Compliance $\tau = 100\%$	Compliance rate $\tau =$ 50%	$\rho = 50\%$ (-10 km/h)
2003–2010 Type 1	46.0	46.0	46.0	46.0	46.0
2003–2010 Type 2 (due to speed)	99.0	85.9	88.5	93.5	90.8
2003–2010 Type 3 | density <24 (*)	140.0	137.8	139.8	139.9	139.8
2003–2010 Type 3 | density >24 (**)	60.0	68.7	60.9	60.3	60.7

(*) 70% of type 3 accidents (**) 30% of type 3 accidents.

Table 12.2. *Observed accidents and predicted accidents by accident type according to speed limit compliance assumptions*

Global predictions speed, density and accidents	Observed	FDs	$\tau = 100\%$	Rate $\tau =$ 50%	$\rho = 50\%$
Speed (km/h)	93.60	88.04	89.33	91.47	90.30
Decrease for the 99 type 2 accidents	0.00	-13.10	-10.50	-5.50	-8.20
Millions vehicle kilometer |density <24 by year	652.30	641.90	651.20	651.90	651.40
Millions vehicle kilometer |density >24 by year	71.50	82.00	72.60	71.90	72.40
% vehicle kilometer| density >24	9.88%	11.32%	10.03%	9.94%	10.00%
% increase in high density vehicle kilometer	0.00%	14.60%	1.50%	0.60%	1.20%
2003–2010 accidents (total)	345.00	338.40	335.20	339.70	337.30

Table 12.3. *Observed and predicted speed and density according to speed limit compliance assumptions*

Empirical speeds are rather low compared with the actual speed limits, but the predicted FDs are lower even by medium density; their use leads to a big decrease in speed, bringing high densities instead of medium densities and thus a high increase in high-density type 3 accidents. Conversely, when the new speed is a function of the speed limit and compliance (in the last three columns), the decrease in speed (and in the number of type 2 accidents) is lower and effective only at high speeds and low densities; the passage from medium to high density is more rare, and the increase in high-density type 3 accidents is very low.

As traffic data covers two years and accident data covers eight years, risks per vehicle kilometer are computed with the assumption that the traffic during the period of 2003–2008 was the same as during the period of 2009–2010.

Risks per 100 million vehicle kilometers	Observed before	Predicted by FDs	Compliance $\tau = 100\%$	Compliance rate $\tau = 50\%$	$\rho = 50\%$ (-10 km/h)
Type 1 risk	0.80	0.80	0.80	0.80	0.80
Type 2 risk	1.71	1.48	1.53	1.62	1.57
Type 3 risk	3..45	3.56	3.46	3.45	3.46
Type 3 risk | density <24/lane per kilometer	2.68	2.68	2.68	2.68	2.68
Type 3 risk | density ≥24 vehicle/lane per kilometer	10.49	10.47	10.48	10.48	10.48
Total risk	5.96	5.84	5.80	5.87	5.83

Table 17.4. *Empirical and predicted risks according to speed limit compliance assumptions*

Type 3 risk for high density is four times greater than that for low density. This result is compared with the work of Golob et al. [GOL 04] who estimated the accident rates (property damage/injury/fatality) per million vehicle miles for eight traffic flow regimes of the FD:

– on the non-congested branch of the FD, the accident rate passes from 1.28 per million vehicle *miles* (light flow) to 1.49 (mixed free flow), then 1.03 (heavy, variable free flow) and lastly 0.55 (approaching capacity);

– on the congested branch of the FD, the accident rate passes from 1.24 (heavy flow at moderate speed) to 2.97, then 3.21 and lastly 5.99 (heavily congested flow).

A direct comparison with [GOL 04] results should take into account differences in units and the fact that his accident database also includes property damage. Albeit this, what is important is the range of the accident rates according to the traffic flow regime, which is very high in both researches.

The accident count decreases from 345 to 338.4 (−1.9%). In the case of full compliance, it would reach 10.5 accidents (3% of the accidents). This decrease might be underestimated, due to conservative approximations:

– use of average speeds, which decrease the link between speed and number of accidents;

– the factor $(V/V_b)^{2,3}$ has been applied here on the average speed and an "average lane" over the two years 2009–2010, and not for each 6-minute period (weighted per vehicle-kilometer) and lane. Given the convexity of the function $(x)^{2,3}$, the application of this function on the average is greater than the average of the applications on every point; thus, $(V/V_b)^{2,3}$ is overestimated (although remaining less than one), leading to an overestimated speed and an underestimation of the decrease in accidents;

– no decrease in the risk of accidents concerning several vehicles at low density, although relative speed should decrease; no decrease in type 1 accidents (due to fatigue), although it is easier to control a slower vehicle;

– the night speed exponent (between 4.1 and 6.1, see Table 12.1) was discarded because it was insignificant and because type 1 accidents were not excluded from the calibration set.

– the rain exponent was not taken into account; the predicted decrease in accidents should even be higher.

Three problems have not been considered here, due to a lack of knowledge for modeling them:

– a possible change in lane assignment and thus in density in certain lanes;

– an increase in motorcycles between the lanes;

– inside a lane, an increase in relative speed if some drivers comply and others not.

Results are highly sensitive to the speed model. This becomes more evident when the two parts of the network are analyzed independently. Variations in speed (and therefore in the number of accidents) are quite different, as shown in Tables 12.5 and 12.6 for the two groups of areas; for motorways in areas A, C and D, which pass from an interurban to an urban status and where the speed limit decreases from 130 to 110 km/h; and for motorways in urban areas E, F, G and I, where the speed limit decreases from 110 to 90 km/h.

130 → 110 km/h	Observed before	Predicted by FDs	Compliance $\tau = 100\%$	Compliance rate $\tau = 50\%$	$\rho = 50\%$ (-10 km/h)
Average speed	102.4	89.7	101.4	101.9	101.4
Type 2 accidents	44.0	32.5	42.9	43.4	43.0
Millions vehicle kilometer, density <24 by year	235.3	228.4	235.3	235.3	235.3
Millions vehicle kilometer, density >24 by year	11.8	18.7	11.8	11.8	11.8
Type 3 accidents, low density	30.8	29.9	30.8	30.8	30.8
Type 3 accidents, high density	13.2	20.9	13.2	13.2	13.2
Type 2 + Type 3 accident decrease		-4.7	-1.1	-0.6	-1.0

Table 12.5. *Observed accidents and predicted accidents, interurban part*

110 → 90 km/h	Observed before	Predicted by FDs	Compliance $\tau = 100\%$	Compliance rate $\tau = 50\%$	$\rho = 50\%$ (-10 km/h)
Average speed	89.1	87.2	83.1	86.1	84.5
Type-2 accidents	55.0	52.4	46.1	50.4	48.2
Millions vehicle kilometer, density <24 by year	417.0	413.5	416.0	416.6	416.2
Millions vehicle kilometer, density >24 by year	59.7	63.3	60.8	60.2	60.6
Type 3, accidents low density	109.2	108.3	108.9	109.1	109.0
Type 3, accidents high density	46.8	49.6	47.6	47.1	47.5
Type 2 + Type 3 accident decrease		-0.7	-8.4	-4.4	-6.3

Table 12.6. *Observed accidents and predicted accidents, urban part*

Because of the non-linearity of the power model, the decrease in type 2 accidents is not "additive", i.e. equal to the sum of decreases obtained when the Power model is applied independently on the two parts (urban and interurban) of the network. A better application would have been to apply the Power model by speed class.

Variations in the number of type 3 accidents on the two parts of the network are not additive either. Let:

– N_u, N_i, $N = N_u + N_i$ be the vehicle kilometers traveled in the urban or interurban parts or the whole network;

– n_u, n_i, $n = n_u + n_i$ be the vehicle kilometers passing from low to high density due to the new speed limit;

– δ_u, δ_i and δ be the differences between the high- and the low-density risks per vehicle kilometer.

The variations in the number of type 3 accidents are respectively $n_u.\delta_u$, $n_i.\delta_i$ and $n.\delta$. However, $(n_u + n_i)\ \delta \neq n_u.\delta_u + n_i.\delta_i$, as δ is an average between δ_i and δ_u, weighted by N_u and N_i and not by n_u and n_i (Tables 12.5 and 12.6).

With the FD model, most decreases in the accident count are obtained in the interurban part (Table 12.5), where speed decreases from 102.4 to 89.7 km/h; in the urban part, when the speed limit passes from 110 to 90 km/h (Table 12.6), the speed and accident decreases are very low. It is quite the opposite with the full compliance model (speed decreases from 89.1 to 83.1 km/h on the urban part. There are two possible explanations: first, low compliance with the 90 km/h speed limit in France, which limits the decrease in accidents (Table 12.6); second, an underestimation of the speed in the FD established for the 110 km/h speed limit would imply an overestimation of the decrease in accidents (Table 12.5) when passing from 130 to 110 km/h and an underestimation of the decrease in accidents when passing from 110 to 90 km/h (Table 12.6). Indeed, the speed in the central lane, in the 110 km/h FD (Figure 12.4(b)), is always less than 100 km/h, which means that following this FD leads to a lower speed than compliance in the 110 km/h speed limit. In the fast lane (Figure 12.4(c)), the FD speed is always less than 110 km/h. The FD speed was not really underestimated, but using for the interurban part of the network, an FD coming from the urban part of the network, where the network configuration (with more ramps) and driver behavior are different, is questionable. Furthermore, is it correct to say that, regarding the urban part of the network, a speed lower than 100 or 110 km/h has incited traffic operators to decrease the speed limit?

12.5. The threefold validation of the approach

The approach is based on traffic models and models linking traffic conditions to accidents. Some assumptions intervene either in the models, in their calibration, or in their use. Validation will concern these three levels.

12.5.1. *Validation of the models*

Two traffic models are used to predict the traffic conditions and two models linking accidents to traffic:

– the first traffic model (prediction of what will be), based on FDs, does not manage a change in the queues from one period to the next. Changes in lane assignment are also ignored. If, when considering the ex-post traffic conditions, these assumptions are not validated, the use of a simulation model is required;

– the second one (prediction of what should be) replaces the actual speed (when higher than the new speed limit) by a function of this limit, based on the compliance rate. This must be applied on individual (and not average) speeds. The impact of aggregation will be computed in a network where individual speeds are available (like the Marius network) in order to validate or invalidate this model;

– the form of the relationship between speed and accident (the power model) is commonly accepted by the scientific community, although confidence in the Exponential model is progressing;

– the accident/density relationship is simplified, thanks to the assumption that there will be no change in the density distribution, conditional to density <24 vehicles/lane (respectively density >24 vehicles/lane). If this is not validated, the calibration and use of another model is required.

12.5.2. *Validation of the calibrations*

New calibrations of the relationships between traffic conditions and accidents (for instance with the ALLEGRO traffic and accident database, on the before period) will validate/invalidate the relationships used:

1) For the first accident type, validating "no relationship with traffic condition" requires, first of all, the calibration of a model such as a logistic model linking the risk of accident to speed and/or density. In the second step, the part of the accident variance explained by this model has to be analyzed. The logistic model will be withdrawn (the assumption that there is no relationship for type 1 accidents is validated.) if this part of the variance is low (10 or 20% of the variance, say) or if the parameters α obtained are close to zero.

2) For the second accident type, a new calibration α of the exponent of the speed power model is necessary.

3) For the third accident type, a new calibration of threshold separating low and high densities is necessary.

Let traffic operators define, by type of accident, a maximum bias (positive or negative) on the predicted accident count. It is possible to convert these biases into positive or negative deviations round parameters α of the relationships linking accident counts with the traffic conditions and, thus, to define an "acceptable" interval round α; the relationships used will be validated if α falls within this interval.

12.5.3. *Verification/validation of the use of the models*

– It is not recommended, in a speed limit reduction framework, to use traffic conditions/accident relationships that are calibrated without any specific speed limit decrease. But road safety measures cannot wait.

– Using the 110 km/h FD which has been calibrated on an urban network for an interurban network is questionable; this urban FD should be compared with the ex-post interurban one.

12.6. Conclusions

A speed limit decrease generally induces a decrease in the road injury and fatality accident count. This chapter presents a method for the *ex-ante* assessment of such a decrease and its application on a French motorway network. Some results on the calibration and the use of relationships linking traffic density or speed to accident counts are given. The envisaged decrease in the speed limit by 20 km/h should lead to a 1.9% decrease in the accident count with the current compliance and to a 3% decrease with full compliance. These figures are a trade-off between a null trend for certain accidents (due to driver or vehicle failure), a decrease in accidents linked to speed and an increase in accidents linked to a high density. However, this prediction is very sensitive to the traffic and safety models, to their calibration and to the conditions of their use.

Breaking down the prediction into the urban and interurban parts of the network shows that most decreases are located on the interurban part of the motorway network (4.7 accidents) where the speed limit decreases from 130 to 110 km/h and the average speed decreases by 13.7 km/h. On the urban part, where the speed limit decreases from 110 to 90 km/h, the predicted speed and accident decreases are much lower (0.7 accidents).

12.7. Appendix: relationships between injury accidents and traffic conditions estimated from the Marius network

12.7.1. *The Marius network of urban motorways near Marseille*

This 75 km long network (by direction) consists of the urban parts of motorways A7, A50, A51 and A55 (in red on the figure) around Marseille. The sections have either two lanes per direction (the lane closed to the hard shoulder, the so-called "slow" lane and the median lane) or three lanes (with a middle lane); 104 traffic stations per direction (one station every 750 meters) are available on the main carriageway and on the ramps. In total, 1.5 billion vehicle kilometers were traveled between June 2009 and May 2010, 5.3% vehicles travel during rain and 15% during nighttime, which is defined between 8 pm and 6 am.

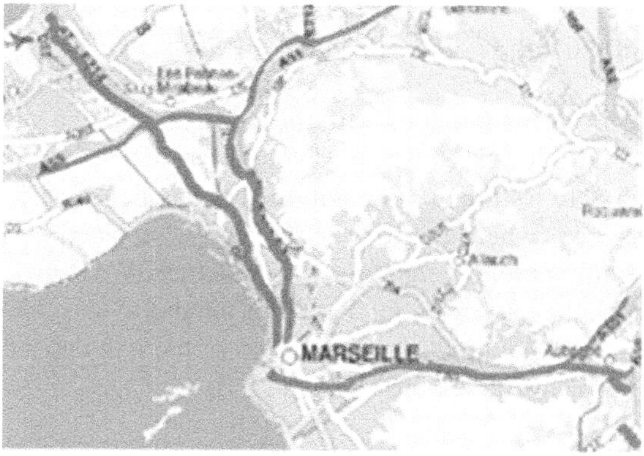

Figure 12.5. *The Marius urban motorway network near Marseille (France). For a color version of the figure, see www.iste.co.uk/jacob/safety.zip*

Three sources of data were used:

– traffic data (time of passage of each vehicle in 100ths of a second, individual speed and length, lane);

– the French accident data file, containing times of accidents, their types, locations and weather condition. In 1 year, 292 injury accidents or fatalities occurred on the network. Traffic data were available for 256 of them;

– accidents due to the driver (fatigue and alcohol) or to a vehicle breakdown are excluded from the calibration;

– weather data, provided by "Météo-France", which recorded 6-minute rain gauge information and hourly human information at a meteorological station located at the Marignane airport, North-West from Marseille.

12.7.2. *The Power and exponential models, logit form*

The Power and exponential models are used here for linking the accident risk to any single variable – speed, occupancy, time headway. We call "POWER $_{\text{Rain}}$" a Power model where a second explanatory variable – the rain occurrence – is introduced; the parameters are estimated on its logit form, where this model reads:

$$POWER_{Rain}: Log[R_i/(1\text{-}R_i)] = \beta + \alpha.Log(V_i) + \gamma.Rain_i + \varepsilon_i \qquad [12.10]$$

where R_i is the accident risk (by vehicle kilometer), V_i is the 6-minute explanatory variable, $Rain_i$ (0 or 1) indicates the rain occurrence and ε_i is the deviation – deviations are assumed to be independent and Gaussian-distributed.

The exponential model, logit form, can be written as follows:

$$EXPORain: Log[Ri/(1\text{-}Ri)] = \beta + \alpha.Vi + \gamma.Rain\ i + \varepsilon \qquad [12.11]$$

When the risk is not monotonous with the traffic variable, a supplementary term (here parabolic) is introduced. However, due to an insufficient number of accidents, it was not possible to estimate four parameters; models are then estimated on datasets excluding rain, where the rain occurrence is replaced by the square of the traffic variable; in that case, the name of the model (POWER or EXPO) is indexed by "NoRain" and by the abbreviation "Para" for parabolic; γ is the coefficient of the square of the indicator; the direction of variation of the risk depends on whether the traffic variable V_i is below/above the value $-\beta/(2.\gamma)$.

$$(POWER_{NoRainPara}): \mathrm{Log}\left(R_i/\left(1-R_i\right)\right) = \beta + \alpha \cdot \mathrm{Log}(V_i) + \gamma \cdot V_i^2 + \varepsilon_i \qquad [12.12]$$

$$(EXPO_{NoRainPara}): \mathrm{Log}\left(R_i/\left(1-R_i\right)\right) = \beta + \alpha.V_i + \gamma \cdot V_i^2 + \varepsilon_i \qquad [12.13]$$

The main relationships useful for assessing the speed limit reduction are presented in the following. The processing is slightly different from Aït-Belkacem [AIT 12], on the same sets of data. A more complete presentation of the relationships can be found in Aron *et al.* [ARO 15]; they include the impacts, on the accident count, of indicators based on a short time headway and on the relative speed.

12.7.3. *Values of the coefficients of significant relationships (single vehicle accidents, daytime)*

Accidents are often due to the behavior of a single driver (speeding), but this does not systematically appear at an aggregated 6-minute level. Moreover, in some contexts (rain and curves), drivers have to decrease their speed; when they do it insufficiently, speeds are lower than on average, but there are more accidents. This makes a tenuous link between average speed and accidents. Despite these limitations, some results have been obtained. A significant relationship between single vehicle accidents and average speed was found for the median lane (only). The Power and exponential models (with a rain coefficient) respectively explain 38 and 39% of the deviance. The $Power_{Rain}$ model exponent is 2.66; the risk increases during rain (γ =1.69); this value would lead to a risk multiplied by $e^{1.69}$ =5.2; however, this is somewhat misleading, because of a concomitant decrease of high speeds during rain. The Power model exponent passes to 7.86 when excluding accidents implying at least one powered two wheeler; the exponent passes to 6.61 when excluding rain conditions and related accidents.

Variable	Lane	Model	β $\sigma(\beta)$	$z(\beta)$	α $\sigma(\alpha)$	$Z(\alpha)$	γ $\sigma(\gamma)$	n	Deviance Null	residual	AIC
Speed	Median	$POWER_{Rain}$	-29.4 (6.1)	0.0%	2.66 (1.32)	4.4%	1.69 (0.00)	42	59.9	37.1	131.5
Speed	Median	$POWER_{Rain}$*	-54.0 (10.3)	0.0%	7.86 (2.21)	0.0%	1.88 (0.00)	33	56.6	27.2	102.4
Speed	Median	$EXPO_{Rain}$	-20.2 (1.4)	<	0.03 (0.01)	2.5%	1.70 (0.00)	42	59.9	36.3	130.7
Speed	Median	$EXPO_{Rain}$*	-25.1 (2.2)	<	0.07 (0.02)	0.0%	1.88 (0.00)	33	56.6	27.3	102.5
Speed	Median	$POWER_{NoRain}$*	-48.2 (11.8)	0.0%	6.61 (2.51)	0.9%		24	26.9	19.8	74.4
Speed	Median	$EXPO_{NoRain}$	-19.6 (1.6)	<	0.02 (0.02)	10.2%		32	28.9	25.8	97.3
Speed	Median	$EXPO_{NoRain}$*	-23.9 (2.5)	<	0.06 (0.02)	0.8%		24	26.9	19.9	74.5
Speed	Middle	$EXPO_{NoRain}$	-12.7 (1.4)	<	-0.03 (0.01)	3.9%		32	19.2	15.0	83.6
Time headway	Slow	$POWER_{Rain}$	-18.6 (0.35)	<	1.15 (0.22)	0.0%	1.25 (0.35)	46	86.2	51.1	154.4

*excluding Power Two Wheelers;
The number of freedom degrees, linked to the number of accidents processed, is indicated in column "n".

Table 12.7. *Significant logistic regressions for daytime, single vehicle accidents, obtained with the software R*

A good correlation does not mean causality: the risk of a single vehicle accident increases with the 6-minute average time headway (see model POWERRain, Table 12.7, last line), but a high average time headway does not cause the risk: in fact, a high time headway is correlated with a high speed, which explains single vehicle accidents.

12.7.4. *Values of the coefficients of significant relationships (multiple vehicles crashes, daytime)*

Whatever the lane, crashes between vehicles increase with occupancy: the coefficients α of the exponential or Power models are positive – see Table 12.8, lines 1 to 4. The percentage of explained deviance is low, but increases up to 40% by the addition of a parabolic term (Table 12.8, line 5); the negative coefficient of the square occupancy means that a very high occupancy decreases injury crashes – maybe because the speed is very low.

Variable	Lane	Model	β $\sigma(\beta)$	$z(\beta)$	α $\sigma(\alpha)$	$Z(\alpha)$	γ $\sigma(\gamma)$	N	Deviance Null	Deviance residual	AIC
Occupancy	Slow	$EXPO5_{Rain\ ramp}$(**)	-16.1 (0.1)	<	14.42 (2.26)	0.0%	0.48 (0.09)	99	144.1	115.9	351.4
Occupancy	Slow	$POWER_{NoRain}$	-13.1 (0.4)	<	1.18 (0.16)	0.0%		83	128.	81.6	281.5
Occupancy	Middle	$POWER_{NoRain}$	-12.8 (0.4)	<	1.32 (0.18)	0.0%		80	142.	91.2	277.8
Occupancy	Median	$EXPO_{NoRain}$	-16.6 (0.2)	<	6.97 (1.32)	0.0%		79	97.2	75 .0	256.4
Occupancy	Slow	$EXPO_{NoRainRampPara}$	-17.5 (0.3)	<	20.03 (4.17)	0.0%	-31.34(0.)	85	126.	75.4	282.9
Time headway	Middle	*EXPORain**	-15.8 (0.2)	<	-0.12 (0.07)	6.5%	0.52 (0.11)	80	63.3	56.5	245.7

*excluding powered two wheelers;

**on the slow lane, accidents related to the access ramp are either included or not (whether the index "Ramp" is mentioned or not).

Table 12.8. *Significant logistic regressions for daytime, crashes between vehicles, obtained with the software R*

The negative correlation between the risk of a crash between vehicles and the average time headway (last line of Table 12.8) might be a causal effect (a too small time headway is risky), or just a (negative) correlation of time headway with the causal variable (occupancy), or a mechanical effect on the accident type: when the time headway is high, accidents involve a single vehicle rather than several.

12.7.5. *Acknowledgments*

We thank the French traffic operators DIR Nord and CETE Méditerranée, who kindly gave us access to their traffic data. Thanks also to our students Y. Saas and I. Ait-Belkacem, who helped us to do the work.

12.8. Bibliography

[ABD 05] ABDEL-ATY M., PANDE A., "Identifying crash propensity using specific traffic speed conditions", *Journal of Safety Research*, vol. 36, no. 1, pp. 97–108, 2005.

[AIT 12] AIT-BELKALCEM I., Evaluation a priori des mesures de modération des vitesses sur le réseau autoroutier de l'agglomération lilloise, Master's dissertation, University Paris-Ouest/IFSTTAR, 2012.

[ARO 13] ARON M., SEIDOWSKY R., DITCHI N., "Traffic indicators and accidents: case of highway network in the South of France TRB", *92nd Annual Meeting*, January 2013 Washington D.C. poster # 13-4638, 2013.

[ARO 15] ARON M., BILLOT R., EL-FAOUZI N.-E. *et al.*, "Traffic indicators, accidents and rain: some relationships calibrated on a French urban motorway network", *18th EWGT*, Delft Transport Research Procedia.

[COH 14] COHEN S., AMIOTTE H., PERUCHON J.E., "Evaluation a priori de l'impact des modifications des vitesses limites sur le réseau autoroutier de l'agglomération lilloise", *Congrès international de l'ATEC-ITS France*, Paris, 2014.

[ELV 13] ELVIK R., "Are-parameterisation of the Power Model of the relationship between the speed of traffic and the number of accidents and accident victims", *Accident Analysis & Prevention*, vol. 50, pp.854–860, 2013.

[GOL 04] GOLOB T.F., RECKER W.W., ALVAREZ V.M., "Freeway safety as a function of the traffic flow", *Accident Analysis & Prevention*, vol. 36, pp. 933–946, 2004.

[HAU 09] HAUER E., "Speed and Safety", *Transportation Research Record*, vol. 2103, pp. 10–17, 2009.

[LOR 05] LORD D., MANAR A., VIZZIOL A., "Modeling crash-flow-density and crash-flow-V/C ratio relationships for rural and urban freeway segments", *Accident Analysis & Prevention*, vol. 37, pp. 185–199, 2005.

[NIL 04] NILSSON G., Traffic safety dimensions and the Power Model to describe the effect of speed on safety, PhD Thesis, Lund University of Technology, 2004.

[PAU 13] PAUW E., DANIELS S., THIERIE M. *et al.*, "Safety effects of reducing the speed limit from 90 km/h to 70 km/h", *Accident Analysis & Prevention*, in press, 2013.

[YEO 13] YEO H., JANG K., SKABARDONIS A. *et al.*, "Impact of traffic states on freeway crash involvement rates", *Accident Analysis & Prevention,* vol. 50, January, pp. 713–723, 2013.

13

Development of a Guideline for the Selection of Vehicle Restraint Systems – Identification of the Key Selection Parameters

"Selection of Appropriate Vehicle Restraint Systems" (SAVERS) is a project funded within the CEDR 2012 Transnational Road Research Programme "Safety". The first work package of the project is aimed at analyzing the existing criteria for identifying the need for the placement of a vehicle restraint system and for the identification of the most appropriate performance class. For this aim, both the existing national standards and guidelines and literature documents have been analyzed in detail. The comparative analysis of 33 national standards and guidelines covering most of Europe and several non-European countries has shown that there are many commonalities, and it is possible to identify the most frequently used parameters with reference to safety barriers. While the majority of the countries have guidelines and/or standards related to safety barriers, there is generally limited guidance for other systems such as crash cushions, transitions and motorcycle protection systems. Lifecycle cost models are usually not included in the standards, but few tools are available worldwide.

13.1. Introduction

Run-off-road (ROR) crashes are extremely severe road accidents that can often result in severe injuries or fatalities. The accident analysis conducted within the RISER project, funded by the EU, concluded in 2005, highlighted that even though only 10% of the total accidents are single vehicle accidents (typically associated

Chapter written by Francesca La TORRE, Ceki ERGINBAS, Robert THOMSON, Giuseppina AMATO, Bine PENGAL, Peter SALEH, Chris BRITTON and Kris REDANT.

with the ROR type accidents), the rate of single vehicle accident events increases to 45% when only fatal accidents are considered.

To reduce the severity of ROR crashes, "forgiving roadsides" need to be designed, and this includes identifying where there is a need for a vehicle restraint system (VRS) and what appropriate VRS should be selected for the specific location and traffic condition.

At the present time, while there are standards covering the testing, evaluation and classification of VRS within Europe (EN1317 and EN12767), their selection, location and installation are based upon national guidelines and standards, often produced by National Road Authorities (NRAs) and/or overseeing organizations. Due to local conditions, these national guidelines vary across Europe.

To allow for the development of a uniform guideline for selecting the most appropriate VRS performance requirements for the given conditions, a number of the Conference of European Directors of Road (CEDR) members have collaborated to fund the research project "Selection of Appropriate Vehicle Restraint Systems" (SAVeRS) within the 2012 Transnational Road Research Programme call "Safety". The aim of the SAVeRS project is to produce a practical and readily understandable VRS guidance document and a user-friendly Web-based tool that will allow the selection of the most appropriate solution in different road and traffic configurations for all types of VRS: safety barriers, crash cushions, terminals and transitions and motorcycle protection devices.

The guidance document and the Web-based tool will have with the following goals:

– ensuring the safety of road users, road workers and third parties;

– optimizing VRS performance in use;

– maximizing VRS serviceable life;

– minimizing VRS whole life costs.

The different solutions considered will be fully compatible with EN1317 and related EN standards (for example the European passive safety standards EN12767, EN40 and EN12899).

To ensure that the SAVeRS project output will be a useable, robust and realistic guidance tool to aid designers in their choice and selection of VRS, a specific evaluation task will be performed to test the effectiveness of the guidelines and of the tool developed in the project. The guideline and tool will be tested for the application to new and current designs and against sites where VRS have been in use for some years and performance information is available.

This chapter gives details of the findings from the first of the work packages within the SAVeRS project i.e. to define the different parameters which can influence the need and selection of VRS, from both a review of national guidelines and standards, and from an associated review of published literature.

13.2. Objectives of the first work package of the SAVeRS project

In order to develop a robust and effective methodology for the appropriate selection of a VRS, it is necessary as a first step to collate, review, fully understand and appreciate current (and proposed future) national guidelines and standards. The aim of work package 1 is:

– to analyze the differing national guidelines and standards to identify, review and categorize information which is currently available relating to the parameters associated with the choice of VRS to develop a single document outlining the approaches taken in each country;

– to collate, review and fully understand international research which has been carried out regarding the parameters considered when selecting a VRS. This may, or may not, be related to the development of guidelines of the NRAs.

In order to achieve these objectives, the work package was split into two distinct tasks, each investigating the areas outlined in the above bulleted list. The subsequent section will outline the aims, methodology and results from these activities.

13.3. Collation and examination of national guidelines and standards

13.3.1. *Methodology*

Standards and guidelines from 33 countries covering most of Europe and several non-European countries were collected in the SAVeRS project (Figure 13.1). It was noted that some countries have adopted and/or adapted guidelines from other countries either neighboring them or with similar infrastructural conditions. More specifically, it was observed that guidelines from the USA were adopted by countries around the Americas, while the German standard is the most widely adopted standard around Europe if countries have not developed their own requirements. Although the majority of European countries have their own dedicated guidelines and standards, they still have common approaches, decision processes, tables and graphs.

Each standard was then read and analyzed in detail to identify the parameters related to:

– the choice of whether to install a VRS or not;

– the selection of VRS performance.

Figure 13.1. *Collection on national standards and guidelines on VRS. For a color version of the figure, see www.iste.co.uk/jacob/safety.zip*

A data matrix was prepared to store and present the identified parameters by country. This was to allow easy identification of those parameters essential for the work within work package 2. As new parameters were identified within each of the items of documentation, it was decided whether to categorize these parameters under "consequence" or "likelihood", such that the elements of risk (the product of consequence and likelihood) could be identified.

Once a complete set of robust parameters had been determined, the data matrix was divided into tabs, identifying those parameters related to the decisions for each VRS type, i.e. roadside safety barriers, median safety barriers, bridge parapets, crash cushions, transitions, terminals and motorcyclist protection systems (MPS). Due to the low level of published guidance for terminals, transitions and MPS, these were removed from the data matrix. For each parameter, it was then determined whether it applied to the decision to install a VRS or to the selection of the performance for the VRS.

After the collation of data from national guidelines, the final matrix was transferred into SPSS, statistical analysis software, which was used to prepare frequency tables for each parameter for each VRS type. These frequency tables were then analyzed to develop evidence-based conclusions on the most used parameters.

13.3.2. *Results*

As each guideline was analyzed, more detailed parameters were detected for some of the countries as compared to others. For example, while some countries only mentioned the "existence of a tree" as a parameter that leads to the decision of barrier installation, some countries went a step further and defined "trunk girth of a single tree".

As one might anticipate, it was observed that the parameters related to the consequences of an accident were used more often for the decision as to whether to install a VRS or not, as these are basically a list of hazards that would necessitate the installation of a VRS to mitigate the danger. Conversely, parameters related to the likelihood of a given type of accident were used more often to determine the level of performance required from the VRS, as these include parameters such as percentage of heavy goods vehicles.

Examination of those parameters most frequently referenced within national guidelines and standards has shown that in terms of the justification for roadside safety barrier, it is the risk to vehicle occupants, travelled speed, road geometry, the existence of risk to third parties and traffic which are the most frequently included parameters (in that order). For median barriers, the same factors are most frequently referenced, but it is the existence of risk to third parties which is mentioned most frequently. When considering those factors which are used for determining the performance requirement of a roadside safety barrier, factors such as the existence of special risk to third parties, traffic and road alignment and/or geometry are considered most frequently (in that order). For median barriers, again, these are factors which are frequently referenced, but with traffic being the most frequently referenced characteristic. When such factors are examined in further detail, it is the presence of embankments and cuttings (and their height and gradient), the presence and proximity of vulnerable road users (such as pedestrians and cyclists), railways, bodies of water and non-deformable roadside obstacles, the average annual daily traffic and actual speeds which are most prominent in determining the need for a roadside safety barrier. When selecting the performance of a roadside barrier, factors such as the presence of structures and railways lines, the presence and proximity of bodies of water and non-deformable roadside obstacles, the average annual daily traffic and actual speeds and the presence of adverse road geometry are most prominent. While factors such as aesthetics and cost are mentioned in some national guidelines and standards, their frequency is low.

For bridge parapets, there is less specific guidance than for roadside and median safety barriers; however, obstructions with a special risk to vehicle occupants and the height of the bridge are the most common factors. When determining the performance of a bridge parapet, factors such as the existence of special risks to third parties (for example railways lines) and obstructions posing a risk to vehicle occupants are referenced most frequently.

Guidance on the need to install crash cushions is very limited; however, it is the presence of a non-deformable hazard which occurs most common (perhaps unsurprisingly). With regard to determining the performance level of the crash cushion, this is limited to the actual speed limit of the road. This is perhaps also as expected, the standards for the testing of crash cushions identifying impact speed as one of the defining parameters for performance. For terminals, transitions and MPS, there is very limited guidance within national guidelines and standards; hence, there is insufficient data upon which to form any justifiable conclusions. It was also observed that, while the majority of the countries have guidelines and/or standards related to roadside and median barriers, there is generally limited guidance for other VRS systems such as crash cushions, transitions and MPS.

13.4. Collation and examination of published literature

13.4.1. *Aim*

The goal of the literature review in the SAVeRS project is to find out how the placement and choice of suitable VRS is approached and assessed by researchers. In fact, the actual decisions taken by NRAs are mostly based on national tradition and acknowledged studies while ongoing research should be able to point out changes driven by the development of the road infrastructures. For the sake of simplicity, the review has been approached by searching how different actors play a role in the decision making of VRS placement. The point of view of policy-makers, vulnerable road users such as cyclists pedestrians and motorcyclists were investigated, and the available analyses of accidents involving safety barriers were assessed looking at both financial implications and safety.

13.4.2. *Methodology*

Most of the studies available in the literature can be divided between in-depth accident analysis and commentaries and assessment of policies.

A large number of studies were undertaken to allow optimization of the number, length and location of VRS installed in the roadside and the central reserve of roads fall in the first category. Most of these studies have utilized collated incident data, although the number, quality and relevance of the incidents investigated should be understood, in all cases. The results of these studies will be reported in the next section.

As powered two wheeler (PTW) safety and the development of MPS have become a specific issue in accidentology, several PTW research projects are dealing as well with crash reconstruction and the expected effects of MPS. Other papers are more general in terms of analyzing common crash circumstances. When it comes to test procedures and standards regarding MPS, several FEMA [MOT 12] initiatives are available in European literature. The EU project Smart VRS [IDI 12], together with results of the first naturalistic riding project [SAL 10], offers deeper insights to various MPS aspects. On the other hand, policy papers usually deal with vulnerable road users and cultural instances such as protection of public health and of the cultural and natural heritage. Pedestrians, cyclists and PTWs fall in the typical "vulnerable road user groups", but these groups are loosely defined without adherence to a strict classification [AVE 05].

Another important issue in roadside crashes is related to removing hazards placed in the clear zone. The decision regarding tree removal in the clear zone and roadside barrier installation is strongly affected by the discussion on the relationship between public health and road infrastructure design. According to Naderi [NAD 03] the US Center for Disease Control and the Surgeon General's office study all the variables which favor active lifestyle living such as pedestrian and bicycle use over the automobile and can improve national health by reducing obesity and related medical care costs. In this context, it is believed that improving the aesthetic aspect of transportation corridors can be beneficial in a double way: by reducing accident frequency and severity and by increasing pedestrian activity.

Boulevard treatment and the introduction of green infrastructure within transportation corridors, however, have presented difficulties in relation to the treatment of the safety zone. In regard to this, although landscaping as a tool to achieve safer roads is socially recognized, researchers are trying to assess quantitatively its effect on driver behavior. Several case studies have been produced: a study on five arterial roads in Toronto between 1992 and 1995 [ROS 98], a comparison between the safety performance of 12 couples of parkways and freeways in four US states [MOK 06], the crash rate before and after landscape improvement in 10 study sites in Texas [MOK 06] and a correlation analysis to identify weak relationships between the quantity of car accidents and some aesthetic properties of road landscape in Lithuania [MAT 13]. The fatal accident rate and the accident cost constituted the dependent variables for comparing the safety performance of parallel sections of selected parkways and freeways in Mok and Landphair [MOK 06]. Most of these studies showed a positive correlation (although sometimes weak as in the case of Matijoaitienė and Navickaitė [MAT 13]) between the aesthetic enhancement of the road landscape and the road collision rate. Also, all the authors concluded that the research results were limited, and more detailed analyses of accidents in relation to landscaping were needed.

The study by Mok *et al.* [MOK 06] is particularly relevant due to the analysis of tree collisions before and after landscape improvements. The research hypotheses of this study were that crash rates significantly decreased after the landscape improvement at study sites and that a decrease in the number of tree collisions occurred after landscape improvements. Of the 61 study sites initially chosen, only 10 could be used to evaluate the effects of landscaping on safety, as the others included additional treatments that can have a direct effect on safety such as pedestrian sidewalk widening, expansion of existing shoulders or installation of bicycling path. Results showed a decrease in crash rate in 8 of 10 study sites; in 2 sites, an increase in crash rate after the landscape treatment was observed. The number of tree collisions showed a decrease of about 70.83% after landscape treatment, but it should be noted that in 9 of 10 locations, only 0–2 crashes were observed in the before period; therefore, the effect of landscaping is not so evident. In the single location where 18 crashes were reported in the before period, only 3 crashes were reported in the after period with a significant reduction in tree collisions. The change was associated with a landscape treatment that occurred in 1992.

It is not possible to select a VRS without considering the financial consequences of the selection. Unfortunately, this is an area that is difficult to quantify due to the difficulty in collecting information covering the relevant costs and benefits for a particular system. There are notable attempts at developing tools to address roadside safety design costs, with the Roadside Assessment Program [RAY 12] in the US being the most ambitious. This is a good example of a method to estimate the costs and benefits associated with different design concepts. A similar evaluation was applied in Sweden in the doctoral thesis of Karim [KAR 08] using lifecycle costs. Similar approaches may be applied in other countries, but no other published articles could be obtained in this review.

13.4.3. *Results*

13.4.3.1. *Roadside hazards*

Running off the road does not necessarily result in injuries. If there is sufficient space available to slow down and stop the vehicle, most incidents will remain without consequences. Most often however, objects or terrain conditions can cause an abrupt slowdown or destabilization of the vehicle leading to injuries for the vehicle occupants. Trees contribute heavily (5–10%) to fatal accidents [SET 02]. Tree crashes are obviously more frequent on roads than on motorways. From all fatal incidents with fixed obstacles on roads (but non motorways) outside urban areas in the Netherlands (2002), trees contributed to 85% of all roadside objects [CRO 04b]. Lighting columns and other isolated supports also represent an important share, varying around 7% of all fatal accidents (STATS19 database, period 1998–2002) [LYN 05]. On motorways, single vehicle ROR accidents involving VRS are responsible for 20–30% of fatal accidents [LYN 05]. Unfortunately, accident statistics generally are not sufficiently detailed to give information about the type of VRS impacted or impact details. Generally speaking, it should be noted that VRS safety effectiveness can be limited in case of an inadequate installation, an inappropriate maintenance or a lack of repair.

Ditches can be considered as a special type of obstacle. Much depends on the cross-section of the ditch (depth and slope). Often, however, entering a ditch will destabilize the vehicle [THO 02]. In some cases, a vehicle that enters a ditch can be guided onto a rigid construction at the end of the ditch. Embankments represent a second type of "special" obstacle. Cut slopes with a gradient below 1:3 are considered to be relatively safe [AAS 11], although some countries allow higher gradients. For example, the Netherlands allow cut slopes with a 1:2 gradient when the transition from horizontal to sloped surface is rounded [CRO 04b]. Based on existing recommendations and numerical simulations, a gradient of 1:3 is the threshold to consider a slope as an obstacle or not [PAR 10]. Higher slopes could lead to a potential rollover risk. For fill slopes, slope and height determine whether this type of obstacle is acceptable or not. For steep slopes, there is a risk that the vehicle leaving the road could lose contact with the surface when it enters the embankment and is "launched". In such cases, the fall height should be reduced to limit possible consequences [CRO 04a, SET 02]. Again, a rounded transition from horizontal surface to sloped surface increases the allowed slope [CRO 04a] in the Netherlands.

13.4.3.2. *Placement of safety barriers*

The available literature on the placement of VRS focuses on safety barriers. Most of the studies make comparative analyses on accident rate and severity on road segments before and after the placement of safety barriers. Information on terminals, transitions and attenuators is scarce. Analysis can be divided between studies on median and studies on roadside barriers. Often these analyses do not discuss the containment level but rather the particular kind of safety barriers used (concrete, steel w-beam, steel cables, etc.) and make a distinction between rigid and flexible barriers. Results are often assessed in terms of containment, accident rates and accident severity.

The effectiveness of two safety barrier types prevalent in the US: the GA-type strong post, corrugated beam safety fence system and a wire rope median barrier have been examined by Alluri *et al.* [ALL 12]. In the case of the strong post system they identified that on a 1,652.3 km length of the barrier there were 8,674 cases in which the safety barrier was impacted. Of these impacts, 94.5% of the impacts resulted in the impacting vehicle being contained by the barrier system. It is important to note that some of the impacts with the barrier were outside of the design parameters of the barrier system (e.g. an impact in which the weight and/or speed and/or angle exceeds the level to which the system has been designed and tested). They noted that compared to roadside safety barrier, median barriers accounted for a slightly higher percentage of incidents in which the barrier was breached by the impacting vehicle. By comparison, when a 162.5 km long installation of wire rope median barrier was examined, it was reported that 549 impacts occurred and of these, containment occurred in 83.6% of cases. In total, 98.1% of cars impacting the barrier and 95.5% of light trucks (whose containment is not part of the design considerations for this type of safety barrier) were contained.

In addition to the in-service performance of the central reserve rope barrier itself, Alluri *et al.* [ALL 12] also examined the before and after effects of the central reserve rope barrier's installation. They found that the installation of the median barrier had reduced the fatal impact rate by 42.2%, the severe injury rate by 20.1% and the minor injury rate by 11.6%. However, the rate of incidents involving possible injury and property damage rose by 53.1 and 88.1% respectively, resulting in an overall impact rate increase of 37.8%. This emphasizes that safety barriers are, in themselves, also hazards and hence their use should only be as a last resort where other safety measures cannot be implemented. Similar results have been found by Bonneson *et al.* [BON 12], Elvik [ELV 95] and Tarko *et al.* [TAR 08] for different type of barriers.

Work by Candappa *et al.* [CAN 09] has also examined the effectiveness of flexible barrier systems, this time on the Australian road network. In this case, a total of 101.6 km of road length was installed with wire rope safety fence. The study compared the impact frequency at road sections before and after treatment. The results indicated that the barriers could be associated with significant reductions in the risk to both casualty and serious casualty rates. These reductions varied from site-to-site, but were, on average, around 76% for all casualty impacts and 77% for serious casualty impacts. The report also states that these figures align closely with work from previous International studies citing, in particular, reports from Sweden and the USA. The report concludes that the introduction of flexible barrier systems, such as wire rope is likely to produce substantial reductions in incident occurrence, in particular in cases of off-road and head-on impacts for both casualty and serious casualty impacts.

The effectiveness of median barriers on the French motorway network was evaluated by Martin and Quincy [MAR 01]. This study showed that the crossover incident was rare, identifying that in 0.5% of car incidents and in 7% of incidents in which the median barrier was struck by a truck, the vehicle was not contained by a safety barrier. The authors also conclude that crossover incidents are more serious than other types of incident with 19% resulting in fatalities and 43% resulting in some level of injury. As a word of warning, the authors state that in recent years in France, extra traffic lanes have been added to existing motorway sections, which has often led to a reduction in central reserve strip width. New motorways are being built with a maximum 5m wide median barrier (to limit ground surface requirements and cost). According to the authors, in these median-strip-width conditions, one possible strategy to reduce the number of median barrier crossings significantly is to place barriers with higher containment capacities (level H2 or over). This same issue is common in most European countries due to limited land availability and most NRAs across Europe therefore require high containment barriers in the medians.

While the safety benefits of VRS are therefore well documented, there are occasions where safety barriers are purposely not installed due in part to the hazard posed by their installation. Instead wide central reserves (9 m) have been used as a lane separator. Davis and Pei [DAV 05] reconstructed five incidents in the USA where a wide central reserve (greater than 9 m) without an installed safety barrier was traversed, causing fatalities on the opposite carriageway. In the USA, a 9 m (30 ft) wide central reserve is thought to be sufficiently wide for 80% of out-of-control drivers to regain control of their vehicle. Davis and Pei [DAV 05] conclude that while the 9 m central reserve was in place, fatal incidents still occurred and could have been mitigated had a normal containment (TL3) safety barrier been in place.

The use of the clear central reserve has also been examined by Donnell *et al.* [DON 02]. In this particular case, the Pennsylvanian Department of Transportation's design policy was reviewed which stated that safety barriers were not required for central reserves with a width of 10 m or more and with an average daily traffic of 20,000 vehicles per day. The report concluded that crossover incidents, while rare, result in fatal injuries in 15% of cases, with 72% incidents resulting in non-fatal injuries. The report also states that on earth-divided roads, crossover incidents decrease as the central reserve width increases (due to the increase in vehicle recovery time). In addition, it was found that crossover incidents occur more frequently downstream of interchange entrance ramps, and that they are more likely to occur during periods of adverse weather (wet or icy) than other types of incident.

Miaou *et al.* [MIA 05] conducted similar work to Donnell *et al.* [DON 02], within the roads of Texas. Their research concentrated further on the modeling of incidents, basing this more directly on the benefit-cost relationship, rather than on the pure safety considerations in the Donnell *et al.* [DON 02] study. They concluded that a positive cost-benefit could be achieved with clear wide central reserves. Similar studies were also developed by Elvik [ELV 95] and Tarko *et al.* [TAR 08] who concluded that median barriers are found to increase incident rate, but reduce incident severity. In general terms, Elvik also concluded that safety barriers and crash cushions reduce both incident rate and severity, adding that safety barriers reduce the chance of sustaining a fatal injury by about 45%, given that an accident has occurred. The chance of sustaining a personal injury is reduced by about 50%.

While the central reserve has been the focus for many papers and research articles, there has been limited associated study for the use of roadside barriers. However, one such study was that of Schneider *et al.* [SCH 09]. Their specific area of study was in locating safety barriers within horizontal curves on rural two-lane highways in Texas.

If a barrier is to be placed it is essential to define the appropriate length of the barrier considering that if a barrier is too short it may allow an errant vehicle to traverse behind the barrier, while a barrier too long in length will present an additional unnecessary hazard to road users. In order to ascertain guidelines for the minimum length of safety barrier to be installed, Tomasch *et al.* [TOM 11] analyzed the run off road crashes derived from a National Austrian crash database. The authors defined the required safety barrier length as the length that allows the vehicle speed to be reduced to an acceptable value upon impact with the obstacle, considering a maximum possible deceleration of 0.3 g behind the barrier. To determine the desired length of a barrier ahead of a hazard they developed a relationship between barrier length and the speed at which vehicles depart the roadway based on the Austrian database. The authors demonstrated that the application of this approach would reduce the number of fatalities among occupants of vehicles striking bridge abutments by approximately 8%.

13.4.3.3. *MPS*

Most ROR crashes in bends occur at radii between 50 and 150 m. Radii relations between subsequent curves (R1/R2 >1) and disharmonic trace, high bendiness and high gradients are critical safety issues for PTW riders. ROR accidents involving PTWs are more likely to occur in left-hand curves than right-hand curves, for countries were standard driving is on the right side of the carriageway [SAL 10]. Most motorcycle collisions with crash barriers occur at angles between 10° and 45° although in the European standard [CEN 12] the prescribed crash test angle is 30°.

Impact of motorcyclists against a fixed object occurred in 4% of the cases in urban areas while variations between 10 and 20% can be observed in rural areas. The most important obstacles referring to particularly severe injuries are trees/poles, roadside barriers and road infrastructure in general. Exposed guardrail posts are the most dangerous aspect of guardrails with respect to motorcyclists. The risk of injury due to hitting a fixed object is related to the impact area and rigidity of the object. Hence, small rigid objects such as support posts are most likely to cause the most severe injuries as they concentrate the impact forces on a small area of the human body [IDI 12].

Several studies as for example Berg *et al.* [BER 05] investigate both sliding and upright crash positions of riders. 2-BE-SAFE [SAL 10] results show the same tendencies, that both crash types are equally relevant, especially from in-depth analysis of Spanish data. Also crashes with the rider still sitting on the bike in an upright position should be considered. For riders remaining in an upright position when impacting a crash barrier, most injuries occur at shallow impact angles, i.e. the rider slides and tumbles into the top of the supporting posts. When a rider impacts with a barrier in an upright position, the motorcyclist is likely to be thrown over the guardrail system if the height of the barriers is too low [DUN 00].

13.5. Conclusions

The review of the existing national standard and guidelines has shown that there are many commonalities and key parameters that are frequently used for identifying if there is a need for a VRS and for selecting the performance of a VRS even though the parameters are typically different: the parameters related to the consequences of an accident were used more often for the decision as to whether to install a VRS, or not since these are basically a list of hazards that would necessitate the installation of a VRS to mitigate the danger. Conversely, parameters related to the likelihood of a given type of accident were used more often to determine the level of performance required from the VRS, since these include parameters such as percentage of heavy goods vehicles.

For the most performance requirement in the most common VRS application (roadside barriers), it is factors such as the presence of structures and railway lines, the presence and proximity of bodies of water and non-deformable roadside obstacles, the average annual daily traffic and actual speeds and the presence of adverse road geometry which are most prominent. While factors such as aesthetics and cost are mentioned in some national guidelines and standards, their frequency is low. While the majority of the countries have guidelines and/or standards related to roadside and median barriers, there is generally limited guidance for other VRS systems such as crash cushions, transitions and MPS.

The available literature on the placement of VRS focuses on safety barriers. Information on terminals, transitions and attenuators is scarce. Most of the studies make comparative analyses on accident rate and severity on road segments before and after the placement of safety barriers. The discussion on median barriers focuses on the effectiveness in reducing accident rate and severity against the option of wider central reserves. In terms of variables for the placement and choice of VRS, these are not generally indicated since, as already stated, most of the studies analyze the same road, for the same traffic and geometrical conditions, before and after a safety treatment. Risk analysis and safety management tools such as "safety-barrier diagrams" are suggested for identifying the risks associated with the installation of safety barrier systems.

Environment and pedestrian protection issues also play a role in the placement of roadside barriers in corridors and rural roads. In regard to this in Europe some countries prefer minimizing the interaction between different road users and between vehicles driving in different directions using central reserve and roadside barriers, while in US especially wide safety zone complemented by landscape and trees are preferred.

Studies on MPS use traffic volume variables such as AADT and percentages of different road users and recommend the installation of MPS in roads with high exposure of PTWs. In terms of road geometry, accident statistics and in-depth analyses point out roads with bends between 50 and 150 m as specific areas of higher risks.

In regard to the influence of cost on the placement of VRS there are useful methodologies that can be exploited for a European economic assessment tool but the main current and future difficulty is to find financial data covering all aspects of a VRS installation and valid beyond a region or national level.

13.6. Acknowledgments

SAVeRS is a project funded within the CEDR Cross-border Transnational Research Programme funded by Belgium/Flanders, Germany, Ireland, Norway, Sweden and United Kingdom. The authors wish to thank the SAVeRS Project Managers, Alastair De Beer (NRA, Ireland) and Anders Hakansson (Trafikverket, Sweden), and Des O'Connor (NRA, Ireland) for his precious help in preparing this paper.

13.7. Follow-up

The SAVeRS was completed in May 2015 and all the deliverables, including the SAVeRS Guideline for the selection of the most appropriate Roadside VRS and the user friendly SAVeRS Tool, are publicly available through the project website (www.saversproject.com).

13.8. Bibliography

[AAS 11] AASHTO, *AASHTO Roadside Design Guide*, US, 2011.

[ALL 12] ALLURI P., HALLEM K., GAN A., *In-Service Performance Evaluation for G4(IS) Type Strong Post W-beam Guardrail Systems and Cable Median Barrier*, Lehman Center for Transportation Research, Florida International University, 2012.

[AVE 05] AVENOSO A., BECKMANN J. (eds), *The Safety of Vulnerable Road Users in the Southern, Eastern and Central European Countries (The "SEC Belt")*, European Transport Safety Council, Brussels, 2005.

[BER 05] BERG F.A., RÜCKER P., KÖNIG J. *et al.*, "Motorcycle impacts into roadside barriers – real-world accident studies, crash tests and simulations carried out in Germany and Australia", *ESV*, 2005.

[BON 12] BONNESON J.A., GEEDIPALLY S., PRATT M.P. *et al.*, Safety Prediction Methodology and Analysis Tool for Freeways and Interchanges, NCHRP Project 17-45, Final Report, May 2012.

[CAN 09] CANDAPPA N., D'ELIA A., CORBEN B. *et al.*, Evaluation of the effectiveness of flexible barriers along victorian roads, Monash University Accident Research Centre, 2009.

[CEN 12] CEN, E. C. F. S., EN1317-8:2012 – Road Restraint Systems. Motorcycle Road Restraint Systems Which Reduce the Impact Severity of Motorcyclist Collisions with Safety Barriers, BSI, 2012.

[CRO 04] CROW, Handboek Veilige Inrichting Van Bermen. Niet-Autosnelwegen Buiten De Bebouwde Kom, Ede, Netherlands, 2004a.

[DAV 05] DAVIS, G., PEI, J., "Bayesian reconstruction of median-crossing crashes and potential effectiveness of cable barriers", *Transportation Research Record N.1908*, pp. 141–147, 2005.

[DON 02] DONNELL E., HARWOOD D., BAUER K. *et al.*, "Cross-median collisions on Pennsylvania interstates and expressways", *Transportation Research Record N. 1784*, pp. 91–99, 2002.

[DUN 00] DUNCAN C., CORBEN B., TRUEDSSON N. *et al.*, Motorcycle and safety barrier crash-testing - feasibility study, Accident Research Centre – Monash University, 2000.

[ELV 95] ELVIK R., "The safety value of guardrails and crash cushions: a meta-analysis of evidence from evaluation studies", *Journal of Accident Analysis and Prevention*, vol. 27, pp. 523–549, 1995.

[IDI 12] IDIADA, Innovative concepts for smart road restraint systems to provide greater safety for vulnerable road users, D 2.4 Final Report SmartRRS, 2012.

[KAR 08] KARIM H., MAGNUSSON R., "Road design for future maintenance problems and possibilities". *Journal of Transportation Engineering,* vol. 134, pp. 523–531, 2008.

[LYN 05] LYNAM D., The travel of errant vehicles after leaving the carriageway, Standards for Highways, 2005.

[MAR 01] MARTIN J.-L., QUINCY R., "Crossover crashes at median strips equipped with barriers on a french motorway network. *Transportation Research Record No. 1758*, pp. 6–12, 2001.

[MAT 13] MATIJOŠAITIENĖ I., NAVICKAITĖ K., Aesthetics and safety of road landscape: are they related? *Journal of Sustainable Architecture and Civil Engineering,* vol. 1, pp. 20–25, 2013.

[MIA 05] MIAOU S.-P., BLIGH R.P., LORD D., "Developing median barrier installation guidelines: a benefit/cost analysis using texas data", *Transportation Research Record No. 1904*, pp. 3–19, 2005.

[MOK 06] MOK J.-H., LANDPHAIR H.C., NADERI J.R., "Landscape improvement impacts on roadside safety in Texas", *Landscape and Urban Planning No. 78*, pp. 263–274, 2006.

[MOK 03] MOK J., LANDPHAIR H.C., "Parkways and freeways: safety performance linked to corridor landscape type", *Transportation Research Board 82nd Annual Meeting*, 2003.

[MOT 12] MOTORCYCLISTS`ASSOCIATIONS, F.F.O.E., New Standards for road – restraint systems for motorcyclists, 2012.

[NAD 03] NADERI J.R., "Landscape design in clear zone: effect of landscape variables on pedestrian health and driver safety", *Transportation Research Record No. 1851*, pp. 119–130, 2003.

[PAR 10] PARDILLO-MAYORA J., "Empirical calibration of a roadside hazardousness index for Spanish two-lane", *Accident Analysis and Prevention,* 2010.

[RAF 03] RAFORD N., RAGLAND D.R., Space syntax: an innovative pedestrian volume modeling tool for pedestrian safety. institute of transportation studies, UC Berkeley, 2003.

[RAY 12] RAY M.H., CARRIGAN C.E., PLAXICO C.A. *et al.*, NHCRP 22–27: Roadside Safety Analysis Program (RSAP) Update, 2012.

[ROS 98] ROSENBLATT J., BAHAR G., "An integrated approach to environmental impact mitigation and safety management: case studies in the municipality of metropolitan toronto", *XIIIth World Meeting of the International Road Federation, Roads/Transportation and the Environment Session*, Toronto, Ontario, Canada, 1998.

[SAL 10] SALEH P., "Interaction between powered two-wheeler accidents and infrastructure", in: FP7E. (ed.), *Project 2-Wheeler Behaviour and Safety, 2-BE-SAFE D1.2,* 2010.

[SCH 09] SCHNEIDER IV, W., SAVOLAINEN P., ZIMMERMAN K., "Driver injury severity resulting from single-vehicle crashes along horizontal curves on rural two-lane highways", *Transportation Research Record No. 2102*, pp. 85–92, 2009.

[SET 02] SETRA SERVICE D'ETUDES SUR LES TRANSPORTS LES ROUTES ET LEURS AMÉNAGEMENTS, Traitement des obstacles latéraux sur les routes principales hors agglomération, France, 2002.

[TAR 08] TARKO A., VILLWOCK N., BLOND N., "Effect of median design on rural freeway safety: flush medians with concrete barriers and depressed medians", *Transportation Research Record No.* 2060, pp. 29–37, 2008.

[THO 02] THOMSON R., "Vehicle impacts in V-ditches", *Transportation Research Record No. 1797*, pp. 82–88, 2002.

[TOM 11] TOMASCH E., SINZ W., HOSCHOPF H. *et al.*, "Required length of guardrails before hazards", *Accident Analysis and Prevention*, vol. 43, pp. 2122–2120, 2011.

14

For the Vision of "Zero Accidents at Intersections": A Challenge between Road Safety and Capacity

Different types of intersection were simulated with different traffic control variants. Simulation results of a wide range of types of intersection and traffic signal phases showed that an additional protected green phase for traffic turning left generally resulted in no relevant restrictions to the capacity of the intersection. In the case of large intersections with a high volume of traffic, a protected signal for traffic turning left increases safety for traffic turning left and oncoming vehicles. In the case of large intersections with a medium volume of traffic and small intersections with a high volume of traffic, "protected" systems for traffic turning left are the most efficient solution both in terms of traffic flow and road safety.

Standardizing these types of regulation at traffic light-controlled intersections helps to reduce conflicts between turning left vehicles and pedestrians and cyclists too. The number and the severity of accidents involving pedestrians and cyclists at such intersections can be reduced significantly.

14.1. Introduction

Intersections in the road network can be constructed and operated in different ways. The combination of the construction and the way in which the traffic is regulated is referred to as the intersection type [ECK 02]. At locations with a high volume of traffic, intersections regulated by right-of-way are as a rule being replaced by intersections regulated by traffic signals. Traffic signals make the road space

Chapter written by Jean Emmanuel BAKABA and Jörg ORTLEPP.

within an intersection available to the competing streams of traffic alternately in a cyclic sequence (phase sequence). The different streams of traffic cross the intersection in sequence.

According to the German Guidelines for Traffic Signals [ROA 10], traffic signals are installed to "increase traffic safety and/or to improve traffic flow quality. [...] The set-up of a traffic signal system has to be considered if accidents which may have been prevented by traffic signal control have occurred repeatedly and if alternative measures (such as speed limits, overtaking prohibitions or constructional crossing aids to pedestrians or cyclists) have proved to be ineffective or not promising". According to section 37 of the German "General Administrative Regulations on the Road Traffic Regulations" (VwV-StVO), situations in which traffic signals are necessary include those where accidents frequently happen because visibility is restricted, and there is no possibility of improving visibility or prohibiting traffic which is crossing or entering the road, and those where there are frequent infringements of right-of-way without this being related to the intersection being difficult to identify or with the right-of-way being difficult to understand. The VwV-StVO also states that the following principles apply to sections 39 through 43: "The free flow of traffic is to be upheld using the means available. In applying this principle, the safety of all road users takes priority over the free flow of traffic".

This study is therefore intended to clarify the impact on traffic caused by different traffic signal control methods used to protect traffic turning left. In particular, it is intended to show the extent to which any claimed negative impact on traffic flow actually occurs and how it can be minimized or offset by an improvement in traffic safety.

14.2. Traffic turning left at signal-controlled intersections

14.2.1. *Non-conflicting and conflicting flows*

"The individual components, as for example the actual layout of the intersection, the division of approaches into lanes, the direction of pedestrians and cyclists and the signalization of the individual traffic streams have to be coordinated in such a way that the preconditions for safe traffic flow are given under all operational conditions and for all traffic loads" (see [ROA 10], section 1.1). Traffic turning left at intersections controlled by traffic signals can be handled in different ways. Traffic streams which do not share any joint conflict areas are to be regarded as non-conflicting streams and can therefore be released together in a single phase. If individual traffic streams share any joint conflict areas, they are termed conflicting traffic streams. They are released separately. Turning traffic that is not subject to special signal control is an exception in this context. Such streams are termed partially conflicting streams in RiLSA.

14.2.2. *Traffic turning left as conflicting streams*

Because traffic turning left shares joint conflict areas with both oncoming traffic and with pedestrians and cyclists traveling in the same direction, they are in principle seen as conflicting flows and must be released in different phases. They are only classified as signal protected if all streams that conflict with them are blocked while they are released. If this approach is adopted, it results in multi-phase controllers in which through traffic and traffic which is turning is released alternately.

14.2.3. *Traffic turning left as partially conflicting streams*

If the release of traffic streams turning left is divided into a "protected" and an "unprotected" period, this is referred to as "temporarily protected release". A distinction is made between "leading green" and "lagging green" which are defined as a time by which the release time for one or more traffic streams starts earlier or finishes later than for other traffic streams released during the same phase [ROA 10]. These times can in principle be programmed with or without visual indication using an auxiliary signal. If such times are indicated, a green arrow is used for the protected release time and an amber flashing signal is used for the release time in which the traffic must pass through oncoming traffic. Leading green with no indication is regarded as unacceptable for reasons of safety, and lagging green with no indication is not recommended, as it is difficult for drivers to estimate the relevant times.

14.2.4. *Selecting cases for investigation*

The following criteria were established for the cases for investigation:

– intersections with single-lane guidance of through traffic in all directions;

– influence of non-motorized users guided in parallel was not considering;

– no significant longitudinal inclination, no unusual lane widths and no unusually large or small turning radii.

14.2.5. *Initial conditions for the calculations*

Simulations were performed to compare the different control variants. The different types of intersection were simulated with different control variants. Each signaling variant was simulated with six different random seed numbers. Initially, the maximum traffic volume without the influence of traffic turning left was determined (basic intersection capacity) and then progressively reduced.

14.2.6. *Results of the simulation calculations*

14.2.6.1. *Large intersection with triangular islands (case 6)*

We shall investigate an intersection with three lanes on all approaches. Traffic turning right flows away freely alongside the triangular island. In the case of signal protection for traffic turning left, this stream of traffic turning right is, however, also protected (exception: "diagonal arrow" with leading green or lagging green).

In the case of simple *two-phase control*, there is no separate protected time for traffic turning left. This means that traffic turning left will tend to flow away during phase changes.

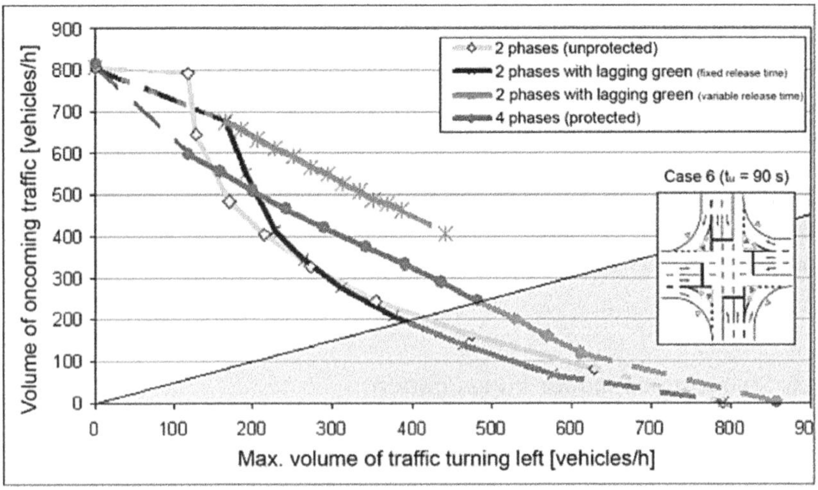

Figure 14.1. *Comparison of the influence of a conflicting vehicle turning left using different control methods (large intersections with triangular islands and a cycle time of 90 seconds). For a color version of the figure, see www.iste.co.uk/jacob/safety.zip*

This shows that up to three vehicles can flow away per cycle at the maximum volume of oncoming traffic. Because of the geometrical size of the intersection, these vehicles will have waited in the interior of the intersection. This results in a maximum volume of traffic turning left of approximately 120 vehicles/h given the maximum volume of traffic in the opposite direction. A significant increase in the volume of traffic turning left can only be observed if the volume of traffic in the opposite direction falls to below 400 to 500 vehicles/h: at a volume of 400 vehicles in the opposite direction, approximately 215 vehicles can flow away to the left i.e. approximately 100 vehicles/h manage to find sufficiently large time gaps in the oncoming traffic. If the volume of through traffic falls still further, even more traffic is able to flow away to the left. However, if the volume of traffic turning left is greater than that of the through traffic stream or oncoming traffic stream, in which case the traffic turning to the right freely in case 6 must also be taken into account, the intersection geometry or the control type will generally be changed, because the traffic turning left would then become the dominant stream and would be assigned more than one lane, for instance, or the intersection would be redesigned. The curves shown in the hatched area are therefore not regarded as significant, in particular not for the unprotected release of traffic turning left.

If a *leading green* or *lagging green* is implemented in a two-phase control system, this time is not available for the oncoming traffic in the course of the cycle. This causes the volume of oncoming traffic to be reduced while at the same time increasing the volume of traffic turning left. The results for leading green and lagging green differ only slightly. If the additional release time for traffic turning left is restricted to a set duration, the potential volume of oncoming traffic falls rapidly to approximately 700 vehicles/h, as the release time must be reduced to 28 seconds.

In the case of *variable leading green/lagging green*, the curves are extended in favor of the volume of traffic turning left, as shown in Figure 14.1. In terms of capacity, the results for leading green are somewhat more favorable than for lagging green, because even when the diagonal green arrow appears, traffic turning left does not move off immediately, whereas in the case of leading green, traffic turning left moves off immediately when the release time begins.

In the case of protected release on all approaches (*four-phase control*), it is assumed that the phase for traffic turning left is activated depending on traffic conditions. The phase is used very often at the high-volume intersection under investigation. If phases for traffic turning left are only demanded sporadically when there is a low volume of traffic turning left, intermediate values of anything up to complete utilization of the release times for oncoming through traffic arise at values below 120 vehicles turning left per hour. Such cases should only occur rarely in reality, because special phases for traffic turning left are regularly incorporated, even if no demand has been registered by the detector, as the possibility cannot be excluded that vehicles have not been detected by the sensor. The more frequently left-turn phases are demanded and granted, the greater the reduction in the volume of oncoming traffic as a result of the changed release times. The difference in total capacity, however, shows that at volumes of traffic turning left significantly greater than 200 vehicles/h, protected release of the traffic turning left using a four-phase control system permits a greater number of oncoming vehicles than is the case if traffic turning left passes through the oncoming traffic unprotected.

Comparison of the control variants reveals the following basic findings:

– at volumes of up to approximately 120 vehicles turning left per hour, the fact that they flow away during the phase transition makes it possible for them to turn without a special phase and thus without restricting the oncoming traffic;

– even a small increase in the number of vehicles turning left (e.g. 200 vehicles turning left per hour) means that protected release using a four-phase control system provides clear benefits with respect to the oncoming traffic capacity;

– the maximum advantage provided by protected release lies in the region of 350 vehicles/h, where the volumes of oncoming traffic and traffic turning left are approximately balanced;

– from the perspective of capacity, it is better to provide for protected release of traffic turning left during a lagging green, provided that this is programed to be variable on demand. This release time cannot, however, be extended without limit, because a minimum green time is reserved for oncoming through traffic, and there may also be traffic turning left in the opposite direction, which should also be granted a release time. For this reason, the curves for these volumes of traffic with a variable release time terminate in the middle of the range of traffic volumes.

Figure 14.2 shows the effects of these programs on the overall capacity of the intersection under the same conditions. This indicates the total volume of traffic that a signal-controlled intersection can handle when certain proportions of this total volume are caused by traffic turning left and when different control systems are installed. In detail, Figure 14.2 indicates the following:

– The maximum total capacity of the intersection under investigation here is just under 7,000 vehicles/h. This is achieved when two phases are used, the proportion of traffic turning left does not exceed 7% and all the traffic turning left is able to flow away during the phase transition.

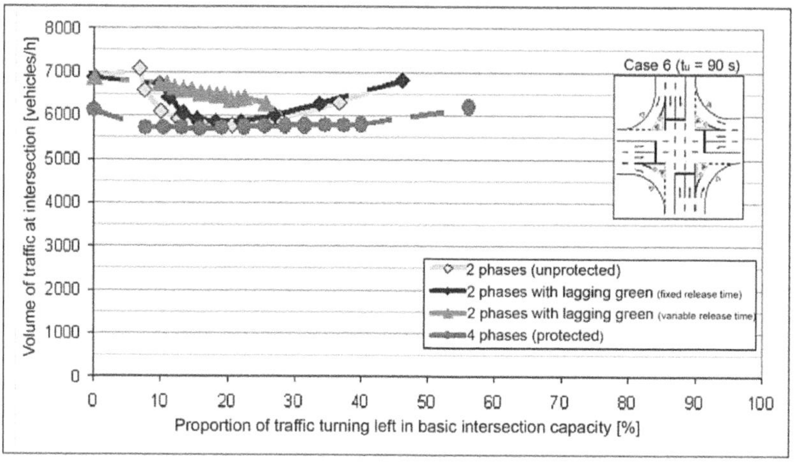

Figure 14.2. *Volume of traffic at an intersection using different control types (large intersections with triangular islands (case 6) with a cycle time of 90 seconds). For a color version of the figure, see www.iste.co.uk/jacob/safety.zip*

– In the case of protected release for the traffic turning left using a four-phase control system, the total capacity is reduced to a little more than 6,000 vehicles/h as a result of the necessary transition times and the need to temporarily block traffic turning right alongside the triangular islands (a protected phase for traffic turning left may only be activated if no conflicting stream is released at the same time). Under certain circumstances, this reduction could be avoided if the triangular island is widened to such an extent that free traffic turning right no longer flows into the intersection with the need to yield, but is instead introduced into the through lane in a merge area.

– As the proportion of traffic turning left increases between 7% up to approximately 18%, the capacities supported by different types of control approach each other and reach comparable values of approximately 5,800 vehicles/h between 20 and 25%.

– If the proportion of traffic turning left continues to increase, a form of control with no protection for traffic turning left provides greater overall capacity, because traffic turning right can flow freely.

– The apparently high capacity provided by control methods with lagging green, above all if this is implemented at a variable duration, only applies on condition that the traffic turning right freely can flow away without being signal-controlled. However, for reasons of safety and capacity, this presupposes the geometrically complex solution already mentioned, in which merge lanes are used to allow this traffic turning right to merge into the main stream without difficulty.

To summarize, this means that the approximately 20% increase in capacity of the intersection under these geometric conditions results from the fact that the opposing streams of traffic turning right must be blocked temporarily during protected release of the traffic turning left. This improvement only arises, however, if the proportion of traffic turning left forms significantly less than 10 or 30% or more of the overall volume of traffic. Under traffic volume conditions that occur frequently, the overall capacities only differ marginally.

14.2.6.2. *Large intersection without triangular islands (case 4)*

A large intersection without triangular islands and with a cycle time of 90 seconds was investigated in order to provide a comparison with case 6 (the corresponding geometry with triangular islands). This case was also calculated using a cycle time of 60 seconds. This also serves to provide a comparison with small intersections with only one lane per approach that tend to be operated with short cycle times.

In this case, the oncoming traffic is made up of through traffic and traffic turning right. In contrast to an intersection with triangular islands, the volume of oncoming traffic is just under 1,600 vehicles/h, as two lanes are available. The difference in transition times results in a slight reduction below twice the volume of traffic from case 6.

From a qualitative perspective, the reductions in traffic volumes using different phase systems show a similar behavior to that in case 6 (with triangular islands), but the quantitative effect is considerably greater.

The reductions in capacity in the case of unprotected release are considerable, because the oncoming traffic occupies two lanes. This means that the probability of sufficiently large time gaps occurring to allow traffic turning left to pass through the oncoming traffic only arises to a sufficient extent when the volume of oncoming traffic is very small (less than 500 vehicles/h). Protected release using a four-phase system, on the other hand, makes it possible to handle between 800 and 1,000 vehicles/h without difficulty when the volume of traffic turning left is high. This results from the fact that traffic turning right can be released simultaneously with traffic turning left originally traveling in a parallel direction.

The results of the simulations show that:

– the maximum total volume of traffic is 6,300 vehicles/h;

– the reduction in the overall capacity of the intersection in the case of unprotected traffic turning left is considerable;

– in the case of streams of traffic turning left protected by a four-phase control system, the overall capacity of the intersection remains at around 6,000 vehicles/h, irrespective of the proportion of traffic turning left.

At a cycle time of 60 seconds, the overall capacity of the intersection is reduced slightly to 6,000 vehicles/h. The capacity is lower as a result of the greater proportion of transition times.

14.2.6.3. *Medium-sized geometrically symmetrical intersection (case 3)*

The intersection under investigation here represents a relatively frequent case in which there are two lanes on each approach. The right-hand lane is for through traffic and traffic turning right and the left-hand lane is for traffic turning left. Calculations were carried out with a cycle time of 90 seconds. Separate protection or release for traffic turning right is not possible because the lane is shared with through traffic, with the result that all oncoming traffic must be blocked if traffic turning left is protected when released.

The results of the simulations show that with a low volume of traffic turning left, up to around 830 vehicles/h can flow away in the oncoming direction. The capacity for traffic turning left to flow away during the phase transition is exceeded at more than 120 vehicles turning left per hour, and further possibilities for traffic turning left are only available if large time gaps in the oncoming stream occur so that it is necessary for the volume of oncoming traffic to drop considerably, e.g. to less than 400 vehicles/h. At these volumes, the capacity of oncoming traffic is regularly considerably more favorable when a four-phase control system or variable lagging green is used.

The overall capacity of the intersection in relation to the proportion of traffic turning left clearly shows the advantage of protected release in the case of larger proportions of traffic turning left in the overall capacity:

– with a proportion of up to 14% of traffic turning left, the two-phase system provides significantly greater capacity than all other types of control. Nevertheless, when the proportion of traffic turning left is expected to be between 20 and 40%, this type of control is particularly unfavorable;

– protected release for traffic turning left (four-phase control) results in an overall capacity of approximately 3,000 vehicles/h for the intersection (which corresponds approximately to half the capacity of multi-lane intersections in case 4);

– a control system with variable lagging green results in the highest capacities. The total capacity is approximately 3,500 vehicles/h. This solution can, however, only be considered when the traffic volumes are asymmetric, as it can only be used for one direction.

Overall, therefore, we see that with larger volumes of traffic turning left, protected release always delivers better results than the two-phase system, even from the perspective of capacity. The proportion of traffic turning left is, however, somewhat higher than with multi-lane intersection approaches, as the volume of traffic in the through streams can reach considerably higher values. The absolute size of the critical streams of traffic turning left is approximately 500 vehicles/h across all approaches in both cases.

14.2.6.4. *Small intersection with one lane on all approaches (case 1)*

This geometry causes the lanes to be used as mixed lanes for all traffic stream types. The result of this is that the traffic behaves in a completely different manner from the cases studied previously, because traffic turning left cannot flow separately from the other streams of traffic, and all other streams of traffic are also affected by traffic traveling in the same direction and turning left. The strategy previously used to simulate full capacity only delivers meaningful results to a certain extent, because the traffic turning left is not subject to any restrictions, and as a result, the traffic streams develop with a virtually equal distribution without further constraints. For this reason, a second series of simulations was calculated in which the volume of traffic turning left was restricted to a maximum proportion of around 10% of the overall volume of traffic on each approach.

The following applies at a volume of traffic turning left and oncoming traffic for both load cases (10 and 33%):

– if it is assumed that there is no traffic turning left, the maximum volume of traffic is 800 vehicles/direction;

– even at low volumes of traffic turning left, the volume of oncoming traffic decreases considerably, irrespective of whether only two phases are used or lagging green of fixed or variable duration are also used;

– if the volume of traffic turning left is reduced while its proportion of the volume of traffic remains the same, the quantity of oncoming vehicles is also reduced. If more traffic turning left is permitted and the proportion is increased up to 33%, the volume of oncoming traffic is necessarily also reduced;

– protected release is only possible if each approach is individually released.

As far as the overall capacity of the intersection is concerned, this means that approximately 3,200 vehicles/h can pass through the intersection if there is no traffic turning left. This value decreases considerably to values below two-thirds of the original level even if there is only a small proportion of traffic turning left without any protected release or with a fixed or variable lagging green.

If each approach is individually released, 1,200 vehicles/h can pass through the intersection. This represents fourfold the volume of traffic on the individual approaches. This value remains constant and is not influenced by the proportion of traffic turning left.

14.2.6.5. *Combination of different approaches at an intersection*

In real life, different approaches are very frequently combined with each other at intersections. In cases such as this, the relevant representation of traffic volumes in the conflict areas can be used to derive the maximum capacities that arise by combining the volumes in the traffic streams. Table 14.1 shows all theoretically possible combinations of the four different approaches that were investigated. Combinations which were investigated are highlighted in color. Other meaningful combinations are indicated.

This results in values which lie between the relevant overall conditions for uniform constructions. These combinations were derived by drawing analogies from the existing results without performing any additional simulations. This means that no new transition times were taken into consideration. The results show that the differences in capacity between protected and unprotected control systems are confirmed for medium to high volumes of streams of traffic turning left. Generally, the variant with protected control delivers considerable benefits for the overall capacity of the intersection. With the exception of small proportions of traffic turning left (up to 10%), two-phase control did not deliver greater capacity in any cases.

<table>
<tr><th colspan="5">Primary Direction</th><th colspan="4">Secondary Direction</th></tr>
<tr><th rowspan="2">Intersection type</th><th colspan="3">Number of Lanes</th><th rowspan="2">Triangular islands</th><th colspan="4">Intersection type</th></tr>
<tr><th>Left-turn</th><th>Through</th><th>Right-turn</th><th>Case 1</th><th>Case 3</th><th>Case 4</th><th>Case 6</th></tr>
<tr><td>Case 1</td><td colspan="3">1</td><td>no</td><td>+</td><td>–</td><td>–</td><td>–</td></tr>
<tr><td>Case 3</td><td>1</td><td colspan="2">1</td><td>no</td><td>+</td><td>+</td><td>–</td><td>–</td></tr>
<tr><td>Case 4</td><td>1</td><td>1</td><td>1</td><td>no</td><td>–</td><td>+</td><td>+</td><td>–</td></tr>
<tr><td>Case 6</td><td>1</td><td>1</td><td>1</td><td>yes</td><td>–</td><td>+</td><td>+</td><td>+</td></tr>
</table>

NOTE.– Intersection types: + sensible; – not sensible; ■ investigated; □ combined

Table 14.1. *Categorization of intersection types*

14.2.6.6. *Protected release of traffic turning left as leading green, lagging green or a combination of both, with fixed and variable durations for each*

For reasons of traffic safety, the lagging green option is generally chosen if the release time is extended in favor of traffic turning left. In other words, when the release time for the oncoming traffic has ended, the oncoming traffic is stopped, and additional time (lagging green) is provided for traffic turning left. If the additional release time is provided before the oncoming traffic is released, this is referred to as leading green. The different effects of various leading green and lagging green variants were tested for case 3 and a cycle time of 90 seconds.

This showed that:

– leading green achieves somewhat higher capacities than lagging green;

– regulation using leading green or lagging green is only possible in one direction of travel. The opposite direction must do without a separate release time for traffic turning left. This therefore only makes sense if the volumes of traffic are asymmetric;

– provision for both directions of travel necessarily leads to a combination of leading green and lagging green. This permits a greater volume of traffic turning left at higher volumes of oncoming traffic. The fact that signal states are difficult to understand must, however, be taken into account;

– variable leading times or lagging times can only be granted limited durations.

14.2.6.7. *Effects of different cycle times*

As a rule, shorter cycle times always cause the loss resulting from phase transitions to be proportionally greater, because the transition times must be assigned values of the same absolute duration. This means that the intersection capacity will generally rise as the cycle time rises. Capacity does not, however, increase linearly with the cycle time, because the time losses have a proportionally smaller impact with longer cycle times.

14.2.6.8. *Possible volumes of traffic turning left when the volume of oncoming traffic is high*

With all multi-lane variants of intersection approaches and low volumes of traffic turning left, it is the case that up to approximately 120 vehicles turning left/h can flow away during the phase transition (at a cycle time of 90 seconds) and that through traffic is therefore not restricted even in the event of unprotected release. Up to this volume of traffic, protected release will always result in reduced capacity for the oncoming traffic, because the time for the special phase must be taken away from the oncoming traffic. The greater the number of cycles per hour (i.e. shorter cycle times), the greater the number of vehicles turning left that can be accommodated during the phase transition. On the other hand, however, the number depends on the geometry of the intersection (number of spaces that the vehicles can occupy in the intersection).

The negative impact on capacity of the special phase for low volumes of traffic turning left is lower if lagging green is used. This is, however, only possible in one direction of travel (asymmetric solution).

As soon as more than 120 vehicles wish to turn left, protected release for these streams sometimes has a considerably higher capacity than the two-phase solution. Setting up a lagging green solution assuming that the volumes of traffic involved are asymmetric) always achieves better results, which can be continued into the realm of higher volumes of traffic turning left by using variable release times. If high volumes of traffic turning left occur in both directions of travel, a four-phase control system must be selected in the majority of cases if leading green is to be avoided (because of the associated safety issues).

The maximum possible volume of traffic turning left in relation to the control method is shown for different geometrical configurations in Figure 14.3. The values shown there have been taken from the graphs and rounded to practical values. The volume of traffic per lane is assumed to be 600 vehicles/h (i.e. 1,200 vehicles/h in the case of two lanes in the opposite direction). It can be seen that only marginally more vehicles can flow away in the event of two-phase control than can flow away during the phase transition. In the case of protected release, this value is sometimes considerably larger, i.e. in particular in cases with multiple lanes, a considerably higher volume of traffic turning left can be handled in the special phase by granting a greater proportion of time. Variable lagging green leads to the greatest number of vehicles turning left flowing away, but it can only be used for one direction of travel.

14.2.6.9. *Overall capacities of intersections with different types of control*

The overall capacity at an intersection reaches approximately 3,300 vehicles/h in the case of two-lane approaches and a little over 6,000 vehicles in the case of three-lane approaches. If traffic turning right can flow away unhindered (alongside triangular islands that are not signal-controlled), this value increases to almost 7,000 vehicles/h.

As soon as a significant number of vehicles turning left arises, the overall capacity of the intersection is reduced, particularly in the case of the unprotected release, at a 15% proportion of traffic turning left (two-lane approaches) or 7% proportion of traffic turning left (three-lane approaches). In contrast, multi-phase systems only show a slight reduction of the intersection capacity, as the proportion of traffic turning left rises. Under favorable conditions, the overall capacity remains virtually constant. Protection of traffic turning left by means of lagging green, particularly if this can be made variable, sometimes results in considerably higher overall capacities than with four-phase control, at least up to the range of a 20% proportion of traffic turning left.

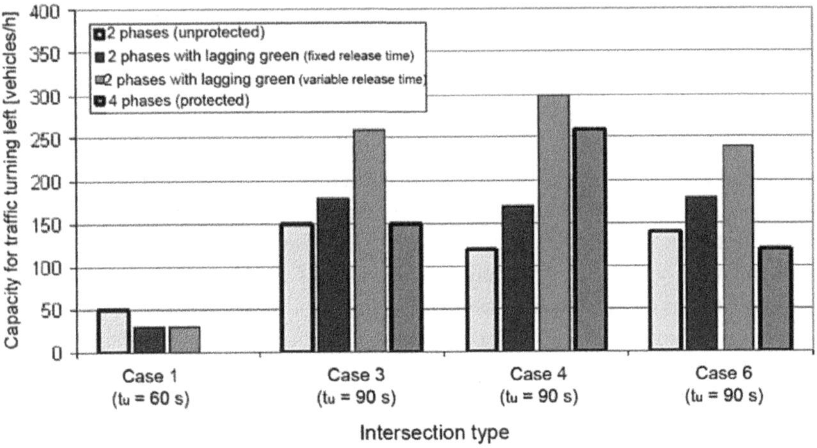

Figure 14.3. *Capacities for traffic turning left at a volume of oncoming traffic of 600 vehicles/(h*lanes) for different control variants for traffic turning left and different intersection types – results of the simulation runs. For a color version of the figure, see www.iste.co.uk/jacob/safety.zip*

Large intersections with traffic turning right freely (case 6) are also special cases: in such cases, only protection with lagging green delivers greater capacity compared with an unprotected two-phase system, whereas the multi-phase systems can only achieve a maximum of the same capacities for the midrange of the proportion of traffic turning left. The reason for this is that unprotected two-phase systems allow traffic turning right to constantly flow freely. This means that this effect only occurs when there are high volumes of traffic turning right. Turning this conclusion around, a large proportion of traffic turning left would also arise, with the result that the values for a low volume of traffic turning left will not be realistic.

14.3. Recommendations

The greater level of safety provided by multi-phase systems means that unprotected two-phase systems should in principle no longer be used. Traffic-related reasons mean that in isolated circumstances, capacity may be restricted as a result of the additional protection for traffic turning left. The following recommendations for use are made:

1) With high volumes of traffic, e.g. in the case of traffic volumes greater than 6,500 vehicles crossing an intersection per hour, it is necessary to create multiple lanes for individual traffic streams. In this case, unprotected release of traffic turning left should never be considered because it is not possible to ensure safety.

2) If the volume of traffic is between 4,000 and 6,500 vehicles/h, three-lane intersection approaches are generally necessary from all directions. Assuming a uniform distribution of traffic relations, the average proportion of traffic turning left is 25%.

3) In the case of overall volumes of traffic of 3,000 to 4,600 vehicles at peak hours, mixed geometric types of intersection will occur. For instance, in the primary direction, the lanes will be divided in a similar way to case 4 (three lanes without triangular islands) and in the secondary direction, two-lane approaches will be used as in case 3. If the volume of traffic is distributed uniformly, i.e. if the proportion of traffic turning left is around 25% of the overall capacity, a four-phase control system will also provide ideal traffic conditions in such cases. Lagging green may also be considered as an option here, provided that the situation can be avoided where traffic turning left passes through the two lanes of oncoming traffic.

4) In the case of traffic volumes between 2,000 and 3,000 vehicles in peak hours, lagging green should be considered if the proportion of traffic turning left is below 20%. If the proportion of traffic turning left is greater, a four-phase control system will always be advantageous.

5) Single-lane approaches should never be planned for traffic volumes between 1,000 and 3,000 vehicles at peak hours. Only a two-phase system of traffic light control would deliver sufficient capacity. Establishing a lane for traffic turning left must be called for in such circumstances (and can also be justified economically).

6) At capacities of up to 1,200 vehicles at peak hours, single-lane approaches with protected release of traffic turning left can be established. It is recommended that these are used up to approximately 1,000 vehicles/h.

14.4. Conclusion

The results of the study can be summarized as follows:

– In the case of large intersections with a high volume of traffic, a separate signal for traffic turning left must always be provided. This considerably increases the safety for traffic turning left and oncoming vehicles, helps to avoid and reduce conflicts between vehicles turning left and pedestrians and cyclists at intersections.

– In the case of large intersections with a medium volume of traffic and small intersections with a high volume of traffic, "protected" systems for traffic turning left are the most efficient solution both in terms of traffic flow and traffic safety.

– Solutions with no separate signal for traffic turning left can only be justified at small intersections where there is no lane for traffic turning left and where the volume of traffic is low.

– It may be necessary to completely prohibit traffic from turning left.

– The costs for converting the traffic signal system to a separate phase for traffic turning left generally lie considerably below the accident costs that can be avoided, and these accidents are generally severe.

If a free flow solution or a roundabout cannot be considered, intersections controlled by traffic signals may have a multi-phase system in order to provide both high capacity and a high level of traffic safety. Only traffic light systems with separated phases also for cyclists and pedestrians can help to improve road safety significantly at intersections. Switching off traffic signals at night has a negative impact on road safety and leads to small energy savings.

14.5. Bibliography

[ALL 01] Allgemeine Verwaltungsvorschrift zur Straßenverkehrsordnung (General Administrative Regulations on the Road Traffic Regulations) (VwV-StVO) dated 26 January 2001 (German Federal Bulletin p 1419, 5206) last amended 2009-07-17 (German Federal Bulletin No. 110 dated 2009-07-29, p. 598), 2001.

[ECK 02] ECKSTEIN K., MEEWES V., Sicherheit von Landstraßen-Knotenpunkten. Mitteilungen Nr. 40 des Institutes für Straßenverkehr (ISK), (Road safety of intersections outside built-up areas) Cologne, 2002.

[ROA 92] ROAD AND TRANSPORTATION RESEARCH ASSOCIATION (FGSV), Richtlinien für Lichtsignalanlagen (RiLSA) (Guidelines for Traffic Signals), Cologne, 1992.

[ROA 10] ROAD AND TRANSPORTATION RESEARCH ASSOCIATION (FGSV), Richtlinien für Lichtsignalanlagen (RiLSA) (Guidelines for Traffic Signals), Cologne, 2010.

[SCH 10] SCHOLZ TH., ORTLEPP J., Verkehrstechnische Auswirkungen der Sonderphase für Linksabbieger an Knotenpunkten mit Lichtsignalanlage (The impact on traffic of the special left-turn phase at signal-controlled intersections), Published by: German Insurers Accident Research (UDV) of the Gesamtverband der Deutschen Versicherungswirtschaft e.V. (GDV, German Insurers' Association), Berlin, 2010.

15

Safety Inspection and Management of the Road Network in Operation

The EU Directive 2008/96/EC on road infrastructure safety management, defining the subject matter and scope, requires "the establishment and implementation of procedures relating to road safety impact assessments, road safety audits, the management of road network safety and safety inspections by the Member States". Starting from the results of the EU research project "Identification of Hazard Locations and Ranking of Measures to Improve Safety on Local Rural Roads" (IASP), a Safety Risk Index (RI) has been formulated as a measure of risk that uses Safety Inspection as a primary source of information associated with design consistency assessment of the horizontal alignment. Moreover, due to its quantitative evaluation, RI can be used to determine the benefits and cost/benefit ratio of different intervention strategies. Then, as there are often situations when the available budget will not be sufficient to undertake all the measures in the road network, an "RI based" optimization algorithm for selecting projects to be included in the budget was defined.

15.1. Introduction

The European Directive 2008/96/CE requires member states to establish and implement procedures related to road safety impact assessments, road safety audits, the management of road network safety and safety inspections. In the Directive, great emphasis is given to road Safety Inspections (SI) as an effective tool for the safety management of existing roads; stating: "Once road sections with a high accident concentration have been treated and remedial measures have been taken, safety inspections as a preventive measure should assume a more important role.

Chapter written by Salvatore CAFISO, Alessandro DI GRAZIANO, Grazia LA CAVA and Giuseppina PAPPALARDO.

Regular inspections are an essential tool for preventing possible dangers for all road users, including vulnerable users". Within the Article 6 "Safety Inspections" it is also recognized that the member states shall carry out safety inspections on existing roads in order to identify the road safety related features and to prevent collisions. These inspections should be performed periodically and by a competent entity. Member states are also encouraged to apply this directive to other national roads, which are not part of the Trans–European Road Network. Two-lane rural roads – that is, the class of highways investigated in this study – are the part of the national road network providing access to land and towns. On local rural roads, low traffic volume results in a relatively small number of crashes per unit length. Moreover, databases on traffic flows and crashes are often characterized by a lack of reliable data. Therefore, for this class of road, traditional approaches to network screening based on crash data can be affected by uncertainty and road safety inspections should undertake a more relevant role.

Despite the significant role of SIs depicted by the Directive, due to the qualitative nature of the process, relevant management issues remain in the use of SI as source of information to prioritize those treatments in road sections which produce the highest accident reduction potential and/or the highest safety benefit-cost ratio.

Based on these considerations, and as a result of the IASP EU research project, a Risk Index (RI) has been formulated to predict the frequency and severity of accidents by using SIs as primary source of information together with design consistency assessment of horizontal alignment [CAF 07]. As the next step in the research, a formal method for selecting treatments to maximize the safety benefits of the investments was defined [CAF 11]. The procedure makes it possible to identify the intervention strategies that produce the greatest safety benefits for variations in the RI, while being compatible with the available annual budget.

The goal of the study presented in this paper was to improve the quality and efficiency of the periodic inspections of the road network by using low cost equipment and specific software tools developed to support inspection tasks, computation of RI and prioritization of intervention alternatives.

This chapter is divided into three main sections presenting:

– road SIs tools in Europe;

– the new software tools designed to improve efficiency and safety during SI and to carry out actual values of RI;

– prioritization and optimization procedures based on RI to identify the intervention strategies that produce the highest accident reduction and/or the highest safety benefit-cost ratio.

15.2. Road safety inspection tools in Europe

Based on the state of the art review, it was concluded that SI is regularly used in at least sixteen European countries (Austria, Belgium, Cyprus, Denmark, Estonia, France, Germany, Hungary, Italy, Latvia, Netherlands, Norway, Poland, Spain and United Kingdom) [NAS 11].

In the following, we report relevant examples of software tools developed to support SI.

15.2.1. *Tool for SI in Austria*

The legal basis for SI in Austria is the RVS guideline 02.02.34, which was published in 2007.

Nast consulting [NAS 11] has developed the software Eves (Electronic Safety Recording System) to carry out road SI. The application fields of the system were divided into two groups, depending on the devices on which the system was supposed to run. For the higher road network, where SI is performed by car, the system has to run on a notebook computer; for the lower road network (e.g. cycle paths, footpaths, etc.), where SI is performed on a bike or on foot, the system has to run on a mobile device.

The system is able to:

– easily register safety relevant events during the inspections;

– automatically locate the gathered events using GPS (Figure 15.1);

– provide the user with a clear user interface to give an overview of the recorded events;

– facilitate post-processing of the recorded events by providing editing by means of video synchronization (Figure 15.1);

– facilitate reporting by automatically generating reports according to pre-defined structures.

15.2.2. *Tool for SI in Norway*

The Norwegian Public Roads Administration published the "Road Safety Audits and Inspections" handbook in 2006. According to this handbook, SI in Norway consists of three steps: Preparation, Inspection and Reporting. In the preparation step a "Vidkon inspection" is carried out by using a Vidkon program (the road section is driven through several times, pictures are taken every 20 meters or a video recording is performed which will be reviewed afterwards in the office, see Figure 15.2). During the Vidkon inspection the description of problems and proposals for remedial measures have to be completed in the report form. The collection of all this data makes the field inspection simpler and quicker. After the completion of the inspections, some supplementary work follows, such as finalizing the road SI report (which is the responsibility of the road safety inspector), reporting that the inspection is completed and the follow-up measures (both tasks for which the project owner is responsible). Next to standard report forms, the additional software Tess is used. This program includes a listing of all documentation and a risk matrix, which can help prioritizing possible intervention measures. It also contains a statistics sheet, which provides all pictures for the entire section [NAS 11].

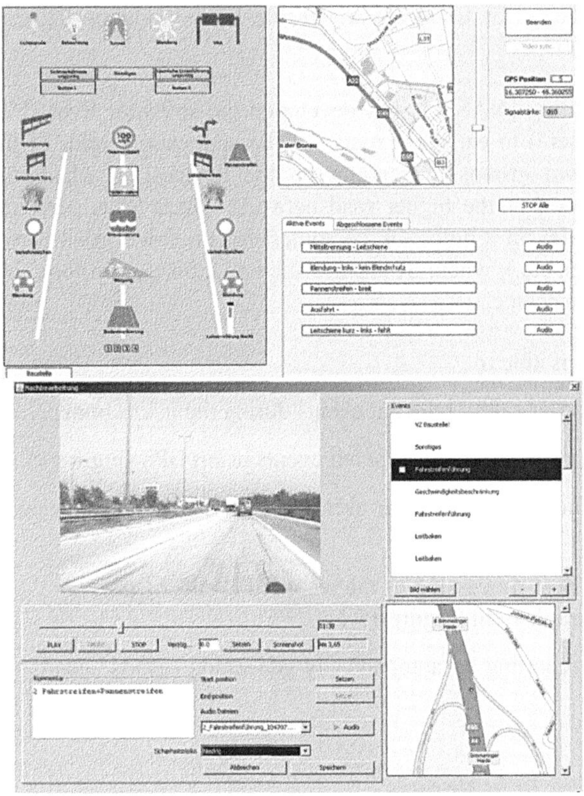

Figure 15.1. *Screenshot of graphical user interface of Eves, a) during inspection b) and post-processing [NAS 11]*

Figure 15.2. *Road Safety Inspection in Norway using Vidkon [NAS 11]*

15.2.3. *Tool for SI in Ireland*

The National Roads Authority published the "Safety Management Procedures for National Roads in Ireland" in 2012, that covers the procedures for management of road safety on the national road infrastructure in Ireland [CUL 12]. It describes the roles and responsibilities of those engaged in the management of road safety and outlines the procedures to be followed to address the requirements of the EU Directive 2008/96/EC on Road Infrastructure Safety Management (RISM) and its transposition into Irish Law under S.I. No. 472 of 2011.

The SI involves a site survey and video footage of the full route during day-time and night-time. The SI team uses an app called "UBIPIX" (http://www.ubipix.com/) to video the route and when it is uploaded, it is GIS located, which will enable the NRA to locate the problem areas easily, and also to review the issues noted (Figure 15.3). From the site visit and video footage the inspection team prepares a report identifying the safety issues. These safety issues are then assessed with respect to the Collision Information and the collision trends and a risk assigned to each safety issue. This report is then issued to the NRA who assesses the safety issues identified and notify other departments, such as road maintenance. For the remaining safety issues the NRA Safety Department will develop recommendations to rectify the safety concerns, which might range from signage and road markings to full re-alignment.

Figure 15.3. *Road Safety Inspection in Ireland using UBIPIX*

This short overview is not exhaustive of the whole scenario of tools available to support the safety inspection process at present, but it makes it possible to point out the main advantages and potential for improvement:

In-field inspection: inspection is a time-consuming process where a lot of information must be collected in a short time by the inspectors; these systems support data recording and GPS/GIS positioning in the road network.

Operator safety: all these tools allow the inspection to be carried out using a vehicle running at low speed, reducing the time spent by inspectors on the road and improving safety.

Reporting: review and team discussion in the office of checklists supported by geo-referenced video and photos improve quality of inspection reports, and facilitate reporting by automatically generating reports according to pre-defined structures.

Selection of treatments: list of documentation and risk matrix help in selecting and prioritizing countermeasures.

Despite the above-mentioned valuable advantages, all the offered tools are able to support the traditional safety inspection tasks, but the process remains subjective and qualitative, finalized to product reports and to identify potential treatments with priority for the sections at highest risk. Instead, a road agency needs specifically, as a final output of the process, a program working with prioritization at road network level considering all the potential treatments in the overall system, to select only those with the highest benefits or benefit-cost ratio, within the available budget.

15.3. Design of new software tools for road inspection

The design of a new system and software tools for Safety Inspection was based on research, practical experience and lessons learned from existing tools; with the aim to maintain their advantages, but also to apply a new comprehensive approach to road network safety management.

15.3.1. *IASP procedure*

In the framework of the European project "Identification of Hazard Locations and Ranking of Measures to Improve Safety on Local Rural Roads" (IASP), a new surrogate safety measure, the safety Risk Index (RI), was defined to supplement crash investigation studies in the safety evaluation of two-lane rural highways, by using data from SIs and design consistency models [CAF 07].

The RI was formulated by combining the risk components related to:

1) the exposure of road users to road hazards, Ef;

2) the probability of a vehicle being involved in a collision, AFf;

3) the resulting consequences should a crash occur, ASf.

$$RI = Ef \times AFf \times ASf \quad [15.1]$$

More specifically, the exposure factor (Ef) measures the exposure of road users to road hazards, depending on the length of the segment under consideration (L) and the average annual daily traffic (AADT).

The accident severity factor (ASf) measures the severity of crashes based on both roadside hazards, identified by a Safety Inspection (SI), and the ratio between the 85th percentile of free flow speed and the posted speed.

The accident frequency factor (AFf) depends on the defects in road features affecting safety, which are assessed by SIs, design consistency and design standards checks:

– SIs relate to the main safety issues that can be effectively detected by means of inspections, and which are consistently present along two-lane rural roads (Accesses, Cross sections, Delineation, Markings, Signs, Pavement, Roadside, Sight distance). Identification and ranking are carried out by a visual survey of the road using a checklist purposely set up for the IASP project. Checklists used in SI procedures are filled in for both directions of the road, with a constant step (e.g. 200 meters) using well defined procedures and criteria for identifying and ranking safety issues with “high level problem”, “low level problem” or “no problem”. The IASP project also defined detailed guidelines to reduce subjectivity in the inspectors’ judgments.

Design consistency and standards checks refer to the safety evaluation of the horizontal alignment based on the design standards for tangents and the Safety Criteria proposed by Lamm [LAM 02] for curves.

The interested reader may refer to Cafiso *et al.* [CAF 07] for background and a more detailed discussion on RI.

15.3.2. *Hardware and software tools for in-field inspection*

As defined in the IASP project, in-field SIs are carried out involving three operators: the driver of the vehicle, and two inspectors positioned in the front and rear seats. The inspectors observe and evaluate road conditions, filling in two different checklists. This task requires relevant expertise and skill of the inspectors. The system presented in this paper frees the operator from having to link the position in the road segment to the checklist to be filled. The inspectors fill the checklists touching the screen of the tablet connected via Bluetooth to the GPS. In this way data are easily collected, stored and linked to the road segment stretch from where they are taken.

The system has the objective to simplify the operator's work, but opens up a whole set of issues that the software should solve without the user’s environment:

1) allows easy data insertion to the user;

2) links inspection data with a unique road section traveling the route in both directions;

3) stores all information to allow the post-inspection review.

In order to achieve an easy and low cost solution, it was decided to create an Android application that could run on every 9/10” tablet on the market. By means of an easy-to-use interface, the application allows the user to insert inspection data for each section of the road using his/her own tablet.

To improve accuracy in vehicle positioning, an external GPS receiver can be connected with the Android application running on the tablet. With respect to this aim, Bluetooth technology, currently presented in many devices, allows a connection to be established between two devices, in a safe and easy way. Finally, to associate images of the road space, frames recorded by a video-camera installed on the vehicle can be geo-referenced and linked to the inspection data through a system of synchronization based on the GPS UTC time used as a metronome for all devices.

The architecture and organization of the system is shown in Figure 15.4. If the GPS module is the "heart" of the system, the Data Acquisition Module is the "brain". This module uses GPS data to support the inspector task via the Graphic User Interface (GUI). The most important part of the infield inspection is the assessment of the safety issues and filling of the checklist for each road section of fixed length composing the road segment under investigation.

For each inspection two checklist forms are available: one for the operator in the front seat and one for the operator in the rear one. The checklist form may be modified before the inspection because it is saved in the form of an XML file, which can be edited at any time.

Once authenticated with the system, the user must choose the checklist form to compile (front seat, rear seat), and then connect the tablet with the GPS receiver, waiting until the calibration procedure can certify the appropriate accuracy. Before starting the inspection it is possible to associate a video-camera to the inspection data. After completing these initialization steps, the system is ready for the inspection execution. Figure 15.5 shows the GUI of the application. It allows the operator to perform each action simply by using touch buttons on the screen. It also shows the status of the GPS accuracy in terms of signal quality. A progress bar gives the operator information about the remaining time to complete the current checklist before starting a new one when entering in the next road section. In this way, the operator can ask the driver to slow down the vehicle speed if he/she has trouble in completing the form-sheet.

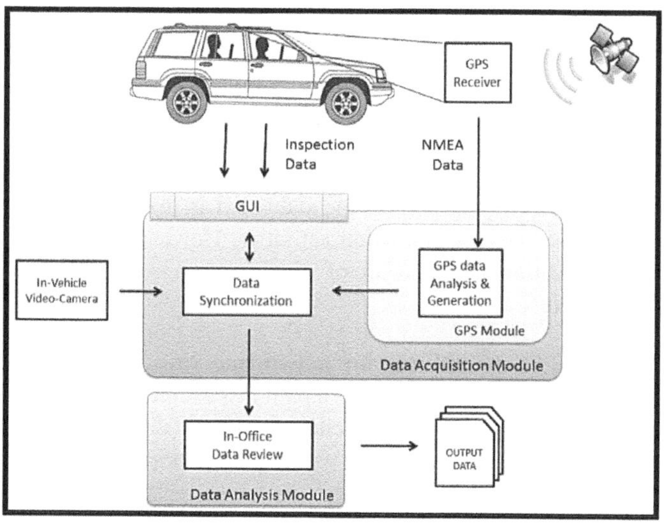

Figure 15.4. *Architecture and organization of the system*

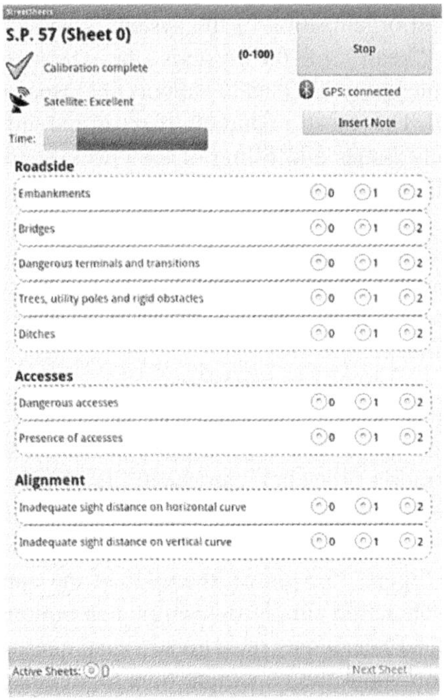

Figure 15.5. *GUI of the application*

Whenever the vehicle enters a new section of the road a new sheet tab is displayed in the lower area of the GUI. The tab color indicates whether the form is under editing (green), has been already filled (black), is empty (red) or only partially filled (yellow). The operator can also move from one form to another to add information or correct any possible errors.

The process of creating the tabs in the status bar is invisible to the operator, so he/she only needs to focus on evaluation activities. However, he/she can constantly keep track of the situation and be aware of which section is under evaluation and the status of all form-sheets.

If the inspection is carried out for a two-lane road, after the forward path inspection, the operator must begin a return path inspection in the opposite direction. Once the acquisition procedure is concluded in both directions, the application merges data obtained in the forward path with those obtained in the return path.

At the end of the infield inspection, the Data Acquisition Module exports all data in the form of text files that can be read and processed by the Data Analysis Module.

15.3.3. *Data Analysis Module for in-office review*

The Data Analysis Module concludes the last part of the inspection process. Once infield data acquisition phase is completed, in the office the Data Analysis Module allows the inspection team to review the checklists and, supported by the video, to fill missed information or to correct errors before drawing up the inspection report.

In this step, the Module puts together data recorded by the front seat operator and the rear seat operator. The user can investigate every single step of the inspection by moving from sheet to sheet and from time to time, with the opportunity to review the corresponding video sequence and, if necessary, modify the values of each checklist (Figure 15.6).

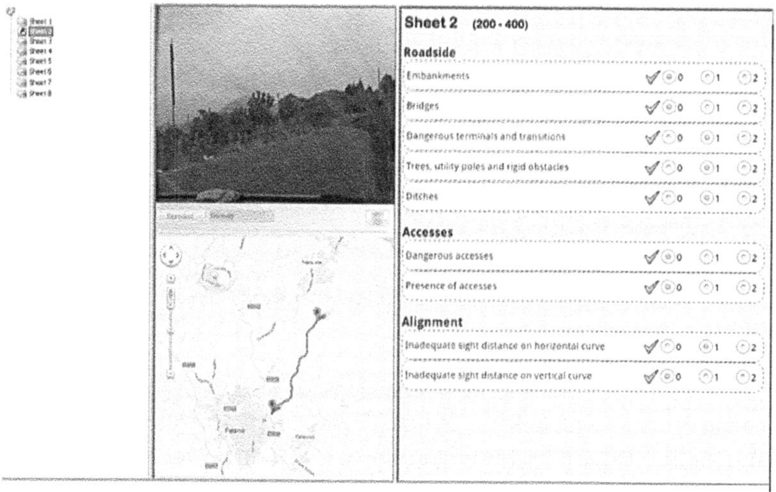

Figure 15.6. *Data Analysis Module for office review*

After completion of the review process, the user can export a file (.xls format) as an output of all the inspection data and use them for SI report and RI evaluation, as well.

15.3.4. *Optimization tool, SAFOPT*

RI is a quantitative measure of risk derived from the SI data acquired with the procedure described in the previous section. In this form RI can be used to define alternative investments that can deliver specified levels of safety. As the correlation between RI and crash estimation exists [CAF 07], it is possible to assume the RI value for each road section as an indicator of accident risk, associating to any reduction of RI, due to safety improvements, a potential reduction in the number of forecast accidents. Therefore, the economic appraisal, which could be performed to compare the benefits of a potential crash countermeasure (m) to its project costs based on crash network screening [HSM 10] can also be applied to safety network screening, considering the reference parameter of the benefits not linked to the reduction in the number of accidents, but to the reduction of RI (ΔRI). For the overall service life, the benefit $\Delta RI_{s,m}$ related to road segments (s) because of intervention (m), represents the total benefit during the analysis period of (T) years:

$$\Delta RI_{s,m} = \sum_T \Delta RI_{s,m,t} \quad [15.2]$$

where

$$\Delta RI_{s,m,t} = RI_{s,0,T} - RI_{s,m,t} \quad [15.3]$$

with

s(1, . ., S), s = index of the examined section and S=total number of sections into which the network is subdivided;

$\Delta RI_{s,m,t}$ = benefit during the lifecycle related to the road segment (s) attributable to intervention (m) at time (t);

$RI_{s,0,T}$ = risk index related to road segment (s) at the end of the lifecycle (T) if no intervention is implemented. $RI_{s,0,T}$ at time (T) is carried out from the first year $RI_{s,0,1}$ derived from SI increased by worsening due to lack of maintenance activities;

$RI_{s,m,t}$ = risk index during the lifecycle related to road segment (s) due to intervention (m) at time (t).

The optimization procedure requires a considerable number of calculations, so the specific software SAFOPT (SAFety OPTimization) has been developed to identify those intervention strategies that make it possible to obtain the maximum benefits within the limits of the annual budget. The whole procedure is detailed in a previous work [CAF 11], and in the following just a synthetic overview will be given. The working logic of SAFOPT is reported in Figure 15.7, which shows the part linked to the calculation of RI, connected to the choice of intervention strategy and finally to the optimization process.

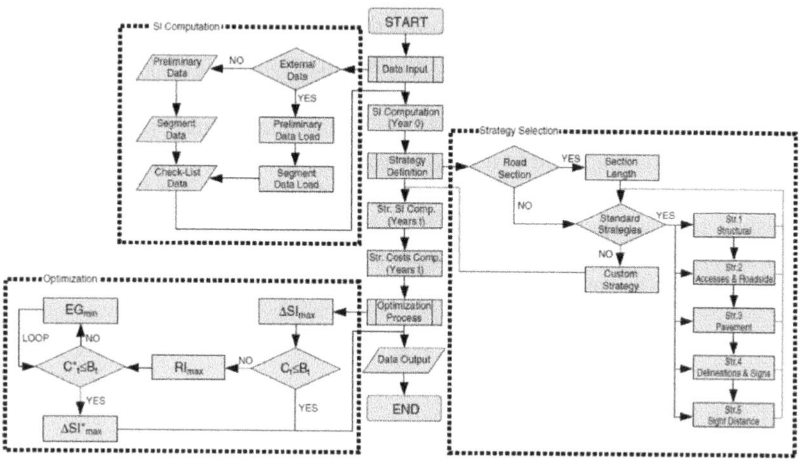

Figure 15.7. *Road network optimization procedure*

The operator can define his own intervention strategy. However, SAFOPT includes some standard solutions:

– strategy 1: structural interventions (cross section, geometry, and markings) carried out to adapt the cross section and geometry of the stretch to the design standards (according to the type of road involved);

– strategy 2: margins and accesses to improve the safety of the road margins regarding a reorganization of the accesses and the elimination of roadside problems;

– strategy 3: paving to re-establish optimal conditions regarding grip and surface evenness, followed by the refurbishment of markings;

– strategy 4: adjustments to delineations and road signs and;

– strategy 5: adoption of measures to improve sight distance by eliminating lateral obstacles with no intervention on the geometry.

The optimization process is carried out by means of a total enumeration algorithm, evaluating the number of years in the analysis period, the budgetary periods (that is, the time duration of allocated budget), the budgetary resource constraints (that is, the budgetary limits defined according to the spending period), and the annual values of ΔRI and costs for each road segment and each strategy adopted. Budget permitting, the best combination of strategies is sought for each spending period to obtain the greatest RI reduction (ΔRI_{max}). When the best combination exceeds the spending limits, an optimization process, based on economic boundary theory, is applied to identify the most efficient strategies to submit to further budget checks [ROB 98]. The result obtained is a combination of strategies (each to be applied to a specific road section in a specific year) that makes it possible to obtain the maximum RI reduction compatible with the spending limits established for each period. By reference to the optimal solution, an intervention program can be drawn up, which makes it possible to identify, year by year, the works to be carried out on each single stretch of the network and their relative costs.

15.4. Case study

A case study is presented to show how the procedure can be applied in Safety Road Network Management, taking into account safety issues in order to justify budget allocations and to select the best program of work. For example, a simple road network was adopted that was composed of about 20 km of two lane rural roads surveyed during the IASP project. Based on the SIs and Design Consistency Evaluations and Design Standards Checks, RIs were computed for the overall network and 28 homogeneous sections were identified. Starting from the "risk conditions", the standard SAFOPT list of strategies was taken into consideration to solve the safety issue, and the optimization process was performed to carry out the best solution for different budget allocations. As a final result, varying the investment budget, it is possible to carry out the corresponding benefit associated to the $\Delta RI_{s,m}$ values (Figure 15.8). Figure 15.8 shows how the present value (PV) of benefits (blue line) and the benefit/cost ratio (dashed red line) vary, increasing the PV of the investments in the cycle life of 10 years considered in the case study. This example is useful to point out which results can be achieved by using the proposed procedure. The study ascertains that each budget allocation gives the maximum benefits (i.e. reduction of crashes) as a result of the optimization process, and different investment costs will produce different benefits.

At program level, if the goal of the road agency is the maximum reduction in the number of crashes, with increasing expenditures the improvement in benefits is predicted and the budget allocation required to reach the maximum benefit (€3.7 M in the example) can be identified. This budget (€1.6 M in the example) is appropriate to solve, in the analysis period of 10 years, any relevant safety issue, and beyond this investment more works do not result in further safety improvements.

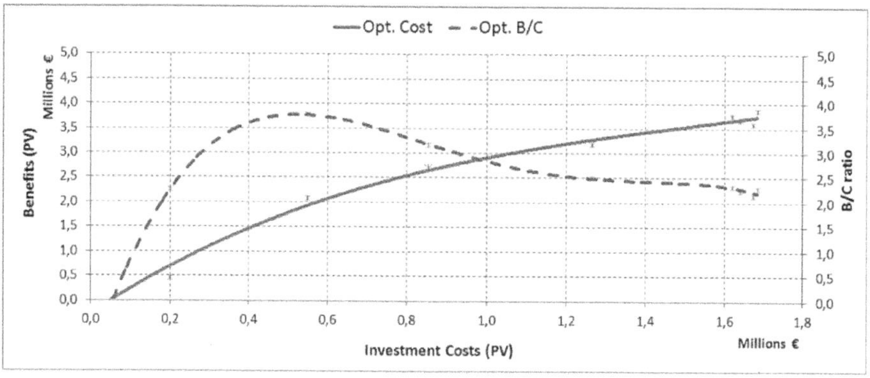

Figure 15.8. *Program level – PV of road network investments vs. benefits and benefit/cost ratio*

If the goal is the benefit/cost ratio, the maximum can be reached with a reduced investment in 10 years (€0.5 million in the example of Figure 15.8). If it is compared with the maximum benefit solution, this investment gives less benefits but leaves the budget (€1.1 M) available for actions in other parts of the road network. Based on the road agency policy, when the 10-year budget is selected, at the project level, the analysis gives the annual investments program in terms of sections to be treated, year and typology of countermeasure to be applied to reach the best result using the available budget. For example, Figure 15.9 reports a summary for an investment cost of €1.3 M in 10 years, with a maximum expenditure per year of €0.30 M, which gives high benefits (€3.2 M) with a B/C ratio of 2.5 (see Figure 15.8).

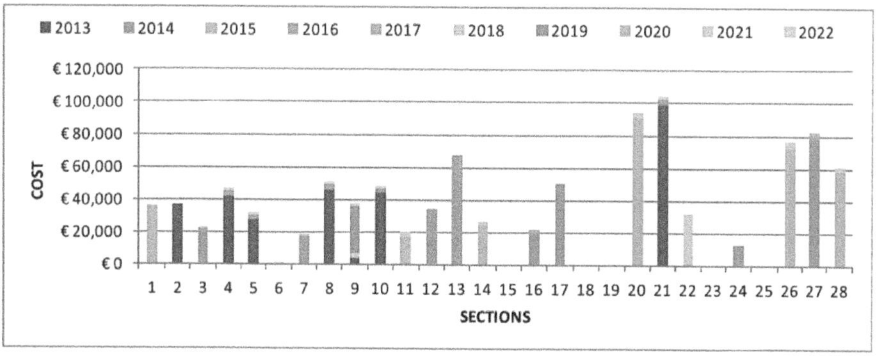

Figure 15.9. *Project level – annual budget allocation in the road network sections. For a color version of the figure, see www.iste.co.uk/jacob/safety.zip*

15.5. Conclusion

The European Directive 2008/96/CE places great emphasis on road Safety Inspections (SIs) as an effective tool for safety management of existing roads; relevant issues remain in the use of SIs as a source of information to prioritize measures and treatment sections with the highest safety benefit-cost ratio. Moreover, from a practical point of view, developing routing inspection on the whole road network gives causes for concern for both safety and cost issues. Some tools already exist and are regularly applied in European countries, but they are finalized to improve the process, which is completed with inspection reporting as the traditional output. To advance the quality and efficiency of the periodic inspections of the road network, a system of software tools supported by low cost equipment have been developed and presented in the paper. The proposed hardware and software system allows easy insertion of data for the user, solves the association problem between data and location in both directions, and after the in-field inspection allows an efficient post-inspection review in the office. Moreover, data collected are suitable to carry out the risk index (RI) defined in the EU IASP project as surrogate measure of safety. This quantitative measure, instead of the qualitative SI report, makes it possible to perform a cost benefit analysis. For road agencies facing the problem of budget constraints when selecting rehabilitation and maintenance projects, the proposed procedure is very effective because it is able to cover the whole process, from the identification of safety deficiencies, with a proactive approach (SIs and design consistency), to the evaluation of the intervention alternatives, providing the highest benefit within the available budget. Using the developed tools it is possible to identify safety issues of existing roads, assess the risks, prioritize the implementation of measures both at route level and at the network level, and aid the road's authority in the allocating budget for Network Safety Management. For practical applications, each road agency can modify the list of roadworks and strategies to take into account local materials and technologies, costs and maintenance needs.

15.6. Bibliography

[CAF 11] CAFISO S., DI GRAZIANO A., "Surrogate safety measures for optimizing investments in local rural road networks", *Transportation research record: Journal of Transportation Research Board*, no. 2237, TRB of the National Academies, Washington D.C., pp. 20–30, 2011.

[CAF 07] CAFISO S., LA CAVA G., MONTELLA A., "Safety index for evaluation of two-lane rural roads", *Transportation Research Record: Journal of Transportation Research Board*, no. 2019, TRB of the National Academies, Washington D.C., pp. 136–145, 2007.

[CUL 12] CULLEN H., "Road safety inspection in the context of the EU road infrastructure safety management directive", *Procedia of Summer School SIIV 2012 – Road Safety Management Theoretical Principles and Practical Application in the Framework of the European Directive 2008/96/CE*, Catania 24–28 September 2012

[HSM 10] FHWA, American Association of State Highway and Transportation Officials, *Highway Safety Manual (HSM)*, First Edition, USA, 2010.

[LAM 02] LAMM R., PSARIANOS B., CAFISO S., "Safety evaluation process of two-lane roads. A 10-year review", *Transportation Research Record: Journal of the Transportation Research Board*, no. 1796, TRB of the National Academies, Washington D.C., pp. 51–59, 2002.

[NAS 11] NAST CONSULTING, *Road Safety Inspection Schemes Review*, ERA-NET ROAD Project, Nr. 823129, 2011.

[ROB 98] ROBINSON R., DANIELSON U., SNAITH M., *Road Maintenance Management: Concepts and Systems*, Macmillian, available at: http://www.ubipix.com/, 1998.

PART 5

ITS and Safety

16

Improving Safety and Mobility of Vulnerable Road Users Through ITS Applications

In recent years, Intelligent Transport System (ITS) applications have assisted in reducing the number of fatalities in Europe. However, Vulnerable Road Users (VRUs) have not benefited as much as vehicle users. The EU-sponsored VRUITS project assesses the safety and mobility impacts of ITS applications for VRUs, assesses the impacts of current and upcoming ITS applications on the safety and mobility of VRUs, identifies how the usability and efficiency of ITS applications can be improved, and recommends which actions have to be taken at a policy level to improve ITS safety and mobility. This paper describes the results of the first phase of the project in which the critical scenarios for VRUs are identified, starting from accident data analysis, and following user needs based on focus groups and expert interviews. On this basis, the most promising ITS applications for VRUs are selected according to their potential to address the specific needs of VRUs.

16.1. Introduction

In recent years, both technological developments and research activities in the fields of Intelligent Transport Systems (ITSs) have primarily focused on motorized transport, to improve on safety and ecological standards by advancing equipment of vehicles and infrastructure. The uptake of ITS applications has assisted in the decrease of road traffic fatalities, particularly among passenger car occupants. However, Vulnerable Road Users (VRUs), such as pedestrians, cyclists, motorcyclists and moped riders have not enjoyed the same decrease in fatalities. Together, they account for 68% of the fatalities in urban areas [CAR 09].

Chapter written by Johan SCHOLLIERS, Daniel BELL, Andrew MORRIS, Alejandra Beatriz GARCÍA MELÉNDEZ and Oscar Martin PEREZ.

Motorcyclists account for 16% of fatalities, which is much higher than their contribution to traffic [CAR 09]. While some projects have considered VRUs from a safety viewpoint, they often aimed at avoiding or mitigating accidents with VRUs by equipping the vehicle and infrastructure. In the vehicle – infrastructure – human approach of ITS research, VRUs and their needs are not an active part of the "human" element in the ITS approach.

What is needed? The VRU must become an active, integrated element in the ITS, addressing safety, mobility and travel comfort needs of VRUs. The VRUITS project, which is sponsored by the European Commission and started on 1.4.2013, aims at actively integrating the "human" element in the ITS approach by focusing on the needs of all relevant stakeholder groups into the development and adaptation process of innovative ITS solutions, aimed at improving traffic safety as well as the general mobility of VRUs. The VRUITS project places the VRU road user in the center, assesses the impact of current and upcoming ITS applications on the safety and mobility of VRUs, identifies how the usability and efficiency of ITS applications can be improved, and recommends which actions have to be taken at a policy level to improve ITS safety and mobility. By applying a multi-disciplinary approach the VRUITS project aims at developing tools to evaluate, field-test and subsequently improve ITS for VRUs.

The first objective of the VRUITS project is to assess societal impacts of selected ITS applications, and to provide recommendations for policy and industry regarding ITS in order to improve the safety and mobility of VRUs. Both ex-ante and ex-post assessments of the applications are performed in order to come to a consolidated set of recommendations.

The second objective is to provide evidence-based recommended practices on how VRUs can be integrated in ITSs and on how HMI designs can be adapted to meet the needs of VRUs, and test these recommendations in field trials.

16.2. Methodology

Based on a comprehensive identification of situations for VRUs that are critical in view of traffic safety, making use of accident data, hospital data and in-depth studies, a taxonomy of the major critical scenarios for VRUs is derived, providing essential input to the development of ITS aimed at VRU safety.

In addition, based on an exploratory research approach applying focus group discussions, expert interviews and a literature review, opinions and needs of all relevant stakeholder groups are integrated into the discussion of VRU and traffic safety. Not only representatives of identified VRU groups but also of national and

European authorities, of infrastructure service providers and ITS related economy are focusing on this approach, resulting in a comprehensive catalog of requirements that need to be considered in the development and application of ITS systems aimed at VRUs.

Based on these findings ITS applications which affect VRU safety and mobility are identified, combining results regarding critical situations for VRUs in traffic and the benefits as well as potential issues and hazards of these systems. As a result, an inventory categorizing and prioritizing the existing and upcoming ITS services targeted at VRUs, also regarding positive or potentially negative impacts on safety and mobility, allows researchers and stakeholders to close existing gaps for new ITS and provides insight into the potential of existing technologies in this field.

The impact of selected services on safety, mobility and comfort is assessed based on accident data, literature and expert analysis. For the safety assessment, the methodology which has been developed for in-vehicle safety systems by Kulmala *et al.* [KUL 10] is adapted towards VRU applications. The methodology covers the three dimensions of road safety – exposure, crash risk and consequence. It not only considers the intended effects, that is the safety factors that the ITS is designed to influence, but also the unintended effects, e.g. the effect of behavioral adaptation [KUL 10], in a comprehensive, systematic and transparent way. In the analyses, the three main factors of traffic safety were covered by nine behavioral mechanisms as first described by [DRA 98]. The method for quantifying the safety effects explicitly takes into account the general accident data available from e.g. the CARE database. The accident data is further subdivided according to different background variables, such as vehicle type, collision type, road type, weather and lighting conditions and location. The inputs for the safety estimates were: expert estimates of effects by mechanism, the total number of accidents/fatalities or injuries from accident data and the share of accidents in each variable category. Some estimates were based on evidence from the literature, others on expert opinion due to lack of data or incomplete data [KUL 10]. For improving the validity and reliability of expert opinion on safety and comfort/mobility impacts, expert judgment models are used, which use the opinions of different experts, as developed by Leden *et al.* [LED 00]: the quantification forms input to a socio-economic impact assessment, taking into account accident trends for target years (2020 and 2030), road exposure trends and estimates on the fleet penetration of the application.

In parallel with the impact assessment, the usability of existing ITS applications for VRUs is assessed. An architecture is developed for the integration of VRUs in the traffic systems of the future, and methods for communication of safety critical information are assessed. The areas for improvement of the HMI for different VRU groups, such as elderly drivers and motorcycles, are identified. Starting from a usability study of current ITS applications, guidelines will be provided on the improvement of the HMI for specific user groups, such as elderly drivers. The performance and usability of different concepts for the communication between road users in safety critical situations will be assessed.

Innovative ITS applications are tested at three test sites, in Helmond (The Netherlands) with an emphasis on cyclists, and in Valladolid and Alcalá de Henares (Spain), with an emphasis on pedestrians. The major aim of the trials is to benchmark the recommended practices regarding usability and integration of VRUs in cooperative traffic systems. Based on the evaluation results, an ex-post impact assessment is made of the applications tested in the trials.

The result of the work in the different areas is combined into a consolidated set of outputs, consisting of recommendations for industry to improve the efficiency and performance of ITS applications, and of policy recommendations.

16.3. Accident data analysis and identification of critical scenarios

Analysis of a range of databases has been conducted to identify a number of scenarios for VRU's. The CARE database has been used as the most representative database for EU accidents and data from national databases from Austria, Finland, Spain, Sweden and the UK have been compared with the CARE data for consistency of results. The main findings from the database are summarized in the following sections.

16.3.1. *Pedestrians*

– In all databases, including the CARE database, it was found that accidents were most likely to occur when the pedestrian was crossing the road remotely from a junction (Figure 16.1).

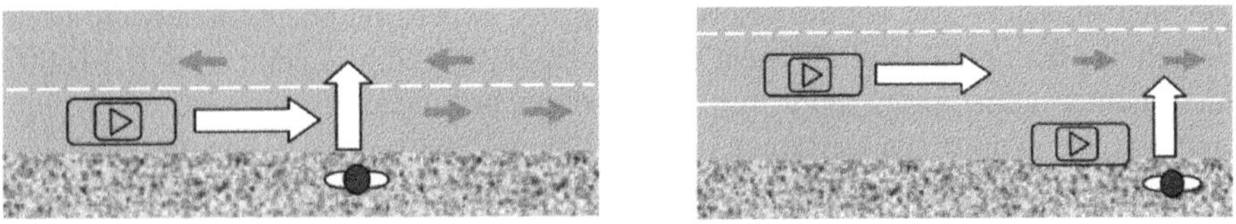

Figure 16.1. *Critical pedestrian scenarios: crossing the street remotely from a junction*

– The accident analyses suggest that in most if not all cases, the environmental conditions are not intuitively detrimental to road-crossing. That is, in the majority of the databases, the accidents tended to occur in fine weather and the road conditions were found to be dry. Also, there was some consistency suggesting that pedestrian accidents tend to occur between 12 pm and 6 pm.

– It was found that in the majority of cases, the accidents tended to occur in urban areas on roads with lower speed limits (50 km/h).

– There was variation in some of the parameters – males are slightly over-represented in the CARE database whereas females are over-represented in some of the individual Member State databases – and there was no consistency regarding the accident month.

– In the majority of databases including CARE, a passenger car was the most frequent collision partner.

– Some important parameters could not be determined. These include vehicle characteristics, vehicle speed pre-collision and pedestrian actions prior to collision.

16.3.2. *Cyclists*

– The majority of cycling accidents in the accident analysis were found to occur at junctions/intersections. One of the most common scenarios involved vehicles pulling out into the path of the on-coming cyclist at an intersection (Figure 16.2(a)).

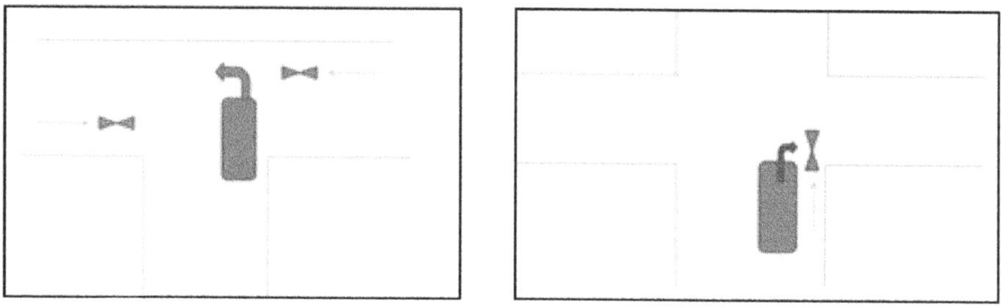

Figure 16.2. *Critical cyclist scenarios: a) vehicle pulling out into the path of an oncoming cyclist at an intersection; b) vehicle turning into the cyclist's path*

– CARE data suggest that the most common scenario involves both the cyclist and the vehicle heading in the same direction, but the vehicle then turns into the cyclist's path (Figure 16.2(b)).

– Overall, males are over-represented in the data.

– The majority of the accidents occur in fine dry weather during daylight hours.

– The majority of the accidents occur in urban areas on roads with relatively low speed limits.

16.3.3. *PTWs*

– The most common scenario in the CARE accident analysis was found to be the PTW being hit by a vehicle (mainly passenger car) initially heading in the same direction and then turning across the path of the PTW (Figure 16.3(a)).

– This was not consistent with the national database analyses, which suggest that the most common scenario involves vehicles pulling out from intersections into the path of the PTW (Figure 16.3(b)).

– Males were far more likely to be involved in PTW accidents compared with females.

– Most accidents occurred within urban environments.

– It is thought that the majority occurred on roads with low speed limits (50 km/h).

– The majority of the accidents occurred in fine and dry weather conditions during daylight hours.

– The majority of the accidents occurred during the summer months (May–September).

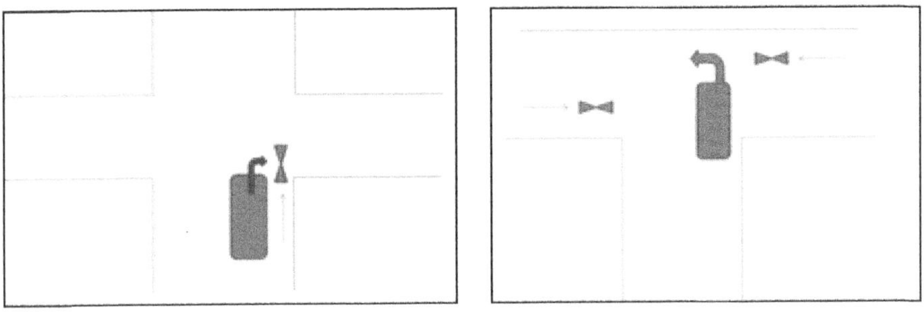

Figure 16.3. *Critical PTW scenarios: a) PTW being hit by a vehicle heading in the same direction and then turning across the path of the PTW; b) vehicles pulling out from intersections into the path of the PTW*

16.4. User needs analysis

Based on focus groups with 143 participants from five different ITS user groups (adults, parents, adolescents, older road users and PTWs) in four partner countries (Spain, Finland, Austria and The Netherlands) and expert interviews with 10 international experts from the fields of ITS technology, policy and VRU representation the following aspects have been comprehensively assessed:

– identification of critical situations for VRUs;

– assessment of needs of different user groups for ITS services & applications;

– identification and prioritization of ITS potential and technology prospects in view of VRU safety and mobility.

Critical situations for VRUs proved to be usually related to high (car) speeds, high complexity and density of traffic, local weather conditions and maintenance of infrastructure.

System knowledge among focus group participants was on a high level with a high share of standard in-vehicle (car) systems, with known infrastructure based ITS mainly regarding traffic lights and traffic signs. Smartphone-based applications for routing and navigation are already known and regularly used by all involved road user groups for pre-trip and on-trip information. Participants in all countries where shown to have experiences on all levels of ITS (mobile applications, in-vehicle and infrastructure) and technologies (informing, intervening and warning) with a high level of experiences, especially among car drivers (Blind Spot Detection, Intelligent Speed Adaptation, GPS, Cruise Control, etc.).

Main identified benefits and advantages of ITS for VRUs are increased "visibility" of VRUs (communication, warning and intervention); increased overall "traffic flow" (automation); "economic" (less fuel consumption) and "ecological" (less CO_2 emissions) aspects; increased "comfort" in traffic (information).

On the other hand, potential adverse effects are a perceived loss of "autonomy, distraction" (sounds, visuals and interaction with HMI), and potential for "overreliance/overconfidence", technical limitations and reliability.

The overall willingness to use ITS for VRUs and the assessment of benefits for traffic safety and general mobility was assessed on a very high level by both experts and focus group participants.

Future technological advancements are mainly expected in view of connecting road user groups (communication between VRUs and vehicles); increasing visibility and vision; standardization of technologies; infrastructural developments and adaptation of legal requirements for broad scale deployment of technologies.

16.5. ITS applications for the critical scenarios and user needs

An inventory of ITS applications targeted to VRUs has been made with a total of 14 systems addressing pedestrians, 34 addressing cyclists, 28 for PTWs and a number of 10 in-vehicle systems which benefit all kinds of VRUs. In order to identify the most promising solutions, a workshop with 40 relevant stakeholders was held in Brussels on 18 September 2013. In this workshop, representatives of VRU groups, national and European authorities, infrastructure service providers and ITS related economy contributed their input to the prioritization process. The participants were divided over three different groups (pedestrians, cyclists or PTWs) according to their expertise/interests, and ITS solutions for two scenarios, which were identified in the accident analysis and user needs as the most important for the specific group, were discussed. More specifically the scenarios were:

– Pedestrians:

- pedestrian crossing the road, occluded or not from a parked car (Figure 16.1),

- support pedestrians at intersections to increase comfort and remove obstacles/barriers;

– Cyclists:

- vehicle on a crossroad, pedal cyclist crossing the road from the right or from the left (Figure 16.2),

- making cycling from location A to location B easier;

– PTWs

- urban junction accidents with cars (Figure 16.3),

- urban single motorcycle accidents on straight roads.

After the discussions, participants were asked to select the three applications which have the most potential for VRU safety and rate these ITS solutions in a questionnaire, according to the following criteria: safety, mobility, technical maturity, deployment potential, acceptance (by VRU, drivers and government authorities), relevance for older people, relevance for people with disabilities, feasibility for children, and usability of system interface. A five-point Likert scale was used for each of them. In addition, they were asked about potential negative side effects of the systems selected.

16.6. Results

Starting from the group discussions and the questionnaires in the workshop, a ranking was made of promising applications. This list was assessed by the project team for coverage of the different user groups, critical scenarios and user needs. A set of the 22 most promising applications were identified and selected for further impact assessment in the VRUITS project. Table 16.1 shows for the selected systems the VRU types for which they apply, the expected impact (either on safety or on mobility and comfort), and the type of ITS (infrastructure, vehicle and VRU as intelligent components of the system), and the maturity of the system. For the maturity level, M indicates that the first systems are appearing on the market. Most of the systems selected are mainly available on a limited number of high-end vehicles or still have a very low penetration level, and still have many possibilities for improvement, e.g. regarding the performance or the suitability for specific user groups. P indicates a prototype, produced as part of a research project, but not yet available on the market. R indicates that the systems are still under research.

The following paragraphs comprise the feedback provided by the experts on the ITS applications.

Blind Spot Detection is considered by lots of experts as a promising ITS for the safety of VRUs, especially in supporting truck drivers when turning right and avoiding cyclists crashes. Besides, VRU acceptance is rated as high, but acceptance by drivers and authorities is lower. They suggest that drivers may worry about the loss of control, and there may also be the problem of drivers relying on this technology rather than checking mirrors. They suggest that if legislation was used to make them mandatory then deployment would be easy. They suggest that the systems should not only detect, but also intervene.

Intelligent Pedestrian Traffic Signals are seen as one of the most promising ITS for pedestrians and cyclists. In this sense, it not only enhances VRU mobility, but also VRUs are less likely to cross during the red light phase. One of the experts claimed that on central urban areas, one of the big challenges for traffic management is to deal with the time shared in pedestrian crossings at junctions, since a green light for pedestrians is a red light for cars. In this sense, having this system available helps to ensure that pedestrians have sufficient time to cross, but also minimizes the negative impact on the vehicle traffic-flow. The acceptance of this system by VRUs is considered to be high; since it responds to the need of pedestrians, especially elderly pedestrians and persons with reduced mobility, to be able to cross the road at their own speed. However, a negative aspect could be that drivers may perceive more frequent stops, but these systems may skip cycles when there is no pedestrian crossing the road (anymore), which is likely to increase acceptance and compliance by the drivers.

ITS applications		*Type of VRU*						*Impact*		*Type of ITS*			*Maturity*
		Pedestrians	Cyclists	PTWs	Disabled	Elderly	Child	Safety	Mobility	Infrastructure	Vehicle	VRU	
1	Blind Spot Detection	X	X	X	X	X	X	X			X		M[1]
2	Intelligent Pedestrians Traffic Signal	X	X		X	X		X	X	X			M
3	ISA (Intelligent Speed Adaptation)	X	X	X	X	X	X	X			X		M
4	Red Light Camera/ Average Speed Camera	X	X	X	X	X		X		X			M
5	Intersection Safety	X	X	X	X	X		X		X	X		P
6	Pedestrian Detection System + Emergency Braking	X	X	X	X	X	X	X			X		M
7	Navigation Systems for VRUs	X	X	X	X	X			X			X	M
8	PTW Oncoming Vehicle Info System			X				X			X	X	P
9	VRU Beacon System	X	X	X	X	X	X	X			X	X	P
10	Cyclist Digital Bicycle Rear-View Mirror		X					X				X	M
11	Roadside Pedestrian Presence	X	X		X	X	X	X		X		X	P
12	Urban Sensing System	X	X	X	X	X		X	X			X	P
13	Automatic Bicycle Identification		X					X	X	X			P
14	Night Vision and Warning	X	X	X	X	X	X	X			X		M
15	Information on Vacancy on Bicycle Racks		X						X	X		X	M
16	Bicycle to Car Communication		X					X			X	X	R

17	Rider Monitoring System			X				X				X	P
18	Crossing Adaptive Lighting	X	X		X	X		X		X			M
19	Infotainment	X	X	X				Neg	X			X	M
20	Real-Time Information Systems for Public Transport	X	X		X	X			X			X	M
21	Road Weather Warning for Pedestrians	X	X		X	X		X	X			X	M
22	Advice System for Elderly Cyclists		X			X		X	X			X	R

M, on market; P, prototype; R, Research.

Table 16.1. *ITS applications for VRUs, selected for further assessment*

Intelligent Speed Adaptation (ISA) is seen as a system that everybody benefits from. VRU acceptance is high, while drivers' acceptance is lower. In this regard, the feasibility of ISA is considered technically high but politically low. As the authorities themselves recognize, there are definitely political issues there – but they go both ways, since on the one hand there is "the car culture" and the drivers' "freedom", but on the other hand the safety and the legal speed limits. They point out the need to start in areas where driving slowly is indisputably a need, such as residential streets and school areas. According to them, a huge business opportunity for ITS providers lies in ITS for enforcement on a huge European market.

For PTWs, the system could be combined with curve warning, in addition to information about the surroundings (accident rates, infrastructure, current traffic flow, etc.). Previous research has however shown that there are negative attitudes towards ISA among rider groups [SIM 07, NOR 11].

Red lights/speed cameras have, according to the experts, a great deployment potential, since pedestrians are struck while crossing a signalized crossing in some cities, even when the pedestrian light is green. This system is simple and cheap and is the only method of ensuring that traffic lights are obeyed. In addition, the payback period is short. Nevertheless, in places where it has been the norm to ignore red lights, rear-end shunts will be a problem in the short term. One of the challenges perceived by one of the participant authorities is "administrative issues": the concern here was "if the authorities can issue fines, mail them, collect and enforce them in due time and with necessary force" so that this measure is effective. Experts suggest that this system should be considered not only as enforcement but also as a preventive action. One of the experts comments that this ITS might be perceived by drivers as a tax measure. To promote a change in attitudes and behavior the installation of red speed cameras is suggested.

Average speed cameras have been used successfully in the UK to reduce drivers' average speed. Experts suggest that their use should be extended, and that they should be combined with in-car information on speed limits. The aforementioned user needs analysis showing that drivers would appreciate having information about the speed limit on a regular basis.

Pedestrian Detection Systems with Emergency Braking is seen by experts as safety-enhancing, with high acceptance by VRUs and a little bit lower for drivers. With regards to deployment potential, the first versions of these systems are already available on high-end vehicles. Regarding negative side effects, experts believe that rejection would occur only if there were many false alarms, which could additionally impose a risk to following traffic by unexpected harsh braking. Another participant says that the biggest risk is delegation of responsibility, and that the driver might compensate by being more reckless. In addition to the detection of pedestrians and cyclists, experts see a good deployment potential in VRU Detection Systems for PTWs, especially to alert the driver of the presence of motorcyclists when entering an intersection.

Trip planning and navigation systems with social networks: according to the experts, this service should be adapted to the needs of VRUs, the elderly and disabled included. As a negative side effect the experts mention an excess of confidence (outdated or erroneous map information), and possible distraction when checking the screen.

VRU Beacon System acceptance might be low for VRUs according to the participants, but it depends on different factors such as power management of the beacon device. However, they consider it is quite feasible for children, and they can imagine it as being standard part of future cell phones. The way the device is introduced into the market should be reflected carefully. Experts suggest the use of tags for kids through infrastructure, such as bus stops, combined with warnings for both the bus driver, to inhibit the start at boarding and unboarding, and for other vehicle drivers to warn of the presence of children. According to the experts, beacons should indicate type of user, but new icons may be difficult to comprehend. Regarding negative side effects, it may be difficult for the system to discriminate between "on boardwalk"/safe and unsafe conditions. In addition, traffic complexity is increased since traffic communication increases.

Oncoming vehicle information systems for PTWs, based on vehicle-to-PTW communication, is seen by experts as a "basic system for a complete ITS-safety system for all road users". However, they think that it should be safe against hackers. It is considered to have deployment potential, as well as being appealing to road authorities. The usability issues are of critical importance, and drivers may choose to switch off/over-ride the system.

Intersection Safety, based on infrastructure-to-vehicle communication, has the potential to enhance the safety and mobility of cyclists, pedestrians and PTW riders. With regards to PTW riders, experts mention the option of combining intersection support systems that detect PTW riders, and flashing lights in the PTW to increase conspicuity.

Urban sensing systems are considered to increase VRU mobility as well as safety. Experts appreciate the feedback to local authorities who can take corrective actions and restore safe and secure conditions. Notwithstanding, they wonder how to check that a reported road hazard is actually true. They suggest using crowd sourcing information and open data to create a virtual real time map of the area including safety data, speed, citizens' views and also to gather data on pedestrians and their movements. This is an issue also raised by PTWs in relation to ITS: they see an opportunity to equip PTWs in order to collect data on accidents.

Rider monitoring systems for PTWs: experts point out that it could help PTWs in case of fatigue, inattention, etc., and they suggest the need for research. Rider monitoring systems on PTW need to be more complex than car systems. Video monitoring for example is hardly feasible, due to helmet and motorcycle clothing. There are very few studies and research works in this field.

16.7. Conclusions

The deployment of ITS technologies has the potential to enhance the safety and mobility of VRUs. However, it is essential for their success that these systems are tailored to the specific needs of road users. For this reason, the current investigation has begun with the identification of the critical scenarios for VRUs, as well as user needs, in order to find the most promising ITS systems to address them, according to a group of experts.

Based on information derived from Focus Group discussions, junctions are one of the most relevant critical situations for VRUs, where VRUs are endangered due to being hardly visible or easily overlooked. Moreover situations where cars overtake cyclists or PTWs are especially assessed as being critical due to high traffic speeds and the perceived reckless behavior of car drivers in some cases.

Data obtained from Focus Group discussions indicated that the visibility of VRUs is generally perceived as a major factor in view of traffic safety, especially in connection with heavy traffic and high speed situations. Correspondingly, technologies and systems enhancing the detectability and visibility of VRUs are considered to have high potential to increase the traffic safety of VRUs.

Previous research has pointed out that ITS technologies capable of distinguishing between different types of VRUs and rapidly deploying tailored countermeasures (such as active braking or airbags) to reduce injuries in the event of a crash offer considerable safety potential [REG 01]. A detailed study on the potential of Pedestrian Detection Systems/Emergency Braking suggests that 40% of fatalities and 25–30% of severe injuries suffered by pedestrians can be avoided [KÄL 09]. Similarly, Blind Spot Detection systems are seen as promising, especially to avoid truck–cyclist crashes.

Cooperative systems, based on simple beacons carried by VRUs, or two-directional devices allowing communication between PTWs and vehicles, are seen as having a great potential. In this sense, the multitude of sensors and communication interfaces that smart phones deliver, together with their rising penetration, could provide a good opportunity to improve the visibility of VRUs in complex/urban environments. Indeed, focus group discussions revealed that smartphone-based applications (for routing and navigation) are regularly used by all involved road user groups.

Regarding PTWs, the potential for vehicle-to-PTW communication systems to address motorcycle conspicuity issues has been previously recognized [BAY 06], though one issue to tackle is the need to reach the necessary penetration to achieve efficiency of cooperative systems. However, while on the one hand systems supporting visibility or communication between PTWs and cars are considered very positive, on the other hand ITS interfering with the riding task or those perceived to take away the autonomy from the rider are regarded very skeptically. Training and education are considered to be of major importance in this group with ITS having mainly adverse effects on riding behavior. This is in line with the research of [BEA 13], who found that riders believe that innovations should focus on protective equipment, rather than systems that prevent crashes; since they believe crash prevention is better addressed through rider training.

In this sense, [HUT 13] found that the social norm and the interface design are powerful predictors of the acceptance of Advanced Rider Assistance Systems (ARASs), while the extent of perceived safety when riding without support did not have any predictive value in their study. Thus, these authors found that the specific social influence of peers confirmed for the acceptance of Advanced Driver Assistance Systems (ADAS) [ARN 08] had proven to have a strong influence on the acceptance of ARAS in their study. This finding revealed the impact that preconceptions regarding the support systems, which might be present in rider circles, may have on the acceptance of the ARAS by an individual. Accordingly, it lends importance to promoting a favorable attitude towards such solutions among motorcycle riders, with the particular aim of avoiding a distinctly negative social norm [HUT 13]. According to PTW's participants in focus group discussions, there are not really specific ITS solutions for motorcycles. Several researchers have reflected this need for assistive systems to be developed specifically for PTWs [BAY 06, BEA 13]. The need for a better understanding of people's attitudes and opinions about ITS applications has been highlighted as an important area for further research [REG 01], especially given the rapid rate at which ITS applications are becoming available. As [VAN 97] stated, it is counterproductive to invest in developing new technologies if the systems are never purchased or if they are purchased but never used. In other words, these systems can only enhance users' safety and mobility if they use them. For this reason, acceptance is a decisive aspect to be considered in the development process of such systems. In the present study on user needs, overall willingness to use ITS for VRUs and the assessment of benefits for traffic safety and general mobility was assessed on a very high level by both experts and focus group participants. Actual acceptance was found to be mainly related to the following factors: price/affordability, usability/implementation, availability, standardization across different platforms and manufacturers, reliability and privacy.

On the other hand, overly high vehicle situational speeds have been repeatedly found to be a very important factor in fatal pedestrian collisions. It is well known that even small decreases in vehicle travelling speeds prevent a large number of pedestrian fatalities. For this reason, Intelligent Speed Adaptation as well as automated speed enforcement have significant potential to reduce the injury consequences of VRUs crashes. Red Light Cameras are seen as a safety enhancing technology, and with great deployment potential. However, promoting a change in attitudes and behavior is suggested in addition to the installation of the Red Light Cameras. In this regard, enforcement should be accompanied by educational measures in order to modify intentions to break the law.

Intelligent Pedestrian Traffic Signals are seen as promising ITS for pedestrians, including the elderly and disabled, and cyclists. It provides pedestrians with the amount of time they need to cross at the speed they are capable of, this being fundamental for all pedestrians, namely the elderly and those with reduced mobility. In this sense, it not only enhances VRU mobility, but also VRUs are less likely to cross during the red light phase.

Starting from the list of applications, the VRUITS project will assess the impact of applications on safety and mobility on VRUs. In a first phase the direct and indirect effects of the functions will be determined, and in a second phase the impacts on safety and mobility and the socio-economic impact will be assessed. The assessment will be the basis for recommendations for policy and industry on measures regarding ITS applications to improve the safety and mobility of VRUs.

16.8. Acknowledgments

This research received funding from the European Union's Seventh Framework Programme (FP7/2007-2013) under grant agreement no. 321 586.

16.9. Bibliography

[ARN 08] ARNDT S., ENGELN A., "Prädiktoren der Akzeptanz von Fahrerassistenzsystemen", in SCHADE J., ENGELN A. (eds), *Fortschritte der Verkehrspsychologie*, DGP-Kongress für Verkehrspsychologie., Deutscher Psychologen Verlag, vol. 45, pp. 313–337, 2008.

[BAY 06] BAYLY M., REGAN M., HOSKING S., *Intelligent Transport Systems and Motorcycle Safety* (Monash University Accident Research Centre Report 260), MUARC, Clayton, Australia, 2006.

[BEA 13] BEANLAND V., LENNÉ M.G., FUESSL E. *et al.*, "Acceptability of rider assisstive systems for powered two-wheelers" *Transportation Research Part F*, vol. 19, pp. 63–76, 2013.

[CAR 09] CARE, EU roads accident database, 2009.

[DRA 98] DRASKÓCZY M., CARSTEN O.M.J., KULMALA R. (eds), "Road Safety Guidelines", CODE Project, Telematics Application Programme, Deliverable B5.2, 1998.

[HUT 13] HUTH H., HELAU G., "Predicting the acceptance of advanced rider assistance systems", *Accident Analysis and prevention*, vol. 50, pp. 51–58, 2013.

[KÄL 09] KÄLLHAMER J *et al.*, Pedestrian Injury Mitigation by autonomous braking, Autoliv Research, paper no. 09-0132, 2009.

[KUL 10] KULMALA R., "Ex-ante assessment of the safety effects of intelligent transport systems", *Accident Analysis and Prevention*, vol. 42, no. 2010, pp 1359–1369, 2010.

[LED 00] LEDEN L., GÅRDER P., PULKKINEN U., "An expert judgement model applied to estimating the safety effect of a bicycle facility", *Accident Analysis and Prevention*, vol. 32, no. 2000, pp 589–599, 2000.

[NOR 11] NORDQVIST M., GREGERSEN N.P., Study on motorcyclist's behaviour and attitude towards road safety 2010, SMC, Borlänge, Sweden, 2011.

[REG 01] REGAN M., OXLEY J., GODLEY S. *et al.*, *Intelligent Transport Systems: Safety and Human Factors Issues*, Royal Automobile Club of Victoria, Victoria, 2001.

[ROS 81] ROSS, H.L., Deterrence of the drinking driver: an international survey (DOT HS-805 820), Department of Transportation, Federal Highway Administration, US, 1981.

[SIM 07] SIMPKIN B., LAI F., CHORLTON K. *et al.*, *Intelligent speed adaptation: results of motorcycle trial*, University of Leeds, UK, 2007.

[VAN 97] VAN DER LAAN J.D., HEINO A., DE WAARD D., "A simple procedure for the assessment of acceptance of advanced transport telematics", *Transportation Research Part C: Emerging Technologies*, vol. 5, no. 1, pp. 1–10, 1997.

17

Experimentation with the PRESERVE VSS and the Score@F System

Security and privacy enhancing technologies are an essential feature for ITS deployment. As the security system generally introduces an overhead, the test and evaluation of a defined security solution is as important as its design. In this context, the PRESERVE project aims to design, implement and test a secure and scalable security system for ITS. In this chapter, we focus on the PRESERVE security tests and precisely on the joint trial tests with the French Score@F FOT. We present the test cases of PRESERVE VSS integrated with Score@F FOT system and evaluate performance indicators. This is the first chapter to describe PRESERVE test cases and Score@F platforms integrating security solution.

17.1. Introduction

Intelligent Transportation Systems (ITS) will enable new cooperative applications, e.g. Road Hazardous Signaling, to improve road safety, traffic efficiency and increase passengers comfort. To achieve this, specialized equipment, the so-called ITS-Stations (ITS-S), will be integrated in vehicles, roadside infrastructure units and central servers. Vehicle to vehicle (V2V) or vehicle to infrastructure (V2I) communications will allow the exchange of Cooperative Awareness Messages (CAM) and Decentralized Environmental Notification Messages (DENM). Based on the exchanged messages, the ITS-enabled applications will provide relevant information to the driver. ITS communications are based on ITS-S reference communication architecture described in ETSI standard EN 302 665 [ETS 10]. The ETSI communication architecture, which is shown in Figure 17.1, consists of four horizontal layers: access, networking/transport, facilities and

Chapter written by Rim MOALLA, Brigitte LONC, Gerard SEGARRA, Marcello LAGUNA, Panagiotis PAPADIMITRATOS, Jonathan PETIT and Houda LABIOD.

applications layers, and two cross entities one for security and the other for management.

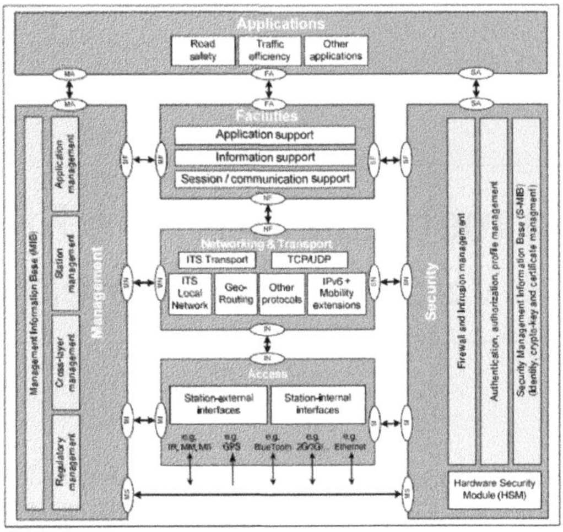

Figure 17.1. *Communication architecture of an ETSI ITS station [ETS 10]*

Without an appropriate design, a compromise of the ITS could have dire consequences: cyber-attacks could make transportation inefficient or put cars and drivers in danger; and ITS could easily give away the whereabouts of cars and thus drivers. That is why security and privacy enhancing technologies are an essential feature for the ITS deployment. This has recently been well understood.

As a result, many European projects dealt with security and privacy of vehicular communications, notably SeVeCom, DRIVE C2X, PRECIOSA and EVITA. Currently, PRESERVE is providing support for on-going Field Operational Tests (FOT) on cooperative ITS. PRESERVE contributes to the security and privacy of vehicular communications and ITS in general. PRESERVE specifies and develops a Vehicular Security Subsystem (VSS) that will be integrated in experimental projects, including, notably, Score@F in France. The Score@F project participates in the preparation of ITS deployment by evaluating standardized application messages (CAM, DENM and I2V messages) in different road environments through selected V2X cooperative driving use cases. Moreover, Score@F implements GeoNetworking protocol and Basic Transport Protocol (BTP) defined by ETSI for vehicular communication based on ITS G5, which is the European profile of IEEE 802.11p. For security and privacy issues, the Score@F system integrates the PRESERVE VSS.

As security generally introduces an overhead (communication and processing and complexity), we have to thoroughly test and evaluate our security system. In particular, we must be certain that the secured ITS remain practical and effective. For this, we have defined a set of performance indicators to evaluate and a test methodology and process to follow that we present in this paper. In fact, we start by internal tests of the PRESERVE VSS, in order to validate the security system *per se*. Then, we integrate the PRESERVE VSS with Score@F use cases, in order to evaluate the effect of security and privacy protection on the communication and networking. Finally, in parallel, we evaluate the effect of security on the performance of the cooperative applications. These tests will be done in two steps: first, in a static environment, and then in a real environment with mobile vehicles. The results of these tests will be the inputs for simulation in order to evaluate the scalability of PRESERVE solution.

The rest of this chapter is organized as follows. Section 17.2 provides an overview of our test methodology. Then, in the following section, we present our defined performance indicators which will be evaluated during joint trial tests with Score@F. In section 17.4 we detail security environments: we present Score@F use cases and platform, we give an overview of PRESERVE security system and we describe test sites. Section 17.5 concludes this chapter.

17.2. Test methodology

In order to determine the overhead introduced by our security system on processing and communication, three different security evaluation/analysis levels are specified. The first analysis level is cryptographic overhead analysis: these tests will evaluate the performance of the crypto-system, e.g. cryptographic delays for signature generation and verification operations. The second level of security evaluation consists of testing and evaluation of delays and overhead relating to the VSS internal processing. For example, for the signature verification process, we test and evaluate three different security policies: verify the signature only, verify the signature and the certificate of the sender and verify the signature and certification chain. Each security evaluation level will be done in two steps: first, in a static environment, and then in a real environment with mobile vehicles.

The third level of security evaluation is evaluation of end to end security overhead such as the time latency introduced by security system from the sender application to the received one. These tests will be done on a real environment with the presence or simulation of attackers. The results of these tests will be the inputs for simulation in order to evaluate the scalability of PRESERVE solution.

17.3. Performance indicators

We defined a set of performance indicators to evaluate our security system.

– Signature Generation Delay *SGD* (ms): the delay for generating one packet signature. This includes calculating a hash (HD) plus performing the actual digital signature generation operation. SGD = HD + SD.

– Signature Verification Delay *SVD (ms)*: the delay for verifying one packet signature. This includes verification of certificate chain plus calculating a hash (HD) plus performing the actual digital signature verification operation.

– Packet Signature Generations per Second *SGPS (1/s)*: for every packet sent, one needs to generate a suitable signature, i.e. SGPS = OPSS. We assume that every packet needs to be signed, which is true at least for CAMs and DENMs, if we do not apply omission schemes.

– Packet Signature Verifications per Seconds *SVPS (1/s)*: for every signed packet received, one needs to verify the signature plus the certificate.

– Pseudonym Change Delay *PCD* (ms): the additional delay introduced when the ITS station switches from one pseudonym certificate to another. It is measured as additional time added to signature generation for a packet.

For performance indicators evaluation we fix points and sensors within the PRESERVE V2X Security System where measurements need to be taken. We also introduce measurement points on the communication stack especially on the Networking and transport layer where VSS is integrated.

17.4. Test environment

17.4.1. *Score@F applications and platform*

Score@F is the French FOT for Cooperative intelligent transportation systems (see [SCO 11, GER 13, JAC 11]). Cooperative ITS systems are based on local wireless communication systems that enable direct two-way communication between vehicles and road infrastructure units (V2I) and between vehicles (V2V). This project aims to develop and validate the standardized ITS communication architecture [ETS 10] and to conduct Field Operational Tests on motorway and CG78 road environments.

SCORE@F project has been assessing applications belonging to three service domains such as represented in Figure 17.2 below. Road safety applications are mainly the signalization of immediate hazards to drivers. Several use cases have been assessed such as road works, stationary vehicles, human presence on motorway, traffic jam ahead and bad weather conditions. A few collision risk warning use cases (electronic emergency brake light, signal violation and wrong way driving) have also been tested in SATORY controlled environment.

Traffic management applications have also been assessed on the request of road operator partners. This includes the collect of traffic data from standard safety messages (CAM & DENM), the provision of the processed data to traffic management center using DATEX II and the immediate feedback of traffic management center through contextual speed and In-Vehicle Signage information to act immediately on the vehicle flows.

Some mobility services have been assessed such as Electronic Hitchhiking enabling a pedestrian to broadcast with his or her smartphone a car-pooling request or the broadcasting of Point of Interest notification facilitating the mobility or achieving the promotion of local businesses. These two categories of applications included some multimedia dissemination (audio, photo and video clip).

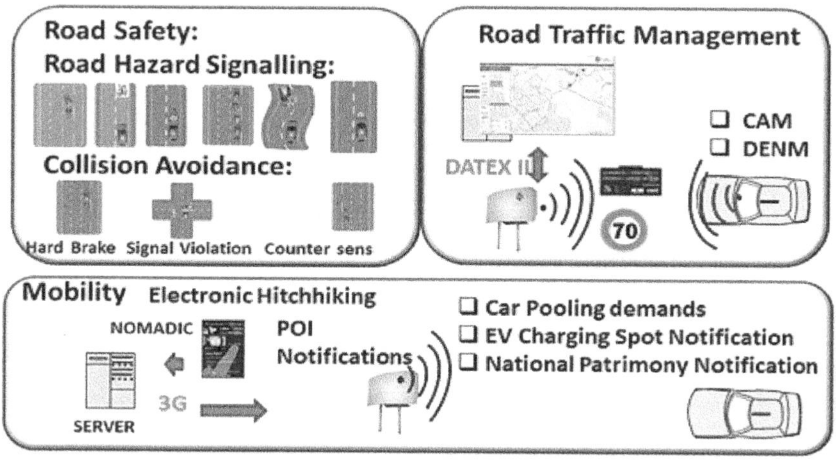

Figure 17.2. *Score@F use cases*

However, the system was tested on two different platforms presented in Figure 17.3. The first platform is formed by two units: an application unit regrouping applications and facilities layers. This application unit is a Nexcom VTC6201. The application unit communicates via Ethernet to the IEEE 802.11p modem, which is a DENSO or COHDA modem. An implementation of the Geo-networking and BTP protocols is integrated on the modem. The application unit retrieves the vehicle data via CAN adapter and vehicle position on the road via camera. The second platform is formed by only Nexcom VTC unit, which integrates an ITRI card as a G5 modem. In both the platforms, an Android tablet is used as a HMI for user field tests.

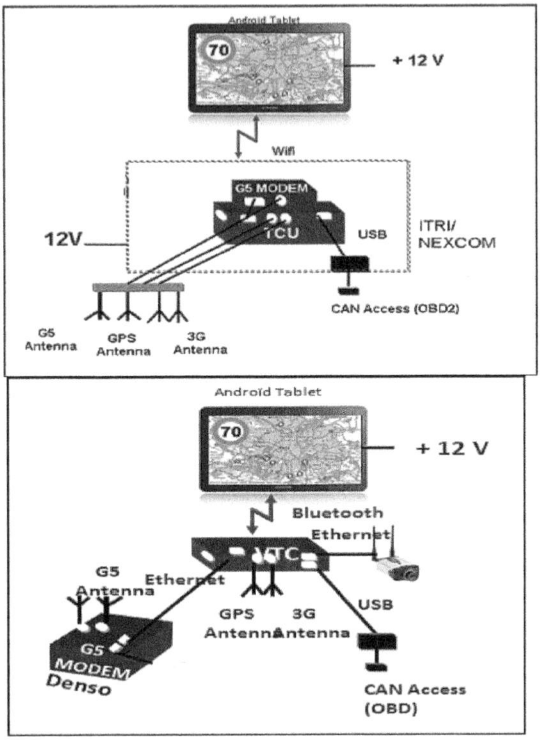

Figure 17.3. *Score@F platforms*

17.4.2. *PRESERVE system*

PRESERVE develops a complete security and privacy solution for V2X communication including an onboard V2X security subsystem (VSS) that will be integrated on ITS-S and an offboard PKI composed of three certificate authorities: RCA, LTCA and PCA. The VSS will integrate results from the SeVeCom, PRECIOSA and EVITA. PRESERVE develops three versions for the VSS: (1) only software security solution, (2) software and FPGA security system and (3) a security ASIC. For the moment, the VSS includes just four modules: secure communication module, pseudonym management module, identity and trust management module and management and configuration. Further modules dealing with privacy enforcing and in-vehicle security will be integrated later. During PRESERVE and Score@F test sessions, we test the VSS software kit and the FPGA-based HSM kit. Regarding PRESERVE PKI, we evaluate PKI performance and test it following test cases detailed in the next section.

17.4.3. *Test site description*

We conducted four joint test sessions; the first two sessions were conducted on a controlled test area which is the Satory track presented in Figure 17.4. This site, owned by the Defense Ministry and managed by Nexter, provides a set of three different tracks from 2 to 4 km, which reproduce a large variety of road situations. The site is equipped with two RSUs covering the whole area and we equipped the two vehicles with the VSS FPGA based version. These tests are conducted only for PRESERVE system validation which is why we tested only signature generation and signature verification functionalities.

Figure 17.4. *Satory test site. For a color version of the figure, see www.iste.co.uk/jacob/safety.zip*

The last two test sessions, held in May and September 2013, were conducted on the Yvelines – Versailles open road site including a section of RD91 of 3.5 KM with five RSU and two vehicles. RSU are set-up from the RD91/RN12 exchanger at Versailles up to the Georges Besse place in Guyancourt, entrance of the Renault Technology Center. This portion, located in a peri-urban/rural area has several interesting configurations. The tested vehicles were equipped with a software-based version of PRESERVE VSS and RSUs were not equipped with PRESERVE. The main objectives of these tests are demonstrating the correct behavior of the whole system in a realistic FOT environment including stations equipped with PRESERVE VSS and others that are unequipped. We integrated the PRESERVE VSS software based on the GeoNetworking stack on both Score@F platforms, one Score@F platform per test session. During these tests, we conducted functional test cases and attack scenarios that we describe in the next section.

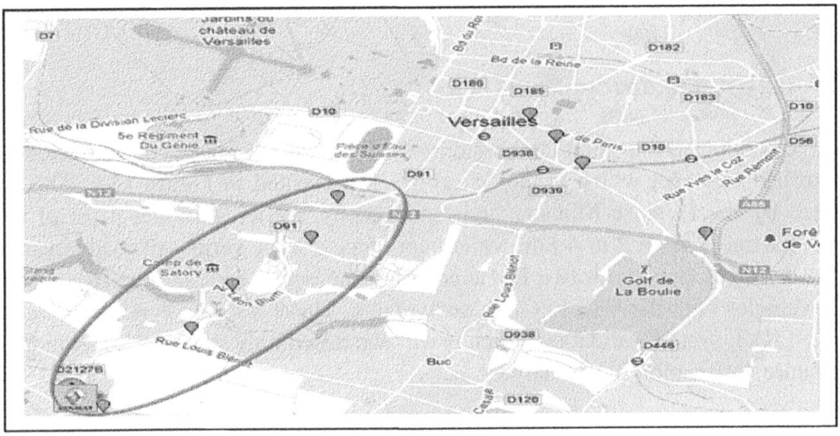

Figure 17.5. *Yvelines – Versailles test site*

17.5. Test case description

We present the use cases tested on PRESERVE and Score@F joint tests. Use cases are split into functional use cases that check the correct functionality and attack use cases that are evaluating behavior of the VSS under certain attacks.

17.5.1. *Functional tests*

In this section, we present functional test cases that we conduct on the last PRESERVE and Score@F joint test session. We give a detailed description of each test case and we indicate which performance indicators are evaluated on each test.

Generation of signed message: on the different test sessions, all messages generated by Score@F use cases and sent over G5 are signed. The purpose of this test case is to evaluate at first the signature generation delay as crypto delay evaluation and next to evaluate packet signature generations per second. In fact the generation of a signed message may require a pseudonym change processing before the signature generation. The packet signature generations per second depends on the VSS processing delay and on the communication (GeoNetworking) stack policy.

Verification of signed message: when the GeoNetworking stack receives a signed packet, it sends it to the VSS for signature verification. The signature verification operation normally includes certificate chain verification. The VSS implements several verification policies. Indeed we can just verify the signature of the received message, we can verify the signature and the certificate of the sender and we can verify the signature and all the certificate chain. We first select the verification policy on the VSS configuration file and then evaluate the signature verification delay for each policy.

Pseudonym certificate change: for privacy requirements, the ITS-Vehicle has to change the pseudonym certificate. The PRESERVE VSS and precisely the pseudonym management module triggers the pseudonym change depending on its pseudonym change policy. Two pseudonym change policies are implemented on PRESERVE VSS; the first one consists of changing pseudonym after a period of time predefined by the administrator. The second policy is based on the number of times a pseudonym is used. We select first the pseudonym change policy on the configuration file of PRESERVE VSS. As prerequisites for this test case: the VSS has a non-revoked/expired long term certificate and has at least two valid pseudonym certificates. When the VSS triggers a pseudonym change all layers of the communication stack have to block message generation. The VSS has to change the pseudonym certificate and the new certificate is preloaded on the VSS. The pseudonym change must be synchronized with the change of identifiers on all communication layers (MAC address, GN ID and station ID). As a result of the pseudonym change test, the VSS has a new pseudonym certificate that is used to sign outgoing messages and the communication stack has a new identifier on each layer. The purpose of this test case is to evaluate the pseudonym change delay, which includes processing delay of communication layers for changing their ID.

Pseudonym certificate refill over 3G: when the number of pseudonym certificates stocked on the VSS reaches a predefined threshold, the VSS has to download a new set of pseudonym certificates. In this stage, we test only the pseudonym certificate refill over 3G but we plan to test the pseudonym certificate refill over G5 in the next test sessions. This test case allows us to evaluate the PRESERVE PKI performance and to validate some VSS functionalities. An adequate PKI implementing three certificates authorities is mandatory for this test case. Moreover, the certificate of each certificate authority has to be stored in the vehicle VSS memory and be valid (not revoked, not expired). The vehicle must have a non-revoked or expired long term certificate and must be equipped with a 3G connector. For pseudonym certificates refill, the VSS generates one or several new ECC key pairs. The generated public keys are included in the signed and encrypted pseudonym certificate request generated by the VSS. This request is then sent over IP/3G to the PCA which validates it in collaboration with the LTCA. If the request is verified, the PCA generates a set of pseudonym certificates and sends it back to the vehicle. The new pseudonym certificates must be securely stored on the VSS. The main objectives of this test are to evaluate the delay for a pseudonym request generation on the vehicle and the processing delay for pseudonym request verification and pseudonym certificates generation by the PKI.

17.5.2. *Attack tests*

Securing the cooperative ITS should not have negative effect on the normal system operation and, therefore, the security functions introduced by PRESERVE should be transparent to the running applications and facilities. The functional tests described in the previous section aims to evaluate the correctness and the performances of those functions under normal conditions, while in this section we describe the test case that includes the presence of an adversary.

An extra payload that carries the security header is added to the messages, and a processing delay is expected for the generation and verification of such a payload. We consider the case where the attacker tries to exploit this delay and attempts to temporarily or indefinitely interrupt or suspend services of an ITS-enabled host. To be able to achieve this goal, the attacker usually saturates the target machine with messages that require computation on the receiver side, so much so that it cannot process the legitimate traffic. Such an attack leads to host "overload", and therefore we address this adversary as the "Overload Attacker".

The adversary saturates the target machine by forcing the consumption of computational resources, such as bandwidth or processor time. When using the security functionalities, those two resources are directly related: the more messages that are received, the more processing time is required. Therefore, the overload attacker needs to send data faster than the receiver is able to process.

To achieve this without deploying an expensive distributed denial-of-service attack over different machines, a single ITS station is deployed using a modified version of the PRESERVE VSS. Instead of running the CPU-intensive cryptographic operations needed to generate a valid signature, the modified version attaches an invalid, i.e. randomly generated, security header. This operation is orders of magnitude faster that the signature generation, and therefore it allows the adversary to overflow the receiver ITS. As a matter of fact, the receiver must still spend processor time to invalidate such a forged header.

During the normal operations, we enable an outsider ITS station to act as the overload attacker with the modified PRESERVE VSS. This device then starts broadcasting invalid messages to other ITS stations at a rate of 1,000 Hz. We then evaluate the impact of such an attack on the system by comparing the measurements of the packet processing time in the normal operations with the ones obtained during the attack. We consider the test successful if the performance degradation affects the applications and facilities only minimally.

17.6. Test results

During tests on the Satory track, we tested and validated the VSS software version and the FPGA based HSM version. The main result of these tests is that the FPGA is successfully integrated with the VSS software subsystem. Regarding tests on real environment, we collected the data for tests analysis. The data acquisition was done using different logging features, in PRESERVE VSS for the API and several modules of the VSS and also in the communication stack. As we completed our tests on 15 September 2013, we continued to analyze logging files for more detailed and end-to-end results. Table 17.1 describes our primary results.

Performances indicators	Results
Signature generation delay (ms)	2,50
Signature verification delay (ms)	33,40
Packet Signature Generations per Second	400
Packet Signature Verification per Second	30

Table 17.1. *Primary experimentation results of PRESERVE security solution within Score@F system*

In PRESERVE technical report 1 [PRE 13], we estimated target performance requirements for VSS, based on previous simulations studies and on load estimations for a standard scenario and a maximum load scenario in urban and highway traffic provided by SIM-TD. This resulted in the requirement of processing 1,000 verifications per second [PRE 13]. These results prove that a software security solution is insufficient for ITS security requirements. A dedicated hardware module for security solution is needed. That is why PRESERVE defines an ASIC for securing V2X systems. The ASIC was tested in 2015 (see [PRE 15]).

17.7. Conclusion

This chapter focuses on the cooperation between the PRESERVE project and the Score@F project, the result of which is the integration of a security system on a standardized ITS communication architecture. We conducted joint tests between PRESERVE and Score@F in order to evaluate the PRESERVE security system. This evaluation is based on our test methodology presented in this chapter and a set of performance indicators. The complete joint test session between the two projects was held in September 2013. The statically analysis of tests logging data of this recent test session is given in the paper. These are the primary results of the integration of PRESERVE VSS into Score@F FOT, and we expected more results as several test sessions were expected in 2014. We expected to have the complete results by the beginning of 2014.

17.8. Acknowledgments

This work was supported by the PRESERVE (PREparing SEcuRe VEhicle-to-X Communication Systems) FP7 European project under grant agreement no. 269994.

17.9. Bibliography

[EHR 13] EHRLICH J., SCOREF: une étape vers le déploiement des systèmes coopératifs en France, Séminaire 'Voiture connectée: un défi de l'Internet des objets', TélécomParisTech, Available at: http://www.telecom-paristech.fr/formation-continue/les-entretiens-de-telecom-paristech/vehicule-connecte-defi-internet-objets/seminaire.html, 4-5 Décembre, 2013.

[ETS 10] ETSI EN 302 665: Intelligent Transport Systems (ITS); Communication Architecture, v1.1.1, September 2010.

[GÉR 13] GÉRARD SÉGARRA, ALAIN SERVEL, Sue Bai: Fault Tolerant C-ITS for Road safety, ITS WC 2013, Tokyo, 2013.

[PRE 13] PRESERVE Technical Report 1: V2X Security Performance Requirements, v1.1, editor: Frank Kargl, 2013

[PRE 15] PRESERVE Deliverable 3.2: FOT Trial2 Results, editor: Carsten Rolfes, 2015.

[SCO 11] SCORE@F PROJECT WEBSITE, https://project.inria.fr/scoref/, 2011.

18

Safety Bus Routing for the Transportation of Pupils to School

This chapter presents a methodology for the planning and operation of students' transportation to and from school. School transportation has a lot of hidden issues that make it difficult to organize and manage (i.e. safety, punctuality, time). This chapter treats the school bus routing problem and proposes routing algorithms that could be applied when planning transportation for students who go to school using school buses. Furthermore, this chapter proposes a routing methodology for pedestrians (from home to the bus stop and school bus and from the bus stop to the school and backwards), in order to provide the children a safer and more efficient way to go to school. For this reason, a "Safe Map" concept, a map with safety parameters integrated to its links and its nodes, is constructed both for the pedestrian and the school bus route, and is presented in this chapter. Also, various solving algorithms are used and compared, presenting a good performance (the genetic algorithm presents up to 12% overestimation of distance in relation to the best solution, with an average overestimation of 5%) and comparing a real case taking into account the safety framework presented in this chapter. These algorithms are developed within the context of the European Project SAFEWAY2SCHOOL.

18.1. Introduction

Going to and from school is a daily transportation necessity for millions of children of age 6–16 within Europe. For example, in Sweden, 250,000 go to school daily, approximately 700,000 in Poland and about 450,000 in Austria [ANU 10]. This transportation is executed by various transport modes (car and bus primarily) and various actors (from the parents themselves to private companies). The likelihood of accidents is a significant issue in this kind of transportation.

Chapter written by Eleni CHALKIA, Josep Maria SALANOVA GRAU, Evangelos BEKIARIS, Georgia AYFANDOPOULOU, Chiara FERARINI and Evangelos MITSAKIS.

According to accident statistics, school bus travel is much safer than car travel to school. This is due to the strict policy applied to the professional transportation of pupils to school, from routing issues to behavioral issues within the bus.

In recent decades, there has been a growing concern about the safety of children during their trip to school. The SAFEWAY2SCHOOL EU Project has developed methods and tools enhancing the safety of children who travel to and from school by bus. These methods and tools focus on bus transport, the consideration of door-to-door safety and the methods and tools which have been developed to provide support both in the planning and monitoring of school bus services. This chapter describes in particular the Safe Route Planning approach and the “Safe Map”.

18.2. The school bus routing problem

The school bus routing problem is a complicated transportation problem which falls into the large category of np-hard problems. The np-hard problem, also referred as the Vehicle Routing Problem (VRP), is one of the most complicated optimization mathematical models and one of the most well-known branches of it is the Traveling Salesman Problem; the school bus routing problem (SBRP) is another important and practical branch of this problem. Three major factors make school bus routing unique:

– efficiency (the total cost to run a school bus);

– effectiveness (how well the demand for service is satisfied);

– equity (fairness of the school bus for each student).

Due to its complexity, the SBRP has been studied by many researchers and lots of studies have this as their main topic. For example, Bowerman *et al.* [BOW 95] have worked on the SBRP and they noted two problems, clustering and routing, with the traditional problem formulation for school bus routing. Their major problem was that the SBRP formulation cannot be solved in a reasonable amount of time, when including all the constraints, since, and this drives us to the second problem, the formulation of a problem involving three routes with 100 students at 20 bus stops would require 3,400 variables and 6,800 constraints. Thus, Bowerman *et al.* [BOW 95] followed the example of all the researchers who have studied the SBRP, including the team behind this chapter, and broke the problem up into sub-problems. Ledesma and Salazar-Gonzalez [RIE 13] and Schittekat *et al.* [SCH 13] proposed a simultaneous solution for the bus stop selection and bus route generation taking into account bounds in the distances traveled by the students, the minimum capacity usage and the maximum number of stops to be visited by each bus. The approach presented in this chapter solves each sub-problem separately, emphasizing the safety issues presented above. Martinez and Viegas [MAR 11] used real data from the

School Bus Service in Lisbon for testing their two-step algorithm, with a clustering first and routing second approach.

Some more simplistic approaches are considered such as target minimizing the cost. As cost, some have considered the operational cost, which is usually defined as the number of buses or combination of the total travel time and bus fleet size. For instance, minimization of the total number of buses is considered as the objective function in studies by Desrosiers *et al.* [DES 80], Swersey and Ballard [SWE 84] and Braca *et al.* [BRA 94]. Alternatively, Gavish and Shlifer [GAV 79] considered as cost the combination of total travel time and number of buses for the objective function. Corberan *et al.* [COR 02] presented an interesting approach, where each objective is considered separately rather than combined with others into a single function to optimize. Conversely, the methodology presented in this chapter focuses on one single objective for minimizing the related costs, but is not limited to time, cost or distance, but it includes the minimization of the exposure of pupils to danger. Desrosiers *et al.* [DES 80] and Braca *et al.* [BRA 94] also added an upper bound on the travel time of a student, and they also considered a time window for the school arrival. Another upper bound constraint defined by Desrosiers *et al.* [DES 80] was the gap between the arrival time at the school and the start time of the school. In more detail, Bennett and Gazis [BEN 72] included in their objective function the total travel time spent by all children. Bodin and Berman [BOD 79], Gavish and Shlifer [GAV 79], Desrosiers *et al.* [DES 84] and Braca *et al.* [BRA 94] added an upper bound on the travel time for each child. Desrosiers *et al.* [DES 80], Swersey and Ballard [SWE 84] and Braca *et al.* [BRA 94] added a time window on the arrival at school. Moreover, Desrosiers *et al.* [DES 80] added an upper bound on the waiting time between school arrival time and school starting time, and upper bounds on the number of students at stops and on routes. Braca *et al.* [BRA 94] imposed an earliest pickup time for children. In addition, they require a minimal number of children to create a route. Paquette *et al.* [PAQ 09] presented an interesting discussion on the quality of the services provided by a dial-a-ride service with reduced mobility, which can easily be applied to the transportation of pupils.

A more comprehensive review of the SBRP can be found in Park and Kim [PAR 10], where the authors summarize the assumptions, constraints and solution methods presented in the literature.

18.3. Methodology for solving the SBRP in SAFEWAY2SCHOOL

From the literature, we have seen that many researchers have tried to solve the SBRP by introducing various constraints and objectives. We also noticed that the complexity and the difficulty in identifying and solving this problem, taking into account all the possible constraints, is great. One major issue though that has not been identified by any of the researchers in the literature is the part of the safety of the children while on board. In SAFEWAY2SCHOOL, we considered safety as one of the major issues that need to be taken into account while trying to confront the SBRTP. Thus, we defined a concept that separates the SBRP into three sub-problems, which are the following:

– the SAFE MAP;

– the pedestrian routing;

– the school bus routing.

18.3.1. *SAFE MAP*

18.3.1.1. *Safety criteria from the literature*

The "Safe Map" concept is the first step of this methodology and it is the most important step in order to define the safety score of each link in the road transportation network where the routing will take place.

While searching the literature, trying to identify the criteria that are important in order to map the safety of the school bus route and also the guidelines that countries have drawn up for evaluating the safety of bus stops were the closest extracts. Two countries have defined safety for the evaluation of bus stops, France and Sweden.

The system proposed by the French National Council of Transport (CNT) sets out seven criteria for evaluating the safety of bus stops in terms of overall infrastructure and three relating to the traffic context. The seven criteria used to "score" the bus stops are:

– path leading to the bus stop (whether it has a pavement, etc.);

– pedestrian crossing (if there is a crossing, its location, etc.);

– perception of the bus stop (presence of vertical and horizontal lighting);

– visibility of the stop for approaching drivers (if it is possible to see the stop in advance (location behind a curve, at a crossing road, etc.);

– size and location of the waiting area (in relation to the number of children, protection from weather, etc.);

– nature and quality of the ground in the waiting zone (e.g. whether these induce children to stay at the stop or wait on the road itself);

– maneuvers imposed on the bus.

For each of these criteria, scores are given on a scale from 0 (safe) to 25 (dangerous). This score is then adapted according to the following traffic-related factors:

– traffic volume:

- the score is multiplied by 2 if there are more than 1,000 vehicles/day,

- the score is multiplied by 5 if there are more than 5,000 vehicles/day;

– likelihood of an accident:

- the score is multiplied by the number of accidents which occurred during the last 5 years within a radius of 200 meters of the bus stop;

– speed:

- the score is multiplied by the legal speed at this location.

Similarly, the model proposed by the Swedish Transportation Agency adopts a similar approach, although it is designed to assess stops in rural areas. It takes into account the risk of accidents and the degree of insecurity for all types of passengers while waiting at bus stops and on their way to/from stops. It distinguishes the following four types of bus stop (taking into account the design and the waiting area).

The risk assessment then takes into account four characteristics of the stop: visibility, lighting, traffic flow and traffic speed (relating to the road on which the stop is sited). The selection of these characteristics is based on the knowledge of previous accidents and the associated risks. The basis for the scoring of different characteristics is more subjective but is linked to factors such as braking distances, visibility distances and other circumstances surrounding accidents. The model has been calibrated through field studies and accident reports as well as subjective assessment. Among the factors which affect the subjective assessment are the type of vehicles (e.g. many large trucks passing), and specific local conditions (e.g. the presence of snow restricting the waiting area, which is an important factor in Sweden).

18.3.2. *Safety criteria used in the presented methodology*

For a school bus operator or local authority who is responsible for school transport services, it is very useful to have a systematic way of assessing the safety level of the road network affected by a school bus service. In order to do this, it is necessary to define criteria which can be used to produce a safety-related "score" for each road section (link).

There are various ways of making such an assessment. For example, a systematic examination can be made of the accident statistics to find out which parts of the network have a record of accidents, in other words, are recognized as "black spots". Another approach is to examine the characteristics of the road and traffic in order to identify the parts of the road network which have the greatest *potential* risk. If the necessary data are available, and also the time to undertake such a survey, the ideal solution would be to combine these two types of approach.

Accident data can be represented on the Safe Map in the relevant geographical location using symbols of different dimensions and color to indicate the number, type and gravity of accidents that have occurred in a given period, e.g. in the last 5 years:

– fatal accident involving a pedestrian: child/adult;

– accident with serious injuries: child/adult;

– accident with minor injuries: child/adult;

– different symbols could be used for accident involving a school bus and for accident involving other types of vehicle.

Among others factors, the potential safety of a section of road is affected by its geometrical characteristics, features such as crossings and intersections, the lighting and signage, and the traffic behavior, as well as various "dynamic" factors relating to weather conditions, e.g. areas prone to ice. Examples of features, which can be responsible for safety risks in rural and urban road environments, are listed below:

– rural areas:

- sharp bends,

- lack of protective barriers on the road shoulder,

- narrow two way roads without a central partition,

- lack of lighting;

– urban areas:

- lack of good visibility for making turns (due to buildings or on street parking),

- fast roundabouts,

- unprotected (left) turns;

– complex intersections.

With regard to traffic conditions, the criteria normally taken into consideration are:

– traffic flow (volume of traffic/hour);

– traffic speed (legal speed limit on a given section).

In order to determine the safety potential of a road, it will therefore be necessary to select the most relevant factors according to the local context. This is because different factors are relevant for different road types (as indicated above for rural and urban areas) and also for different regions and countries. For example, in Sweden the risk of snow and ice makes certain roads dangerous. Good road lighting is also important because of short days in winter. On the other hand, on many country roads in Italy the presence of roadside barriers is a critical safety factor where there are steep drops, as well as a central barrier when the roads are narrow.

18.3.2.1. *Method proposed for calculating the risk level of the school bus route*

Taking into account all the aforementioned criteria, we have identified a methodology on how to evaluate the safety aspect of a route link that follows the steps below:

– select the most relevant risk criteria with regard to the local situation;

– assign each criterion a weight from 0 to 5 (where 5 indicates a strong influence and 0 no influence) to reflect the relative impact of this criterion on safety (using local or national statistics as guidelines);

– for each road section, assign a score from 1 to 5, on the basis of a survey or available data;

– multiply the two values (score and weight) to obtain a total risk value for each criterion;

– to obtain the overall score, sum the values for each risk criterion and divide by the total value of weights. The result is the road safety index.

The resulting overall score can then be translated into three risk categories as shown in Table 18.1.

Final score (S)	Risk category	Link with the routing and mitigation strategy
$0 \leq S \leq 1$	Green	This road is considered relatively safe
$1 < S \leq 3$	Yellow	This road is considered borderline
$3 < S \leq 5$	Red	This road involves a significant safety risk

Table 18.1. *Risk categories*

18.3.3. *Pedestrian routing*

In real life pedestrian routing is a problem that really concerns the families of the students. It really affects only the children of a certain age, since the very small ones are picked up from their front door.

On the contrary to the actual school bus routing methodology, the literature for pedestrian routing is quite poor. Additionally, some literature only refers to the allocation of the students at the bus stops, based upon the distance criterion, but without referring at all to the route that is going to be taken by the child.

Schittekat *et al.* [SCH 06] developed a holistic methodology for school routing that also included the allocation of the children at the bus stops. They determined a set of potential stops, so that each student has at least one stop that he/she can walk to. Afterwards, they determined the school bus routes so that all students are picked up at a stop they are allowed to use, while making sure that the capacity of the buses is not exceeded. In this way it was possible that a stop is not used by the bus, as long as the student is allowed to go to another stop.

In the process that we propose, the allocation of the children at the bus stop is based on the distance criterion, similar to that proposed by Schittekat *et al.* [SCH 06], but the final allocation of the students at the bus stops is based only on the safety criterion. The safety of the route that the students will take is going to be calculated by using the pedestrian safety criteria of the previous section.

Thus, on the contrary to most vehicle routing formulations, in which a set of stops is given and only the school bus routes need to be determined in order to visit each stop, we will present the pedestrian routing problem as a standalone issue, that is not affected by vehicle routing, but is only effected by the children's safety. In the SAFEWAY2SCHOOL pedestrian problem we identify a set of potential stops based upon the bus stop safety criteria, find the set of stops each child can visit according to distance criterion and finally determine for each student to which stop should move, so that the safety of the walking route is maximum.

The algorithm that is going to be used for the final allocation of the students to the bus stops is going to be a variation of Dijkstra's algorithm that is used to solve the shortest path problem. In our case we will use Dijkstra's algorithm in order to solve the safest path problem.

Dijkstra's algorithm for a given source vertex (node) in the graph, finds the path with the lowest cost (i.e. the shortest or safest path) between that vertex and every other vertex. It can also be used for finding costs of shortest paths from a single vertex to a single destination vertex by stopping the algorithm once the shortest path to the destination vertex has been determined.

According to Dijkstra, the node at which we are starting is called the initial node and the distance of node Y is the distance from the initial node to Y. Dijkstra's algorithm will assign some initial distance values and will try to improve them step by step. On the contrary, SAFEWAY2SCHOOL pedestrian algorithm will assign safety values to each link.

1) Assign to every node a safety value: set it to zero for our initial node and to infinity for all other nodes.

2) Mark all nodes as unvisited. Set initial node as current.

3) For current node, consider all its unvisited neighbors and calculate their tentative safety.

4) When we are done considering all neighbors of the current node, mark it as visited. A visited node will not be checked ever again; its safety recorded now is final and minimal.

5) The next current node will be the node with the highest safety in the unvisited set.

If all nodes have been visited, finish. Otherwise, set the unvisited node with the maximum safety (from the initial node, considering all nodes in graph) as the next "current node" and continue from step 3.

18.3.4. *School bus routing*

The "Safe Map" is now the basis for the routing algorithms to be implemented. First the pedestrian algorithm is implemented, which follows Dijkstra's algorithm framework that is used to solve the shortest path problem, which is used in order to solve the safest path problem. When the children are assigned to the bus stops, the final branch of the problem follows, which is their route to the school, on the bus.

As should be clear from the previous discussion, most papers on the routing of school buses have focused on capturing as many aspects of reality as possible, building intricate multi-objective models with many complex constraints. On the other hand, we focus on a basic version of this problem to develop efficient methods that take the safety issue of the problem into account. Our problem formulation has only a single school, one type of student and one type of bus, with fixed capacity and has been based upon the multi-trip elementary shortest path problem with resource constraints developed by Akca *et al.* [AKC 10].

– We will try to optimize the following two vehicle routing criteria:

- the total time travelled by each student,

- the safest path taken from the school bus.

– Assumptions:

- the route starts and ends at the school (depot),

- the cost of the route is proportional to the safety of the route and the time travelled from each child,

- travel times between each stop are known and accurate,

- demands at each stop are known,

- loading time per child is constant for every child (=1 minute per child),

- the route will be executed only if there are more than 10 children on board,

- if the bus stops at a bus stop, the children will be boarded,

- each stop is served only once,

- each stop is served by one route.

– Constraints:

- the bus capacity is not exceeded by the number of children on board,

- each child should not be in the bus for more than one hour.

Constants:

M = set of nodes;

B = set of links;

I = set of customers;

Δ = set of routes;

a = weight of the time factor;

b = weight of the safety factor.

Variables:

Sik: the safety mark of arc (i, k);

xik,δ: binary variable (0–1) of arc (i, k) for route δ;

yik,δ: flow at the arc (i, k) in route δ;

LV: capacity of the school bus;

LT: time limit of the route;

Tik: travel time of the arc (i, k);

Di: demand of node *i*.

Mathematical programming formulation:

$$Min \sum_{\delta \in \Delta} \sum_{(i,k) \in B} x_{ik,\delta}(aT_{ik} + bS_{ik}) \quad [18.1]$$

$$\sum_{\delta \in \Delta} \sum_{k \in M} x_{ik,\delta} = 1 \qquad \forall i \in I, \forall \delta \in \Delta \quad [18.2]$$

$$\sum_{k \in M} x_{sk,\delta} = 1 \qquad \forall \delta \in \Delta \quad [18.3]$$

$$\sum_{k \in M} x_{ks,\delta} = 1 \qquad \forall \delta \in \Delta \quad [18.4]$$

$$y_{ik,\delta} - L^V x_{ik,\delta} \leq 0 \qquad \forall (i,k) \in B, \forall \delta \in \Delta \quad [18.5]$$

$$\sum_{k \in M} y_{ik,\delta} - \sum_{k \in M} y_{ki,\delta} + D_i \sum_{k \in M} x_{ik,\delta} = 0 \qquad \forall i \in I, \forall \delta \in \Delta \quad [18.6]$$

$$\sum_{(i,k) \in B} T_{ik,} x_{ik,\delta} \leq L^T \qquad \forall \delta \in \Delta \quad [18.7]$$

$$\sum_{i \in I} D_i \sum_{k \in M} x_{ik,\delta} \geq 10 \qquad \forall \delta \in \Delta \quad [18.8]$$

$$\sum_{i \in I} D_i \sum_{k \in M} x_{ik,\delta} \geq 10 \qquad \forall \delta \in \Delta \quad [18.9]$$

$$x_{ik,\delta} \in \{0,1\} \qquad \forall (i,k) \in B, \forall \delta \in \Delta \quad [18.10]$$

$$y_{ik,\delta} \geq 0 \qquad \forall (i,k) \in B, \forall \delta \in \Delta \quad [18.11]$$

where $x_{ik,\delta} = 1$, if arc (i, k) is used in route δ and 0 otherwise; and $y_{ik,\delta} = 1$, if equal to the amount of flow on arc (i, k) for route δ for all (i, k) that belong to B.

The objective equation [18.1] indicates that the total cost of all routes, which is the weighted sum of the arcs safety and travel time, should be minimized. Equation [18.2] specifies that the customer node i may be visited only once. Equation [18.3] requires that the vehicle should enter and leave a customer node an equal number of times in the same route. Equations [18.4] and [18.5] define that each route starts and ends at the school (depot). Equation [18.6] guarantees that the vehicle capacity is not exceeded in any route. Equation [18.7] requires conservation of flow of each route at each customer node. Equation [18.8] limits the total time of a vehicle's route to the time limit. Equation [18.9] puts a lower bound in the number of students of each route. Equations [18.10] and [18.11] are the integrality and non-negativity requirements on the variables.

18.4. Application to Thessaloniki

The algorithms that are initially proposed to be tested and compared for the Thessaloniki case study are as follows:

– savings algorithm;

– greedy algorithm;

– genetic algorithm.

A set of benchmarks have been used for testing the efficiency of the algorithms, as proposed in Marinakis [MAR 09] and Bielli *et al.* [BIE 11]. A set of equivalent VRP benchmarks [TSP 95] have been used for evaluating the performance of the three proposed algorithms. The results are given in Table 18.2.

Algorithm	Minimum (%)	Average (%)	Maximum (%)
Greedy	13	82	190
Savings	1	12	21
Genetic	1	5	12

Table 18.2. *Overestimation of the algorithms in relation to the best solution*

It can be observed that the savings and the genetic algorithms present the best fit, with an average overestimation of 5% and 12% respectively, while the greedy algorithm presents much higher values (more than 80%), as expected. Data related to safety are missing from the instances library, but they have been created for the Thessaloniki use case and the results are presented below.

The above methodology has been applied to the city of Thessaloniki. The road network has been divided into two levels as shown in Figure 18.1, a primary network for the bus routing with 3,000 links and 1,000 nodes, and a secondary network for the pedestrian routing with 140,000 links and 50,000 nodes.

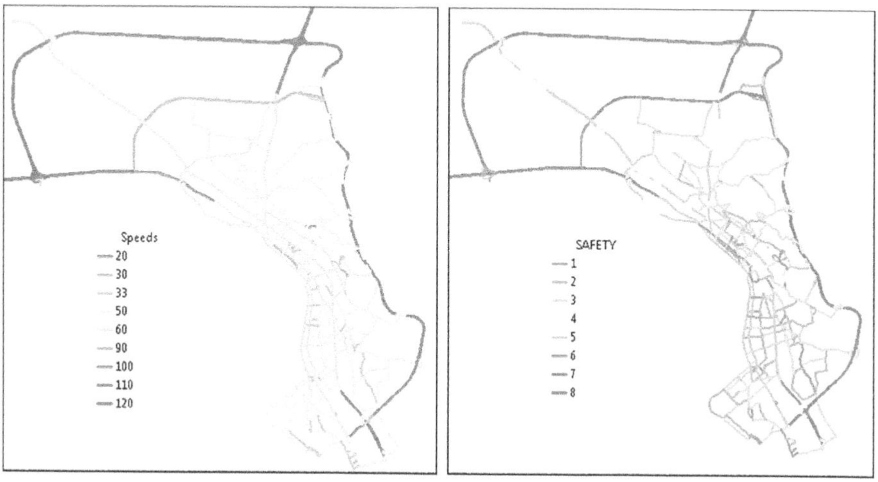

Figure 18.1. *Thessaloniki speed and safety maps. For a color version of the figure, see www.iste.co.uk/jacob/safety.zip*

The data of the links used for the calculation of the scores and the routing are basically length, travel time and safety, as shown in Figure 18.2. Three algorithms have been programmed in MATLAB, two heuristics (greedy and savings) and one meta-heuristic (genetic algorithm). Two routing engines have been used, both based on the Dijkstra algorithm, one with turning restrictions (LB) and one without turning restrictions (NB). The problem to solve is composed of 60 children, 29 bus stops, 1 school and 1 bus.

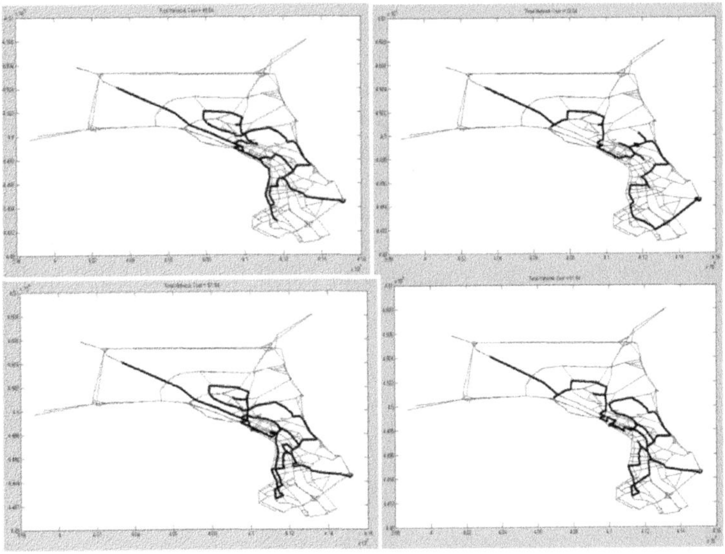

Figure 18.2. *Solutions obtained by the greedy NB (up-left), genetics NB (up-right), greedy LB (down-left) and genetics LB (down-right). For a color version of the figure, see www.iste.co.uk/jacob/safety.zip*

The solutions obtained by the different solving algorithms and routing engines are presented in Figure 18.3, while the numerical values of the case study are depicted in Figure 18.4.

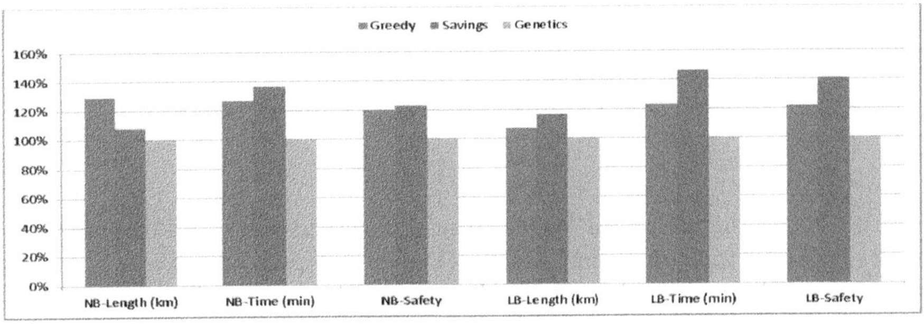

Figure 18.3. *Comparison of the performance of the solutions obtained by the various algorithms. For a color version of the figure, see www.iste.co.uk/jacob/safety.zip*

In all cases the genetic algorithm performed better, presenting a reduction of 10–20% in most of the indicators in relation to the other algorithms. In the safety aspect, which is the most important, the reduction that has been achieved ranges from 20–40% in comparison with the greedy and the saving algorithms. The NB algorithms run much faster, they need 1/3 of the iterations in comparison with the LB, and they provide a good solution eight times faster than the LB, but they are not realistic since turn restrictions exist in real life, especially for buses.

18.5. Conclusions

This chapter is an extract of the wider research that was realized within the SAFEWAY2SCHOOL EU Project, concerning the safety and security of children's transportation to and from school, following a holistic approach that covers the transportation from door (home) to door (school).

In SAFEWAY2SCHOOL we have followed an approach that first of all splits the SBRP in to three separated ones and also focused at the safety perspective of it.

Thus, three individual problems emerged which needed separate development in order to get the greater picture of the SBRP solved. These three problems are the "Safe Map", "Pedestrian Routing" and "School Bus Routing".

The "Safe Map" is the first step to be followed and it focuses on the identification of the different safety levels of the links of a map of an area, where the routing will take place. This is a preliminary step, which is very important to do, since in this way, we can construct a map which has all its links and nodes characterized with their safety level. Different criteria have been chosen and evaluated, like the traffic flow, the speed, the lighting of the road, the railway crossing, etc., using a rank from 0 (safest) to 5 (dangerous). Using these criteria the safety of the component accrues and is inserted into the map.

The route, from the bus stop to the school, took into account two very important issues. First of all, the bus should follow the safest route and also each child should not be in the bus for more than 1 hour. The network was initially characterized by its safety, using the "Safe Map" approach, and then the algorithms were implemented using savings, greedy and genetic algorithms (the algorithms have been validated by testing their performance in a set of benchmark problems, presenting average overestimations of 12, 82 and 5% respectively). The algorithms have been applied to a real case using the safety framework presented herein. In all cases the genetic algorithm performed better, presenting a reduction of 10–20% in most of the indicators in relation to the other algorithms. In the safety aspect, which is the most important, the reduction that has been achieved ranges from 20–40% in comparison with the greedy and the saving algorithms.

18.6. Acknowledgments

The research presented in this chapter was conducted within the context of the SAFEWAY2SCHOOL (SAFEWAY2SCHOOL) project "Integrated system for safe transportation of children to school", funded by the Research Directorate General of the European Union under the Seventh Framework Programmer for Research and Technological Development.

18.7. Bibliography

[AKC 10] AKCA Z., RALPHS T.K., BERGER R.T., "Solution methods for the multi-trip elementary shortest path problem with resource constraints", *Optimization Online the Mathematical Programming Society*, 2010

[ANU 10] ANUND A., LARSSON J., DUKIC T. *et al.*, SAFEWAY2SCHOOL Deliverable 1.4 "Accident Analysis", 2010.

[BEN 72] BENNETT B., GAZIS D., "School bus routing by computer", *Transportation Research*, vol. 6 pp. 317–326, 1972.

[BIE 11] BIELLI M., BIELLI A., ROSSI R., "Trends in models and algorithms for fleet management", *Procedia Social and Behavioral Sciences*, vol. 20, pp. 4–18, 2011.

[BOD 79] BODIN L., BERMAN L., "Routing and scheduling of school buses by computer", *Transportation Science*, vol. 13, pp. 113–129, 1979.

[BOW 95] BOWERMAN R., BRENT H., PAUL C., A multi-objective optimization approach to urban school bus routing: formulation and solution method, *Transportation Research Part A: Policy and Practice*, vol. 29A, no. 2, pp. 107–123, 1995.

[BRA 94] BRACA, J., BRAMEL J., POSER B. *et al.*, A computerized approach to the New York City school bus routing problem, Graduate School of Business, Columbia University, New York, 1994.

[COR 02] CORBERAN A., FERNANDEZ E., LAGUNA M. *et al.*, "Heuristic solutions to the problem of routing school buses with multiple objectives", *Journal of the Operational Research Society*, vol. 53, pp. 427–435, 2002.

[DES 80] DESROSIERS J., FERLAND J., ROUSSEAU J.-M. *et al.*, "An overview of a school busing system", *Scientific Management of Transport Systems*, vol. 9, pp. 235–243, 1980.

[GAV 79] GAVISH B., SHLIFER E., "An approach for solving a class of transportation scheduling problems", *European Journal of Operational Research*, 1979.

[MAR 09] MARINAKIS Y., "Metaheuristic algorithms for the vehicle routing problem", *Encyclopedia of Optimization, Part 13*, Springer-Verlag, pp. 2055–2061, 2009.

[MAR 11] MARTÍNEZ LUIS M., VIEGAS JOSÉ M., "Design and deployment of an innovative school bus service in Lisbon", *Procedia – Social and Behavioral Sciences*, vol. 20, pp. 120–130, 2011.

[PAR 10] PARK J., KIM B., "The school bus routing problem: a review", *European Journal of Operational Research*, vol. 202, no. 2, pp. 311–319, 2010.

[PAQ 09] PAQUETTE *et al.*, "Quality of service in dial-a-ride operations", *Computers & Industrial Engineering*, vol. 56, no. 4, pp. 1721–1734, 2009.

[RIE 13] RIERA-LEDESMA J., SALAZAR-GONZÁLEZ J.J., "A column generation approach for a school bus routing problem with resource constraints", *Computers & Operations Research*, vol. 40, no. 2, pp. 566–583, February 2013.

[SCH 06] SCHITTEKAT P., SEVAUX M., SORENSEN K., "A mathematical formulation for a school bus routing problem", *IEEE Proceedings of the 2006 International Conference on Service Systems and Service Management*, 2006.

[SCH 13] SCHITTEKAT P. *et al.*, "A metaheuristic for the school bus routing problem with bus stop selection", *European Journal of Operational Research*, vol. 229, no. 2, pp. 518–528, 1 September 2013.

[SWE 84] SWERSEY A.J., BALLARD W., "Scheduling school buses", *Management Science*, vol. 30, no. 7, pp. 844–853, 1984.

[TSP 95] TSPLIB 1995: http://comopt.ifi.uni-heidelberg.de/software/TSPLIB95/index.html.

19

Spreading Awareness of Traffic Safety through Web Application

Public awareness of the state of traffic safety is of crucial importance on the way towards its improvement. Although the data concerning incidents and accidents in traffic and their participants have been publicly available in Slovenia since 1994, the tabular-only format has, until recently, prevented the broader public having a clear view into the topic. To overcome this shortcoming we have developed a web-based application that puts the events from the traffic accident database onto an interactive map. The map display is customizable and it includes the possibility of filtering the data according to a user-specified criteria including the display of potential conflict points. This way it is possible to examine the traffic safety status for an arbitrary area or a particular set of roads or list the accidents matching to a particular group of participants. Using the application the broadest public can identify the problems in traffic safety on various levels, from strictly local to nation-wide and pass them to the decision-making authorities for consideration.

19.1. Introduction

Access to information on traffic accidents on Slovenian roads is interesting for both experts in the field and the lay public. Appropriate presentation of past events helps increase general awareness of the importance of traffic safety and enables analyses to be performed, which can prove beneficial in the planning of measures necessary to remedy the situation [KIM 96, ELV 08].

Since 1994, the Slovenian Police has been collecting the traffic accident data and publishing them for the past few years on its official website (Policija). However, no tool has been developed yet, which would enable a publicly available visualization

Chapter written by Miha AMBROŽ, Jernej KORINŠEK and Ivan PREBIL.

of data from the entire period on a geoinformation base layer. For management of a uniform traffic safety database, which includes data about the past traffic events as well as other publicly available data about road conditions and performed research studies and measurements, a software tool [PRE 10] has been developed, which is intended for both traffic experts and operators of road infrastructure. Due to a large amount of data as well as requirements regarding support software and data security, this tool is not intended for a broader public. However, the work described in this chapter has helped the authors to bridge this gap and provide a publicly accessible website for displaying traffic accident data on an interactive map.

19.2. Current state of traffic accident data in Slovenia

19.2.1. *The Traffic Accident Database*

The Traffic Accident Database (known by the Slovenian acronym BPN for "*Baza Prometnih Nezgod*") is managed by the Police and is updated several times a year. It includes data about events and event participants arranged in separate tables (*tbl_BPN_N_1994–2012* and *tbl_BPN_U_1994–2012*, respectively). The event table includes data about the event type, its location (including the data about the road where a specific event happened), participant types and worst consequences for participants. The participant table includes data about the participant's age, gender and driving license holdership, participant's role in the event, severity of injuries resulting from the event and participant's psycho-physical condition (based on performed tests for alcohol and drug intoxication). Each participant from the participant table is connected to a specific event in the event table by means of a uniform key. Predefined field values for certain database fields are stored in auxiliary tables. Figure 19.1 is a schematic presentation of the structure of the Traffic Accident Database. Table 19.1 shows a summary of the most relevant data present in the database.

The analysis of BPN data has shown that the quality of data on specific events and thereby the possibility of an appropriate geoinformation display varies significantly. Therefore, preparation of data was based on algorithms, which were used to identify events the reliable geolocation of which could not be determined. If possible, the correct geolocation of these events was determined and their amended data entered into BPN. All events which could not be arranged in this matter were eliminated from BPN. Also eliminated were the events for which only an official note was entered by the Police, without any classifications of the injuries to participants. Consequently, out of 730,150 events, 472,305 remained in the database as of the last update.

19.2.2. *Infrastructure data*

Roads in Slovenia are generally divided into two categories – state-managed roads (roughly 6,000 km in total length), the database of which is maintained by the Slovenian Roads Agency, and municipal roads (32,000 km) which are managed by municipalities. The Slovenian traffic network (Figure 19.2) also includes 1,228 km of railways, which are included in the map for the purpose of analysis of accidents occurring on road-railway crossings.

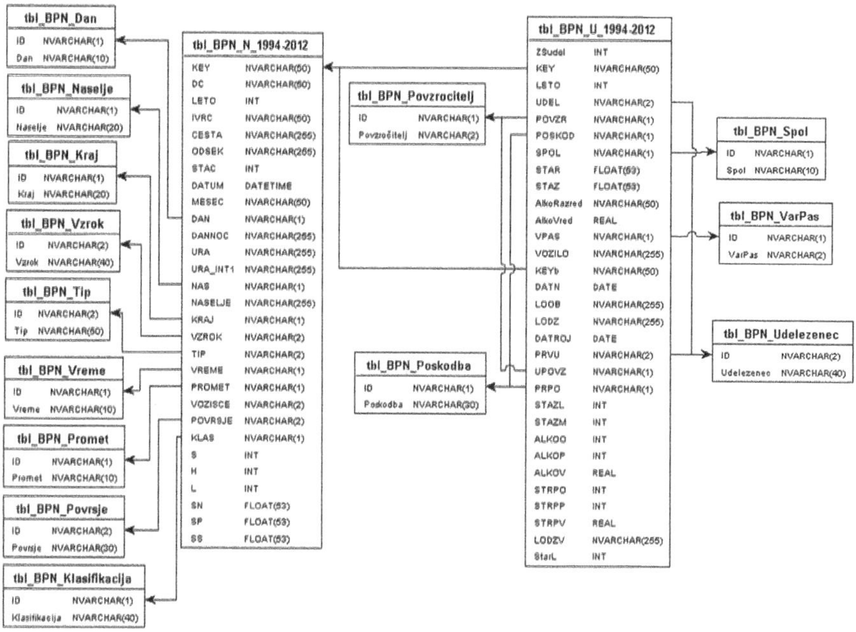

Figure 19.1. *Structure of the Traffic Accident Database*

Also obtained were publicly available data about settlement, municipality and region boundaries as well as data about public infrastructure (roads, railways and waterways), building register with house numbers and topographical numbers. Also included are maps in different scales and orthophotographic images. These data are collected and made publicly available by the Surveying and Mapping Authority of the Republic of Slovenia (GURS) and are updated regularly. In addition to the data, provided from the authorities, an overlay of roads from the OpenStreetMap database (OpenStreetMap) is also imported. This overlay combines the roads from the separate municipal and state-managed roads and includes additional data used for road categorization and conflict point identification. Any inconsistencies found in the existing OpenStreetMap data are promptly corrected and submitted back to the OpenStreetMap servers.

Event table		Participant table	
Field	Description	Field	Description
KEY	Unique event identifier	KEY	Unique participant identifier
LETO	Year of occurrence	LETO	Year of occurrence
CESTA	Road designation (if available)	UDEL	Participant type
ODSEK	Road section designation (if available)	POVZR	Participant is accident inducer (true/false)
STAC	Running stationing (if available)	POSKOD	Type of sustained injury
DATUM	Date of occurrence	SPOL	Gender
MESEC	Month of occurrence	STAR	Age
DAN	Day of week of occurrence	STAZ	Driving license holdership period
DANNOC	Time of day (allowed values: day/night)	VPAS	Usage of protective measures (true/false)
URA	Hour of occurrence	DATN	Date of accident
NASELJE	Settlement (if available)	DATROJ	Date of birth
KRAJ	Locality (if available)	ALKOO	Breath alcohol test commissioned (true/false)
VZROK	Accident cause (one of 12 predefined values)	ALKOP	Breath alcohol test positive (true/false)
TIP	Accident type (one of 11 predefined values)	ALKOV	Value of breath alcohol test
VREME	Weather conditions (one of 8 predefined values)	STROP	Blood alcohol test commissioned (true/false)
PROMET	Traffic intensity (one of 5 predefined values)	STRPP	Blood alcohol test positive (true/false)
POVRSJE	State of road surface (one of 9 predefined values)	STRPV	Value of blood alcohol test
KLAS	Injury classification (one of 5 predefined values)		
Xkoos	x GC coordinate of accident location		
Ykoos	y GC coordinate of accident location		

Table 19.1. *Descriptions of relevant data fields in Traffic Accident Database*

19.2.3. *System problems with the current dataset*

State-managed roads in Slovenia (including the motorways managed by the Slovenian Motorway Company) are marked with unique identifiers and divided into sections each of which also has its own unique identifier. The road sections have a defined direction and running stationing in meters. Physical stationing markings (Figure 19.3) are generally available along the roadway in 500 m intervals. As a result, the geolocations of the events on the state-managed roads are readily determinable.

Figure 19.2. *Road and railway network in Slovenia (red – state-managed roads, grey – municipal roads, black – railways). For a color version of the figure, see www.iste.co.uk/jacob/safety.zip*

Figure 19.3. *Marking on a state-managed road*

Municipal roads, however, are not marked with systematic identifiers and neither do they have a unified stationing system nor any physical markings along the roadway. To determine the location of events on municipal roads, the Police usually use nearby buildings and their house numbers which can be several meters or in some cases even further away from the actual location of the accident (Figure 19.4). Furthermore, the buildings used to locate the accidents on the same microlocation are often chosen inconsistently between the events. All this makes the exact geolocation of events on municipal roads difficult to determine and may lead to wrong conclusions in regard to dangerous locations within the road network. The problem is even graver as most roads in the urban areas fall into the municipal road category. The current geolocating procedure thus effectively prevents the precise microlocation of events in areas such as larger intersections or on multi-lane alleys. This is further complicated by missing the exact definitions of accident type and cause [POL 12]. The improved procedure of cataloguing the future events on municipal roads should therefore include geolocation by a GPS device by the scene investigation team and clearer definitions of accident types and causes.

Rather than being discarded, however, the problematic data is included in the database and presented on the geoinformation overlay in the web application in a distinct color. This makes the users aware of the problems and may help in implementing the proposed changes in the data collection procedures in the future.

Figure 19.4. *Questionable locations of traffic accidents attached to objects with house numbers (yellow markers), actual locations of traffic accidents on the roads (green markers). For a color version of the figure, see www.iste.co.uk/jacob/safety.zip*

19.3. Identification of conflict points

In order to provide the users of the application with an insight into possible conflict points on the traffic network and to facilitate the matching of the falsely located traffic accidents to those points a system of road categorization and crossing point ranking has been devised. The roads are categorized according to the categorization scheme used by the Slovenian Road Agency. Each category of road has a weight factor "*r_weight*" assigned to it, based on the value of its "*highway*" tag in the OpenStreetMap database (OpenStreetMap).

The conflict points are assumed to be on every crossing of two or more paths (roads, cycleways and footways) and on crossing of those with railways. These points are called "*crossing points*" and to each of them a significance factor "*cp_weight*" is assigned based on the number "*leg_count*" and the categories of the ways that cross in it, as follows:

$$cp\_weight = \sum_{i=1}^{leg\_count} r\_weight_i$$

Each traffic way junction has at least one crossing point, while those more complex ones may have two or more. As none of the datasets used include the accurate data about the junctions, a heuristic algorithm has been devised to find and group the crossing points into blobs that form larger junctions as displayed in Figure 19.5. The blob centroids are also calculated and can be displayed as a separate overlay on the map.

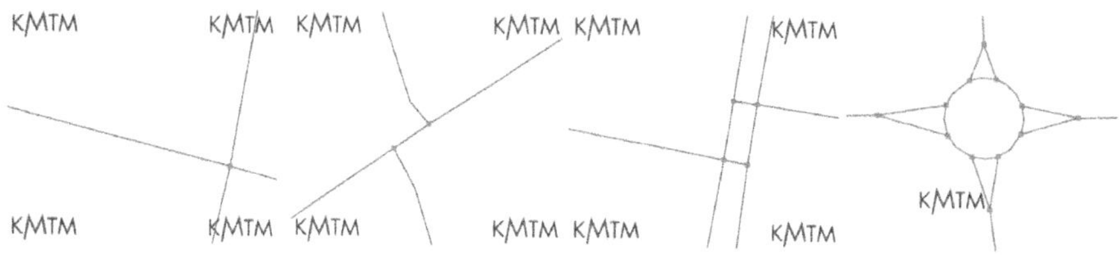

Figure 19.5. *Some possible arrangements of crossing points in junctions consisting of roads*

The significance of the conflict points identified this way is presented as two separate overlays (one for crossing points and another for junction blobs) that include symbols colored and numbered based on the value of the *cp_weight* parameter. Figure 19.6 shows a cut-out of such a display with active OpenStreetMap and Traffic Accidents overlays.

Figure 19.6. *Display of identified crossing points and junctions with their significance (colour and size of marker) and number of ways crossing (number in the marker) in a residential area of a city (past accidents displayed for reference). For a color version of the figure, see www.iste.co.uk/jacob/safety.zip*

19.4. Application structure

The main request in the development of a web application was to make it widely applicable and accessible. Also important was the use of development tools and software components which will enable simple updates of databases and functionality upgrades. Consequently, for the software and application development open-source tools were applied (Apache, PHP, PostgreSQL, PostGIS, Geoserver and OpenLayers). Figure 19.7 is a schematic presentation of the application structure.

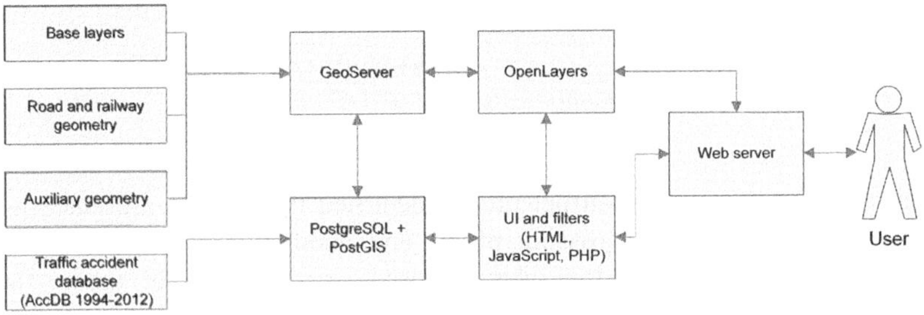

Figure 19.7. *Web application structure*

19.4.1. *Server components*

The server software consists of open-source software solutions adapted to specific application needs. The data is stored in a PostgreSQL relational database. The web server consists of the Apache2 HTTP web server with added Apache Tomcat7 servlet container and PHP "hypertext preprocessor" and is extended with the software for storage and display of geoinformation base layers and data overlays – the GeoServer "web mapping" service for serving map data. The database is supplemented with the PostGIS spatial database package. Other support packages are also installed for spatial data preparation and processing (GDAL, JAI, etc.).

19.4.2. *User interface*

The user interface is implemented using OpenLayers open-source tool with some modifications on the client side and the PHP open-source tool on the server side. It consists of two main parts: an interactive map and a search criteria entry form.

The interactive map (Figure 19.8) enables the user to view a selected area of the Slovenian territory. The user chooses either a map or an orthographic image as the geoinformation base layer. Through a selector pullout the user can also choose to display various data overlays (boundaries of settlements, municipalities, regions, various road types, railways, waterways, buildings and topographic nomenclature as well as traffic accidents) as shown in Figure 19.9. The map is equipped with standard map navigation tools. Under the map there is a status bar showing the current scale and the number of displayed traffic accidents according to chosen criteria.

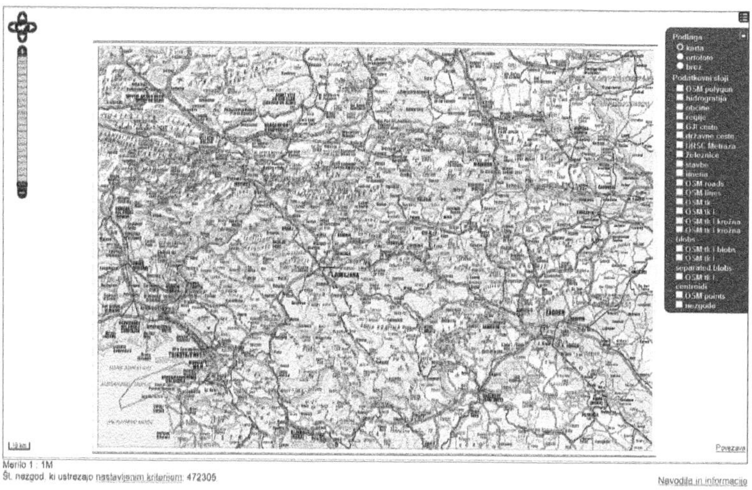

Figure 19.8. *Interactive map with display setting tools and overlay pullout*

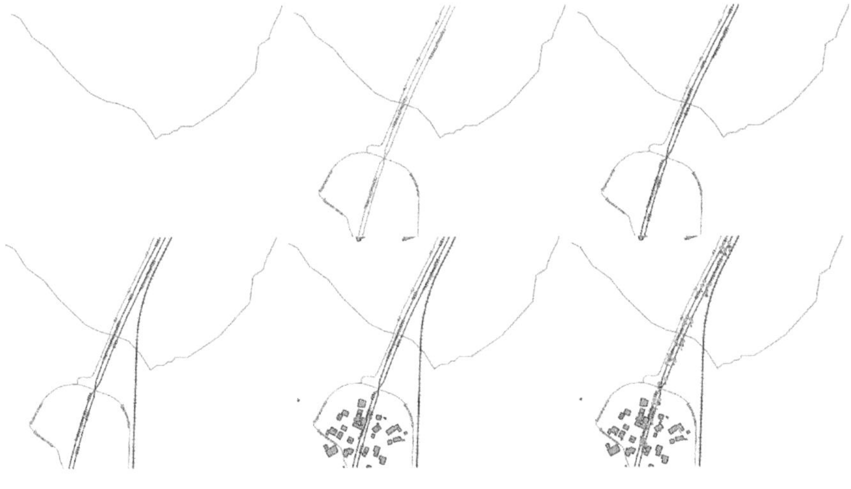

Figure 19.9. *Adjusting the display by adding data overlays (added overlays from top-left across to bottom-right: municipal boundaries, municipal roads, state-managed roads, railways, buildings, traffic accidents)*

The search criteria entry form (Figure 19.10) allows for adjustments of the traffic accident display in accordance with the user-set criteria. The user can choose from time and spatial data (period, region, municipality, time of accident, weather, road and traffic conditions), typological data (injury classification, type and cause of accident), data about the road type and systematization as well as data regarding the type of accident participants. Once the criteria are set, the user confirms the selected options, after which the application display is refreshed in accordance with the selected criteria.

Accidents are presented on the map with accident identifiers, each consisting of a symbol (yellow dot with red outline) and a letter representing the most severe accident that occurred on the location. For each location on the map, the user can access the data associated with it. A mouse click on each accident identifier on the map provides access to data on all accidents at a specific location which are in accordance with the selected criteria. Also available are data related to the road a specific accident occurred on and data on topographic entities at the location. The data is shown in an information cloud pop-up (Figure 19.11), which also includes a link to the participant data for the selected accident. A mouse click on the link opens a new window with relevant data on all accident participants (Figure 19.12).

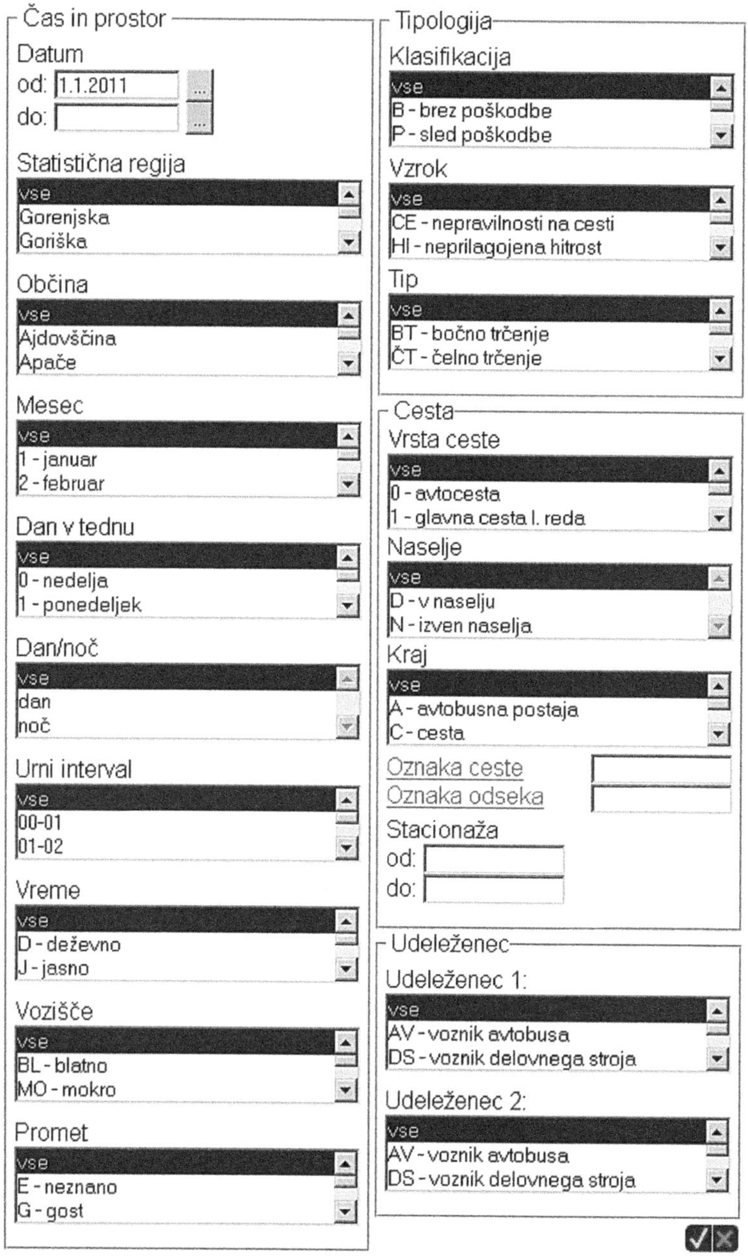

Figure 19.10. *Search criteria entry form*

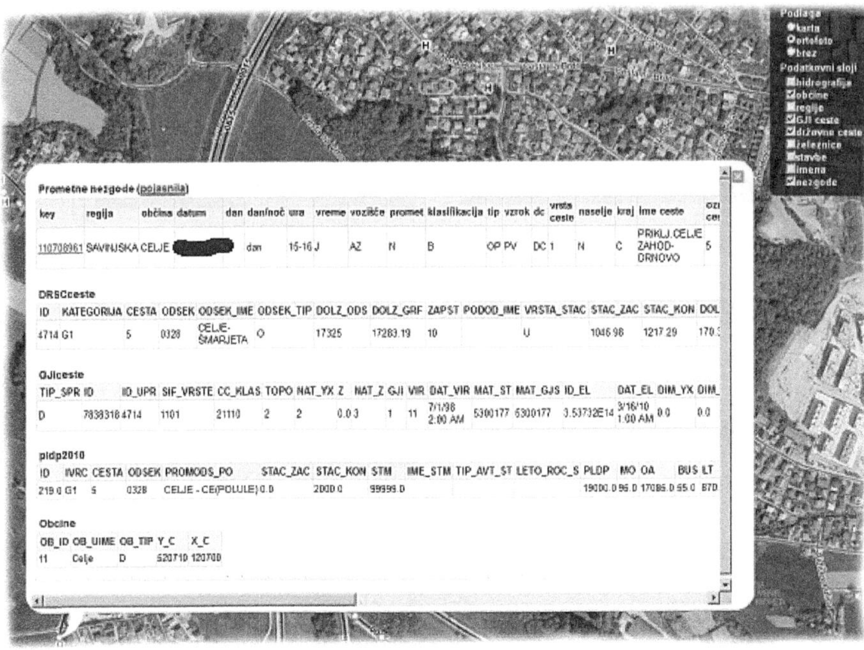

Figure 19.11. *Cloud pop-up with data on the selected spot on the map*

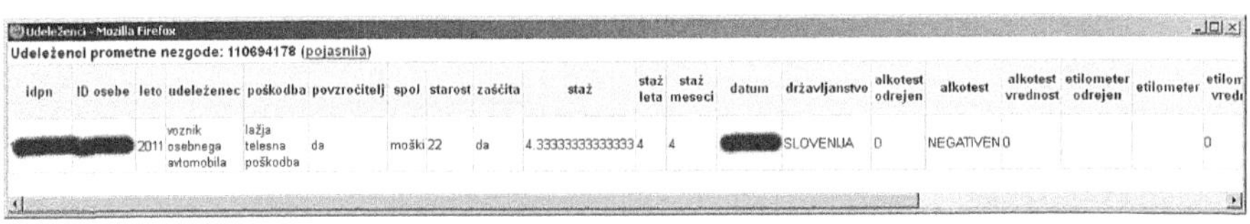

Figure 19.12. *Window with data on accident participants*

19.5. Use of the web application

The application is used to enable various displays of traffic accidents determined by geographic location, time and other criteria. For each displayed accident, the user interface can also be used to access all relevant data about a specific accident and its participants. All data are shown as standard HTML files and can therefore be "copied and pasted" in all software tools that support such formats (text editors, spreadsheets, presentation software and HTML editors).

Publishing the data and sharing them among users can easily be performed by means of a permanent link to the current display which is always located at the bottom of the map (Figure 19.13). The link can be used on other websites or sent via email. The permanent link retains all the display settings and active filters in order to provide the exact same display as presented to the user upon activating the link.

Application users have access to detailed guidelines for use of application – the online help – and explanations of parameter descriptions and values in BPN. As a reference, a table of road sections on state-managed roads is also added.

Figure 19.13. *Link to the current map display with applied filters*

19.5.1. *Examples*

Figure 19.14 shows two example displays prepared by the application of filters. The upper example shows all accidents on the A1 motorway on the ortophoto base layer. The lower example shows all accidents involving cyclists on road sections 1181, 1182 and 1183 with included data overlays (without a base layer).

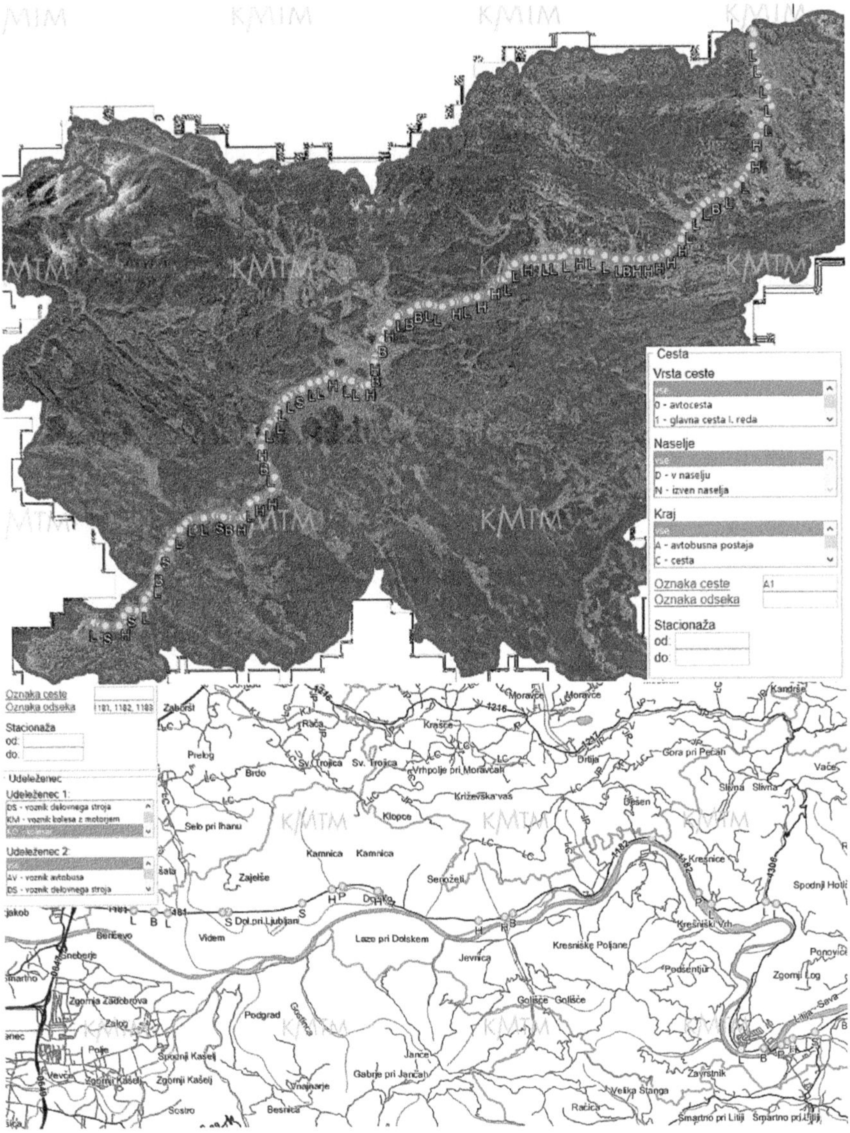

Figure 19.14. *Two examples of filtered traffic accident displays (top: filtered by road, bottom: filtered by road section and type of participant). For a color version of the figure, see www.iste.co.uk/jacob/safety.zip*

19.6. Conclusion

The newly developed web application is intended for members of a broader public who wish to participate in the identification and improvement of the conditions at "black spots" in the Slovenian road network. The application enables various displays of data about traffic accidents that have taken place in the Slovenian road network since 1994.

The web application is based on web standards. To function, it only requires a modern web browser and does not include any additional software requirements. The user interface is designed to be clear, universal and easily comprehensible to the user and is set out in the Slovenian language. Its main component is the interactive map which enables the display of any arbitrary segment of the Slovenian territory and the data associated with it. The search criteria entry form allows for adjustments of the traffic accident display in accordance with the user-set criteria. Any display of the data prepared this way can be accessed through a direct permanent link.

The server component of the application is based on open-source software solutions, due to which the system is flexible and allows for simple data update as well as addition of functionality and possible expansions of the existing dataset.

Traffic accident data are obtained from a database managed by the Police and appropriately consolidated and amended upon each database update. Geographical data, which enable the display of geoinformation base layers in the form of maps and orthographic imagery as well as topographic and transportation infrastructure data overlays, are obtained from publicly available sources and also updated as necessary. The traffic accident data in the display are presented with their geolocations as recorded by the police. This includes potentially mislocated events – mainly those on the municipal roads that are assigned to roadside objects by their house numbers.

The application is publicly accessible and available to everyone willing to obtain insight into the topic. Although primarily targeted at traffic safety professionals and decision makers at different levels, it has also been positively received by traffic accident analysts and scientists from various fields. The recorded use of the application is ranging from presenting road "black spots" to local communities to traffic accident statistics compilations performed by authorities and various interest groups.

By employing the principle of public participation, the quality of the data can be monitored and in some cases improved by the users. Since the system has been set up, the public feedback has been used in requests for improvements of data collection and recording procedures as well as to pursue targeted measures for improving safety on specific spots.

19.7. Acknowledgments

The application was partly financed by the Slovenian Traffic Safety Agency and is hosted on their website.

19.8. Bibliography

[AVP 13] AVP (2013) – *Zemljevid prometnih nesreč od leta 1994/Interactive map of traffic accidents since 1994*, http://nezgode.avp-rs.si/, (visited 25. 8. 2015), 2013.

[ELV 08] ELVIK R., "A survey of operational definitions of hazardous road locations in some European countries", *Accident Analysis and Prevention*, vol. 40, pp. 1830–1835, 2008.

[GUR 12] GURS – Republika Slovenija, Ministrstvo za infrastrukturo in prostor, Geodetska uprava Republike Slovenije, Availbale at: http://www.gu.gov.si/en/ (visited 18. 3. 2013), 2012.

[KIM 96] KIM K., LEVINE N., "Using GIS to improve highway safety", *Computers, Environment and Urban Systems*, vol. 20, no. 4/5, pp. 289–302, 1996.

[POL 12] POLICIJA, Republika Slovenija, Ministrstvo za notranje zadeve, Policija – Prometna varnost – statistika, http://www.policija.si/index.php/statistika/prometna-varnost/ (visited 25. 8. 2015), 2012.

[PRE 10] PREBIL I., AMBROŽ M., ŠUŠTERŠIČ G. *et al.*, *Presojanje cestno prometnih razmer na osnovi simulacij in meritev dinamike vožnje v področju nevarnih cestnih odsekov*, projektna naloga: končno poročilo. Ljubljana: RS MP, Direkcija Republike Slovenije za ceste (in Slovenian), 2010.

Geoserver, http://geoserver.org/ (visited 25. 8. 2015).

OpenLayers: Free Maps for the Web, http://openlayers.org/ (visited 25. 8. 2015).

OPENSTREETMAP, Available at: http://www.openstreetmap.org/ (visited 25. 8. 2015).

PHP: Hypertext Preprocessor, http://www.php.net/ (visited 25. 8. 2015).

PostGIS, http://postgis.refractions.net/ (visited 25. 8. 2015).

PostgreSQL: The world's most advanced open source database, Available at: http://www.postgresql.org (visited 25. 8. 2015).

The Apache Software foundation, http://www.apache.org/ (visited 25. 8. 2015).

PART 6

Railway Safety

20

Overview of Freight Train Derailments in the EU: Causes, Impacts, Prevention and Mitigation Measures

The European Railway Agency (ERA) has estimated that over 500 freight train derailments are happening every year on open lines, costing the EU27 Member States more than 200 million Euros per year. Given the projected freight market growth in Europe, it is essential to ensure the reduction of the number of freight train derailments. The chapter summarizes the most relevant results achieved within the first part of the European FP7 funded project D-RAIL, which aims to make significant improvements in the detection and prevention of derailments, and the mitigation of their subsequent effects. It presents a comprehensive overview of recent mainline freight train derailments in Europe (with a major focus on their causes and impacts), and an extensive investigation on supervision and monitoring systems and their efficiency. This work also provides conclusions and recommendations to the safety authorities and European key organizations, responsible for accident reporting methodologies and impact analysis.

20.1. Introduction

The FP7 project D-RAIL, which aims to make significant improvements in the detection and prevention of freight train derailments, and the mitigation of their subsequent effects has started with a review and analysis of recent freight train derailments to evaluate root causes and the severity of impact.

Chapter written by Cristian ULIANOV, François DEFOSSEZ, Gordana VASIĆ FRANKLIN and Mark ROBINSON.

The European Railway Agency (ERA) has estimated that open line freight train derailments cost the EU27 Member States more than 200 million Euros per year [ERA 09]. Previous research [DNV 11] suggests 3.7 fatalities per year, on average, from incidents that involve the release of dangerous goods, and 0.2 fatalities per year from other derailments. Since 2003, RSSB has commissioned four research projects on behalf of the rail industry in Great Britain in relation to both passenger and freight derailments [RSS 03, RSS 04, RSS 06, RSS 07, RSS 08]. One of the findings was that the majority of derailments were due to the track being non-compliant with standards. UIC has a collection of reports, by ERRI [UIC 64-83], where the analysis was conducted on how track twist, curve, wheel-loads and combined causes influence derailments.

ERA looked at the short- and medium-term measures for the prevention and mitigation of freight train derailments [ERA 12], but D-RAIL focused on long-term measures, starting with an analysis of recent derailments in Europe, Russia and the USA. The understanding of the main causes of derailments is important for finding measures to prevent them in the future. Analyzing the impacts of accidents is an important factor to consider and recognize before any derailment detection and mitigation measures are proposed in order to ensure that they are affordable and could be practically adopted by relevant stakeholders.

Another objective of the research [DRA 13] was to identify and evaluate the monitoring and inspection technologies that can be used to check crucial parameters of the vehicle-track system and warn of the potential for train derailments. More importantly, this work has identified technology gaps and areas where existing technologies need to be further developed. Only technologies that target the major derailment causes identified earlier in D-RAIL [DRA 12a, DRA 12b] have been considered for the study.

20.2. Research methodology

Data about derailments exist both in European and national databases, either public or confidential. These belong to a variety of organizations and are presented in a number of formats differing in structure, information under which criteria is reported, definition of causes of accidents, etc. The main sources of information for this research included:

1) ERA's safety database, ERADIS;

2) European derailment database created within the previous study by Det Norske Veritas [DNV 11];

3) European UIC safety database;

4) Great Britain's Safety Management Information System (GB SMIS) administered by the Rail Safety and Standards Board (RSSB);

5) Rail Accident Investigation Branch (RAIB) investigation reports;

6) Austrian safety database and investigation reports;

7) France's SNCF safety database;

8) Germany's DB safety and event database.

In order to provide a comprehensive review, analysis and comparison, it was necessary to harmonize the data into the same format. The classification of derailment causes used in D-RAIL [DRA 12a] was slightly modified and adapted. D-RAIL decided on a "cause" categorization and created a new database. A representative from each country was responsible for identifying the appropriate derailment causes for the data for that country, and where necessary new causes were added to the harmonized classification system. Given the low number of derailments compared to the number of different causes, the statistical error is large for the less common causes. Despite the heterogeneity of the data, merging multiple data sets with different thresholds increases the size of the data set and reduces the statistical error when ranking based on numbers of derailments.

The analysis was conducted for freight train derailments on main lines. This excluded a large number of derailments occurring in shunting yards and which do not incur significant costs (and may not even be reported). Data from the period 2005–2010 were selected as being the most complete. Accident data were collected to capture crucial aspects of derailments, including environment, infrastructure, rolling stock and operation, and create a key platform for the entire project and subsequent analysis of freight derailments. The database also includes information about the number of derailments, derailment causes, location, costs and monitoring systems.

The D-RAIL database [DRA 12a] enabled the identification of key causes, which produce most damage and losses, as well as which causes were the most common. Some data was collected on derailments with multiple causes, but not enough for meaningful analysis.

In the next phase [DRA 13], the investigation on monitoring systems employed the following methodology:

1) Identification of prevention and mitigation technologies. Collection of data (specifications, requirements, conditions, principles, measuring capabilities/ parameters, etc.).

2) Analysis of existing and emerging technologies: classification, relation between measured parameters and derailment causes/mechanisms, efficiency of systems, gap analysis and potential for improvement, etc.

3) Assessment of existing technologies against defined criteria, and with respect to:

i) efficiency;

ii) potential for improvement.

20.3. Results and discussion

20.3.1. *Derailment trends*

In order to compare data from different regions, Figure 20.1 shows the number of derailments per billion tonne-km, in the period 2005–2010. Freight derailments are in decline in Great Britain and Austria, but there is no clear trend for France or Europe as a whole (based on the databases from UIC and from DNV's study for ERA).

20.3.2. *Analysis of derailment causes*

Derailment often happens as a result of a combination of factors. The GB's Rail Accident Investigation Branch (RAIB), for example, categorizes causes as:

1) immediate cause;

2) causal and contributory factors;

3) underlying causes.

It is not always easy to decide what the main cause was, and many accident databases only record the immediate cause. Therefore, the study considered only the main causes; where data regarding multiple causes was available this was shared between categories as appropriate. Figure 20.2 presents the breakdown of derailment causes into major categories. Within Europe, rolling stock causes are responsible for 38% of major freight train derailments, followed by infrastructure (34%) and operations (22%). The spread between countries can be significant because of differences in operations, the nature of the infrastructure, the types of rolling stock, etc.

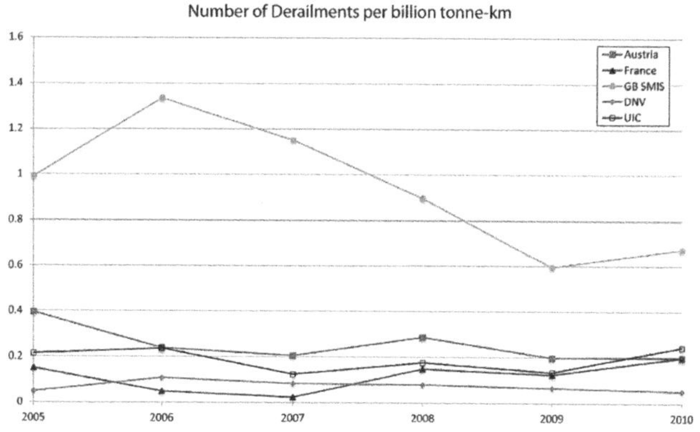

Figure 20.1. *Number of derailments per billion tonne-km for three selected European countries and from two European datasets over six years (2005–2010). For a color version of the figure, see www.iste.co.uk/jacob/safety.zip*

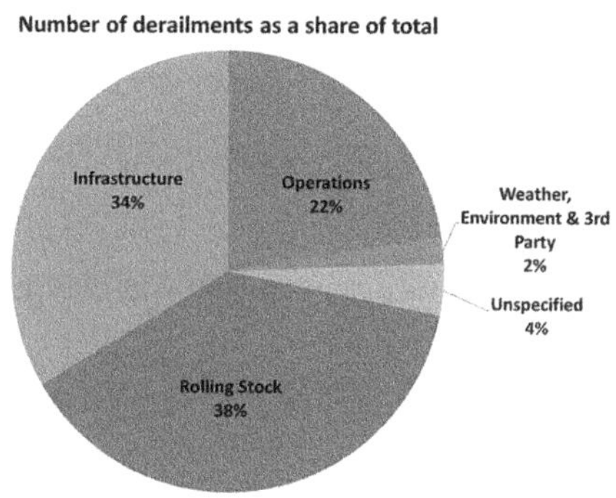

Figure 20.2. *Breakdown of number of derailments in Europe into major cause categories (harmonized data from new D-RAIL database). For a color version of the figure, see www.iste.co.uk/jacob/safety.zip*

This is true both within Europe and outside. For example, the USA's major category is infrastructure at 50% and the dominant cause within that is rail failure. For both the USA and Russia, structural failure of the track superstructure is more likely to be the cause of derailment than track geometry failure. In Europe, by contrast, track geometry failure is more likely to be the cause of derailment. The dominant derailment causes within the infrastructure category are: excessive track gauge, track height/cant failure, rail failures, switch component structural failure and excessive track twist. There is also significant variation between different countries in Europe. For example, within the infrastructure category during the period considered, no derailments in GB were caused by rail breaks and excessive track twist was the major cause; in France and Austria, rail breaks were a major cause of derailments but no derailments were caused by excessive track twist.

Rolling stock derailment causes were more difficult to classify, although the following four groups could be identified: hot axle box and axle journal rupture, failure of bogie structure and supports, suspension failure and wheel rupture. GB has had a higher percentage of derailments as a result of brake failures, the majority of which have been due to handbrakes left applied to wagons when moving on the mainline. Of course, "handbrakes left applied" can be regarded as a human failure rather than a rolling stock component failure. Human factors are a major element in the derailment causal chain, and one that is not reported consistently; the GB SMIS database has a major focus on human factors, and D-RAIL has categorized such causes as "operations."

Other dominant operations causes include: wagon wrongly loaded, point switched to wrong position, other mishandling of train including driver caused SPAD, brake shoe or other object left under train.

Overall, the ranking of major derailment causes in Europe is [DRA 12a]:

1) axle failures (including bearing, axle and journal failures and hot axle boxes);

2) excessive track gauge;

3) wheel failure (all types – surface damage and wheel/tyre breaks);

4) skew loading;

5) excessive track twist;

6) track height/cant failure;

7) rail failures (breaks and profile);

8) suspension failure.

When analyzing the above causes with representatives of infrastructure managers, they were surprised that switches and crossings (S&C) are not on the list. This is perhaps a reporting issue, since the complex geometry of the switch can trigger a derailment that ultimately has a different cause; for example, an over-speeding train, if the driver accelerates before the end of the train clears the S&C. But also, if the points have the wrong setting, or are changed while the train occupies them, then these have separate causes within the operations category. Other causes, surprisingly not included in the list, were over speeding and under speeding.

20.3.3. *Impact analysis*

The economic impact of freight train derailments has also been assessed. Cost threshold is one of the criteria for reporting derailments upstream to the European central statistical and safety and databases EUROSTAT and ERADIS (ERA's database), and these categorize derailments differently. EUROSTAT records only significant accidents, which are defined as accidents either causing fatalities or with total damages amounting to costs of over € 150,000. ERA uses the term severe to indicate a derailment with a mechanical impact such that it would have the potential to cause a leakage in a dangerous good wagon, and three general categories of severity:

1) "severe":

i) *"immediate severe"*: immediately identified as severe;

ii) *"not immediate severe"*: undetected for some time and potentially end up being classified as severe;

2) "not severe": detected quickly and the train is brought safely to a stop.

The distribution of total derailment costs from collected derailments in the period 2005–2010 is presented in Figure 20.3. A large number of derailments fall below the € 150,000 threshold for reporting to EUROSTAT. There are a significant number of derailments within the 1-2M€ range, but above that the data become very scattered.

The average costs of derailments were calculated as follows:

1) If considering all derailments, the average cost per derailment is € 802,361. However, this is significantly skewed by maximum value in the dataset.

2) If considering just the derailments in the range € 100,000 – 2,000,000, i.e. excluding "low-cost" derailments and extreme events (these thresholds were selected in respect to data in Figure 20.3), the average cost per derailment is € 608,353.

Based on the approach in the second bullet point above, the value of € 600,000 as an average cost of a "severe" derailment is suggested for analysis.

Most European databases do not provide detailed information of derailment cost split into categories, but only the total derailment costs. Even where information on cost components exists, each source has a different way of calculating, naming and dividing the costs. Disruption cost is difficult to estimate and ranges from only 15% of the direct cost in one data set to 150% based on ERA's cost benefit analysis model. It would be reasonable to assume that the indirect costs, if unknown, are equal to the direct costs.

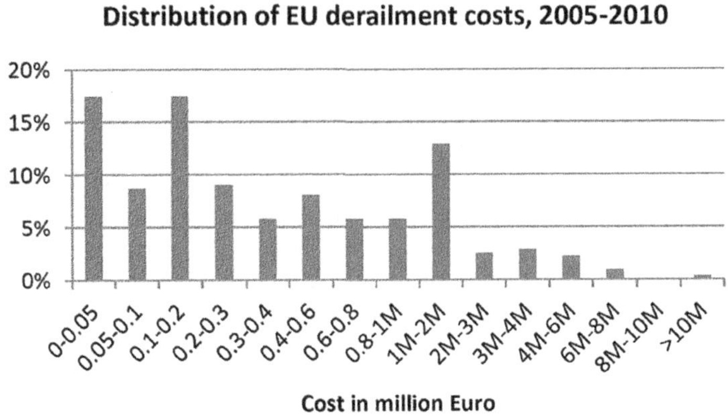

Figure 20.3. *Distribution of total derailment costs of selected derailments in the period 2005–2010*

A number of observations can be made [DRA 12b] for modeling derailment costs:

1) There are 500 derailments per year, of which 7% (35 derailments) involve dangerous goods.

2) There are, on average, two fatalities per year and three serious injuries per year, at costs of € 1.5 M per fatality and € 0.2 M per serious injury, so the human cost is € 3.6 M per year. This is equivalent to a human cost of € 7,200 per derailment.

3) Environmental clean-up costs are negligible excepting the 7% of derailments involving dangerous goods. If the minimum cost per dangerous goods derailment (€ 250,000) is assumed here, this is equivalent to € 17,500 per derailment.

Based on this, the human cost and environmental cost add a fixed cost of € 24,700 per derailment, independent of the type of derailment. However, this is an average value, and could be thought of as, for example, six severe derailments per year, each incurring costs of € 2 M (rather than 500 derailments per year, each incurring the cost of € 24,700 per derailment).

In data collection, the costs were separated into two major groups:

1) *direct costs*, meaning just railway asset costs of infrastructure and rolling stock that are damaged during or after a derailment;

2) *indirect costs*, including e.g. disruption cost (delay minutes, etc.), fatalities and injuries costs, legal and litigation costs, third party damage, environmental

(could include post-accident clean-up operation, etc.), attendance of emergency services, public dangers (hazardous cargo), loss of cargo and freight, etc.

The analysis of the cost-impact data from the study shows an average breakdown of direct costs in: infrastructure 78% and rolling stock 22%. The cost benefit analyses by ERA [ERA 09] and DNV [DNV 11] suggest splits of 75%/25% and 80%/20%, respectively [DRA 12b], so a breakdown of direct costs into 80% infrastructure and 20% rolling stock would be a reasonable approach.

For calculating the total impact in cases where only direct costs are known, the direct cost should be multiplied by a factor *f*, i.e.:

$$C_{TOTAL} = f * (C_{INF} + C_{RS}). \quad [20.1]$$

where C_{INF} is the direct cost to the infrastructure, C_{RS} is the direct cost to the rolling stock and C_{TOTAL} is the total cost (the 80%/20% split above corresponds to C_{INF}/C_{RS}).

ERA's cost benefit analysis model gives a factor of 2.5 (data for the USA indicate this factor to be in the range 1.8–2). Indirect costs were not well documented so conclusions of their share in the total cost cannot be made with certainty. Analysis of the data provided by infrastructure managers in the D-RAIL project suggests that this factor may be much lower (only 1.33), but this factor probably varies considerably between countries.

20.3.4. *Overview of prevention and monitoring systems for reducing the occurrence of derailments*

The study covered both wayside and vehicle mounted systems. The capabilities of these technologies for measuring parameters and/or faults have been identified and analyzed. A brief summary is presented in the following sections.

20.3.4.1. *Wayside systems*

1) Wheel impact load detector – Axle load checkpoint (Q, Y and Y/Q)

Wheel impact load detectors (WILD) and axle load checkpoints (ALC) can reveal overloading or unbalanced loading of vehicles and wheel defects such as out-of-round wheels and wheel flats, which increase the risk of rail breaks and the consequent potential for derailment. For derailment prevention, trackside inspection of vehicle forces extends to analysis of curving behavior and running instability as well. Checkpoints placed on curves can measure the horizontal as well as vertical track forces. There are a large number of different ALC devices in use around the world. ALC provide information on the running behavior of passing trains. On-track installations allow for grinding, tamping and other maintenance operations without the need to dismount equipment. The operation of WILD/ALC devices is automatic – the equipment measures, diagnoses and sends information to operator.

2) Hot box, wheels and brakes detection – infrared

Bearing housing, wheel and brake disc temperature measuring devices (known as hot axle box and hot wheel detection units HABD/HWD) are installed along the line at appropriate distances and designed to monitor any overheating, which could lead to train accidents. They can also detect "cold wheels" if brakes are not applied. These devices can be installed either individually (e.g. just HABD) or together. They are widely used in Europe. If the temperature passes the threshold, an alarm is triggered and the dispatcher receives the information of the value of the measured temperature and the place, and can take the appropriate actions.

3) Acoustic bearing defect detection

Acoustic bearing defect detectors record sound produced primarily by bearings, but also by flat wheels or dragging equipment, as the wagon passes. The system can then analyze the sound to detect bearing failure in the early stages before a significant defect develops. This information is important both for helping with bearing maintenance management (by following the bearing fault trend), and for safety (emergency withdrawal from service if needed).

Acoustic detection of wheel and bearing is ideal for detecting and managing faults of a fleet of vehicles (individually identified using, e.g. RFID), and fewer actual detectors would be necessary than hot axle box detectors. In addition, bearing faults can be detected much earlier, reducing costs and impact. Acoustic bearing detection is not as widely used as HABDs. For railway networks used by multiple fleets with some unknown traffic, HABD remain a necessity.

Both acoustic and HABD can detect the following bearing failures: cracked/broken cup; damaged cone and damaged roller.

4) Wheel profile and diameter systems – laser-based wear measurement systems

Optical wheel profile and diameter systems (WPDS) operate with high-speed video cameras and a combination of lasers which are directed towards the wheels. They can provide automatically and in real-time laser-based profile and wear condition measurement of each wheel when the wagon passes over the measurement site. These systems can prevent derailments by detecting wheel profile deviations and major wheel parameters at an early stage.

Endowed with efficient algorithms and powerful computers, these systems generate in a very short period of time a computer model of each wheel, and then compute the measurements for flange thickness, flange height, rim thickness, flange angle, diameter and other main wheel parameters. These measurements can detect wheel faults such as: thin flange, tall flange, vertical flange, thin rim, out of gauge, out of round, hollow wear, etc. The collected measurements are then transmitted to a database, which is used to monitor the condition of wheels and to adapt maintenance to actual wear. Laser-based technology provides the highest level of accuracy and repeatability for railway wheel measurements on moving trains, at both low and high speeds.

5) Wheel tread condition monitoring detectors

Wheel tread condition monitoring detectors are wayside automatic devices that are capable of detecting discontinuities on the running surface of the wheel. Such defects include cracks (surface-breaking and/or subsurface, thermal), wheel flats, dents, shelling, spalling and metal build-up at the outer profile edge.

The technologies used are non-destructive and could be:

1) ultrasonic scanning;

2) high definition cameras and *ad-hoc* illuminating system;

3) electromagnetic scanning.

20.3.4.2. *Vehicle-mounted systems*

1) Track geometry measurement systems

Track geometry measurement systems (TGMS) capture precise data on some critical geometric features, such as: track gauge, cross level, longitudinal levels (both rails simultaneously), alignment (both rails simultaneously), curvature and radius, cant, twist, change of gradient; position and distance.

2) Rail profile measurement systems

Non-contact optical technologies can be used to obtain the full rail profile at speeds up to 350 km/h. Rail profile measurement systems (RPMS) can use high level laser and video technology to provide accurate and instantaneous report on the rail profile condition. The video cameras capture full cross-sectional rail profiles from the base to the top of the rail surface to allow useable and accurate measurements. Such a system can measure the following parameters: rail profile; vertical and horizontal wear; rail inclination and head width. By comparison with a reference profile, parameters out of tolerance can be identified immediately.

3) Rail surface inspection systems

The current rail surface defects are abrasion (due to skidding, ballast imprints, burns), linear defects, cracks and progressive fractures, shelling, broken welds or head checks. The inspection of the rail surface is mainly a process using optical systems in high definition. These non-contact, optical systems allow acquisition, storage and analysis in real-time of the recorded images. Moreover, associated with relevant measurement systems and advanced image processing algorithms, rail surface inspection systems can classify defects according to their properties and positions. This type of system can generally be used up to 200 km/h. Other systems based on magnetic eddy current principle are also designed to inspect the surface and near the surface of the rail.

4) Rail internal structure inspection systems

Rail internal defects are: transversal and longitudinal defects, internal flaws and broken welds. In order to detect and classify such defects, ultrasonic rail flaw detection systems have been designed. Ultrasonic signals are sent inside the rail, under different angles and using different techniques (pulse-echo or pitch-catch) depending on the kind of defects sought. The different ways to organize the probe arrangement also allow different inspection capabilities. A coupler is needed to send the waves from the transducers into the rail, so a sled filled with fluid (water or other) is sliding or rolling in front of the probes. Formerly handmade, the ultrasonic inspection has been automated to make inspections all along the track.

5) Head check inspection systems

If "headchecks" (defects occurring at the gauge corner of the rail) are not treated, the surface cracking will typically grow in length and depth leading to significant reduction of the rail service life. Different technologies are used to detect such a defect:

1) Systems using machine vision allow automatic detection and precise localization of 0.15 mm headcheck defects at 120 km/h enabling early-stage detection of gauge corner cracking when grinding is still a possible solution.

2) Some systems use state of the art magnetic technology to detect and measure rail surface cracking and material loss. The measurement technology provides repeatable, high-resolution measurements of rail surface cracks over a wide range of operating speeds. The detailed measurements can be analyzed to provide meaningful indices of rail damage that can be used to assist in planning rail rectification.

20.3.5. *Assessment of existing prevention and monitoring systems*

The assessment of the relevant above systems was carried out using a matrix, which was developed [DRA 13] based on the top derailment causes presented in section 3.2. The assessment employed *nine criteria* (defined in Table 20.1), and each technique was evaluated against these, and ranked on *three levels* (high, moderate and low), with respect to their relevance to each criterion. Finally, considering all criteria, each system was ranked using quantitative indicators (from 1 to 10), indicating its efficiency in preventing or reducing the risks for derailment, for each of considered causes.

Assessment Criteria	Definition
System resilience	Resilience against external factors (weather condition, human factors, maintenance activities, etc.).
Technology platform	Issues relating to integration in the overall system (practical installation, user friendliness).
Standards, engineering acceptance requirements	Compliance to existing standards and overall engineering requirements.
Cost	General estimation of initial costs – investment/system price and installation.
Operational limits	Influences relating to operational aspects (disturbance, regular traffic disruptions, etc.).
Cross border interoperability	Capability to be used on different networks (particularly within the EU, according to TSIs).
Diagnostic alerts and data communication	Capability to capture, analyze and communicate measured parameters (degree of automation, reliability of data transfer, etc.).
Measurement effectiveness	Capability and accuracy (effectiveness) to capture key derailment parameters.
Derailment prevention efficiency	Efficiency of preventing derailments, considering both the reduction of derailments, occurrence and implementation (i.e., spacing of wayside detectors, or frequency of rail bound inspections).

Table 20.1. *Definition of assessment criteria*

The assessment [DRA 13] was completed and validated by rail experts with relevant expertise in the area, both project partners and stakeholders representatives consulted for this purpose. Furthermore, the potential for improvement was analyzed and discussed, including recommendations for possible directions. Considering both their efficiency and, especially, their potential for further improvements, it was observed that the most promising systems are those relevant to *wheel failures, skew loading* and *spring and suspension failures*.

20.3.6. *Gap analysis and overview of emerging technologies*

Based on the assessment presented in the previous section, and measuring capabilities shown in section 3.4, a gap analysis has revealed the limitations of existing monitoring technologies and their potential for improvement. The analysis has also considered emerging technologies and state-of-the-art techniques employed by systems developers. A summary of observations and results is presented below.

20.3.6.1. *Axle, bearing and journal failure detection*

Hot axle box detectors are capable of detecting bearing/journal temperature when it overheats and passes the threshold. The main problem is that the train has to be stopped, so this system does not give the early warning. Acoustic bearing detection systems solve this problem, as they can detect defect at early stages. However, some bearing parameters (such loose backing ring or loose/damaged seal) still cannot be detected with automatic technologies and workshop inspection is required.

Axle shafts and journals can develop fatigue cracks. Crack initiation can start from a scratch or corrosion. Axle related derailments are generally associated with the fracture of the axle under a vehicle, usually under conditions of high dynamic loading. These types of failures can be detected in the workshop by NDT inspection, e.g. ultrasound. There is potential for automated technologies dealing with axle defects.

20.3.6.2. *Wheel failure detection*

Wheel-related derailments are generally associated with the fracture of the wheel under a vehicle, usually under conditions of high dynamic loading. High temperature/overheating related failures are associated with the change in metallurgical properties of the wheel and the formation of thermally induced cracks.

The current wheel monitoring practices are described in section 20.2, but gaps exist in automatic detection and/or measuring of the following: cracked/broken flange, thermal flange crack, cracked/broken/shattered/spread rim, cracked/broken plate, thermal plate crack, hole in plate, loose wheel, wheel gouge, subsurface defect, etc.

20.3.6.3. *Suspension failure detection*

Spring and suspension failures are generally associated with improper loading or failure of the spring and suspension elements under load. They can include broken springs and other suspension components, failed elastomeric elements, etc. These failures could be detected indirectly by axle load check points, because they will create high impact forces. The inspection in the workshop is the only direct one capable of detecting such failures, so there is a gap for automatically detecting suspension failures.

20.3.6.4. *Skew loading detection*

Wagon loading derailments are generally associated either with improperly loaded wagons that generate undesirable dynamic behavior, or with wagons which generate excessive loading that can result in failure of a wagon structural component or in failure of a track component.

These have been checked with axle load check points and vehicle profile gates. If the vehicle exceeds the limits it has to be excluded from traffic. The technologies exist, but there is a gap in their implementation – ideally, wagon faults should be prevented or detected earlier, and skew loading identified at the point of loading.

20.3.6.5. *Track geometry detection*

Inspection of the infrastructure is mainly done by track geometry measurement vehicles. Excessive track gauge and track twist are the principal defects that lead to derailments. The trend is towards compact optical (non-contact) systems to be installed on nearly any vehicle and at high speeds. This will make it possible to use shorter inspection intervals without affecting track availability.

To evaluate the system integrity of the track and geometry, measurements should be made close to (or under) a test wheel with controllable vertical and lateral loads. Ideally, the track should be tested with loads that reach (or exceed) the maximum load expected from the traffic. Such a measurement requires special vehicles. From a track engineering point of view, it would be beneficial to have the geometry data for track with and without the additional controlled loading, measured by the same vehicle. This is a potential area for improvement, regarding both measurement and analysis. Vertical track alignment should especially be measured with an applied load and a high sample rate to cover short defects such as dipped rail joints, and this is another area for improvement.

20.3.6.6. *Rail failure detection*

Rail defect detection is particularly important for derailment prevention. Ultrasonic testing is well established and powerful. There is scope for making measurement systems more compact and cheaper, and also for improving the analysis software (which often requires a skilled user).

Defect types where inspection technologies need development include:

1) rail foot corrosion pit induced cracks;

2) size of internal rail cracks;

3) increased inspection speed for rail defect detection is needed.

20.3.6.7. *Emerging technologies, prototypes*

Ongoing research in this area is bringing new technologies and applications that are either in a conceptual or prototype phase and might be relevant for the future. Some emerging technologies are:

1) wayside crack detection:

   i) laser-based ultrasonic cracked axle detection,

   ii) automated ultrasonic-based cracked wheel detection;

2) train fault detection system (TFDS) image acquisition;

3) acoustic-based inboard bearing defect detection;

4) displacement sensor-based bogie hunting detection;

5) optical monitoring of loading.

20.4. Conclusions and recommendations

The first significant conclusion, apparent even during the initial collection of data, is that every country has different systems for collecting and management of data on derailments, and different thresholds. This creates difficulties in merging and comparing datasets. Different organizations and countries place the emphasis on different aspects; human factors and multiple causes need to be considered, otherwise the record of the causes and contributory factors of a derailment can be misleading.

European regulation should create the framework so that each Member State uses an equivalent reporting procedures and templates to enable fully compatible and comparable databases. A unified safety reporting across Europe, using the same economic principles and cost components, would be a great benefit.

Derailment costs are also complex, in part because the economic impact of a derailment extends far beyond the asset costs. There are delay minutes, environmental impact, societal impact, and so on, and different countries will evaluate these in different ways. Typically the reported costs are simply the direct costs, of which 80% pertain to infrastructure and 20% to rolling stock; the indirect costs, in the absence of real numbers, can be assumed to be approximately the same as the direct costs.

Overall, the existing situation in Europe regarding the prevention of freight train derailments through implementation of automatic inspection and monitoring systems is acceptable. However, considering the latest technical innovations and developments, there is a significant potential for improvement. There are still a large number of derailments happening with the associated costs, and there is always the potential for disaster if dangerous goods trains derail.

The survey of existing technologies in use in Europe and the assessment of their performances and efficiency in capturing key derailment parameters have enabled some conclusions and recommendations concerning both technical aspects and regulation:

1) There is a huge implementation cost associated with automatic inspection and monitoring systems, which is a great barrier for their adequate use.

2) The development of new technologies and the improvement of existing ones for suspension checks and detection of subsurface cracks in wheels and rails are essential.

3) Rail defect detection would need higher inspection speeds to improve track access.

4) More emphasis should be given to preventing a need to stop the train and exclude it from service if a critical fault that can lead to derailment is detected.

5) There is a huge amount of data being collected; development of algorithms for real-time processing is essential for quick identification of dangerous defects.

6) Mechanisms for cross-border data communication and interoperability need to be improved.

20.5. Acknowledgments

This work was funded by the European Commission within the FP7 project D-RAIL "fig. Development of the future rail freight system to reduce the occurrences and impact of derailment".

20.6. Bibliography

[DRA 12a] D-RAIL, Summary Report and Database of Derailments Incidents. Deliverable D1.1, 2012.

[DRA 12b] D-RAIL, Report on Derailment Economic Impact Assessment. Deliverable D1.2, 2012.

[DRA 13] D-RAIL, Survey and Assessment of Existing Inspection & Monitoring Systems. Deliverable D4.1, 2013

[DNV 11] Det Norske Veritas (DNV), Assessment of freight train derailment risk reduction measures: B2 – Risk model and potential effectiveness of measures, Report for European Railway Agency, Report No: BA000777/07, Rev: 02, 2011.

[ERA 09] EUROPEAN RAILWAY AGENCY (ERA), Study: Impact assessment on the use of derailment detection devices in the EU railway system, Ref. ERA/REP/03-2009/SAF, 2009.

[ERA 12] EUROPEAN RAILWAY AGENCY (ERA), Prevention and mitigation of freight train derailments at freight train derailments at short and medium term, Final report ERA/REP/02-2012/SAF, 2012.

[INT 64-83] INTERNATIONAL UNION OF RAILWAYS (UIC), ERRI: Project B 55: Prevention of derailment of goods wagons on distorted tracks, Reports: RP1-1964, RP3-1966, RP4-1970, RP5-1973, RP6-1975, RP7-1978, DT 78 – 1981, RP 8 – 1983, 1964–1983.

[RSS 03] RAIL SAFETY AND STANDARDS BOARD (RSSB), Derailment mitigation – Categorisation of past derailments (T078), Risk Solutions, 2003.

[RSS 04] RAIL SAFETY AND STANDARDS BOARD (RSSB), Development of a prototype model for managing derailment risk due to track faults (T207), Risk Solutions, 2004.

[RSS 06] RAIL SAFETY AND STANDARDS BOARD (RSSB), Cost-effective reduction of derailment risk (T357), Risk Solutions, 2006.

[RSS 07] RAIL SAFETY AND STANDARDS BOARD (RSSB), Developing the incident cost model (T682), Risk Solutions, 2007.

[RSS 08] RAIL SAFETY AND STANDARDS BOARD (RSSB), Identification of existing and new technologies for wheelset condition monitoring (T607), Risk Solutions, 2008.

21

A Risk Assessment Tool for Public Transportation

As security and security perception become increasingly important in the public transportation-environment, operators and authorities need to decide which security measures need to be adopted. Although such decisions need to be taken soon, they represent a significant additional cost to the already strained public budgets. That is why security investments need to be correlated with a set of identified risks, which can be defined as the product of the likelihood and impact of specific threats. A clear overview of the existing threats is necessary before any investment decision is taken. The Risk Assessment methodology developed within the SECUR-ED project is the latest tool designed to specifically help Public Transport Operators identify their risks, define their priorities concerning security investments and, consequently, enable them to decrease the risks to a "tolerable" level. This methodology has already been implemented within several metro/bus networks and the non-confidential results will be presented here.

21.1. Security – a growing concern for Public Transport operators

Given the escalation in violent activities throughout the world, security in Public Transport (PT) is a salient topic for both EU decision makers and sector representatives. Given the fact that many of the 500 million EU citizens use public transport services daily, security also becomes a matter of perception.

In order to enhance the security levels, the most obvious step undertaken by the Public Transport Operators (PTOs) and local/national authorities is to increase the spending in security-related technologies, with CCTV being the preferred choice, and manpower. These investments are not cheap – security technologies, more security staff and the necessary training required in either cases – but are necessary

---

Chapter written by Andrea SOEHNCHEN and Mihai BARCANESCU.

in order to protect the passengers, the staff and the assets. Yet PTOs also need to take into account the commercial aspect of their activities. Part of the rationale behind such investments is that the security enforcements will maintain or even boost PT use, due to the feeling of security induced. Given the current evolution of violence, such costly investments need to be made sooner rather than later.

Nevertheless, while aiming to address both the security matters and the security perceptions, PTOs need to balance these investments. An increase of security measures and personnel does not automatically translate into the enhanced protection that is being sought, but it may trigger the perception that PT systems are more insecure than they actually are.

That is why security investments require a comprehensive approach, in order to strike the desired balance between costs and effectiveness. It is commonly agreed that security investments must be directly linked to the existing (or perceived) risks, which can be defined as the product of the likelihood of the threats and their impact. The term "threats" in security applies to a wide array of unlawful activities, from low and medium crimes, such as graffiti, aggression or theft, to major attacks such as bombings or arson. A realistic knowledge of the risks involved is a necessity before taking any managerial decisions concerning security.

Consequently, this chapter presents a tool developed by security experts which allows the PTOs to properly assess their security situation and take the best possible measures in order to mitigate the challenges: the risk assessment.

It is based on the COUNTERACT risk assessment methodology, updated according to ISO31000:2009(E) [MAT 09].

21.2. The risk assessment procedure

21.2.1. *A tool developed to mitigate security risks*

There are several methodologies concerning risk assessment in PT, a widely used one being the methodology developed in COUNTERACT, a project funded by the European Union through Framework Programmer 6 (FP6). The product, an improved and broadly accepted risk assessment methodology, has proven to be very successful and fit to be the standard practice in the PT context.

Notwithstanding the good results achieved through the implementation of this practice, new security challenges have been emerging due to the growing population of cities, a fact which directly leads to an increase of threats from all sorts of criminal activities. These aspects, as well as the advancements in security-related technologies, have encouraged the EU stakeholders to start their research of upgrading the existing risk assessment methodology. This has been achieved with the help of EU funding through SECUR-ED, an FP7 project, the largest EU-funded project in the field of land transport security [HAM 13].

The risk assessment developed by the SECUR-ED partners is typically dedicated to the specific environment of PTOs, having been designed as a tool to help the operators and local authorities to quote their risks and define their priorities concerning the security investments. The aim is to decrease the associated risks by solving the detected vulnerabilities and mitigating the challenges.

PTOs are facing challenges concerning the security of customers, staff and assets, such as infrastructure and installations. There are crimes and offenses that target or take place in the PT network, from petty offenses to major attacks. The frequency of these security incidents is also important, since the "day-to-day" offenses will encourage the spread of criminality, while at the same time diminishing the feeling of security among the users. In addition, the operators must also be ready to face unforeseen incidents that, without specifically targeting PT, can have a major impact on operations and result in significant losses, such as fires or explosions. Moreover, these negative consequences can be magnified in the case of intermodal and interchange stations, which are neuralgic nodes of urban public transport. The consequences of such incidents range from serious personal injury and damage to property to disrupted operations, loss of customers' trust, loss of revenue and competitive advantages.

PTOs need therefore to prevent or reduce the impact of such incidents. The main rationale behind performing and applying a risk assessment is to spot the relevant security gaps in order to decide what measures may be appropriate to improve the overall security level.

21.2.2. *Definition and terminology*

Within the SECUR-ED project, the following terminology has been defined:

Security is here defined as the absence of intentional threats. Thus, the term "security" encompasses all measures, actions or systems aiming at preventing intentional threats from compromising the well-being of passengers and staff or the integrity of infrastructure, rolling stock and installations.

In the security context, a threat is the expression of intention (or perception of a possible intention) to provoke a security incident, i.e. to harm or injure, damage equipment or infrastructure, disrupt operations etc. Security threats may materialize into security incidents that are a concern for safety.

The security risk is the degree of exposure to a threat. The risk increases with the potential impact and the probability of a threat materializing. Risk is measured in escalating categories.

Vulnerability describes a weakness, e.g. in physical structures, personnel protection systems, processes or other areas that, in the presence of hazards, may compromise safety or, in the case of security threats, be exploited by perpetrators.

Conducting a risk assessment is the initial step in the broader approach to security issues known as the risk management process. The risk management process rests on five main pillars, as shown in Figure 21.1 below. The risk assessment itself is made up of the three middle pillars. Each of these pillars is, in turn, made up of different steps that help to concentrate and coordinate the effort of the PTO in order to obtain the highest possible degree of protection. Conducting a risk assessment is the most important prevention measure to be taken by PTOs as part of their risk management strategy.

21.2.3. *Preliminary steps in conducting risk assessment*

Establishing the context is the preliminary step before the actual risk assessment procedure (see Figure 21.1, point 1). The build-up of this context must take into consideration both the internal and external factors.

Regarding the internal factors, the PTO must first define the parameters according to which the entire undertaking will be analyzed. First of all, the PTO needs to identify its assets, both the tangible (e.g. passengers, staff, vehicles, facilities) and intangible (e.g. reputation, trust, business objectives, etc.) in order to have a clear picture of what needs to be protected. The potential risks that the PTO faces must then be defined. Based on this information, the PTO can thus decide the goals and objectives of the risk assessment, as well as the total amount of resources that need to be dedicated for conducting the risk assessment.

The external factors that need to be taken into consideration encompass the following aspects: the legal framework applicable to PT and security, the legacy systems, other relevant stakeholders, the criminal activities in the city/region where the PT is operating, etc.

It is important to notice that a risk assessment in PT security should be done in a qualitative way. A quantitative assessment based on mathematical formulas and calculations is not possible due to a lack of statistics (especially concerning terrorist incidents).

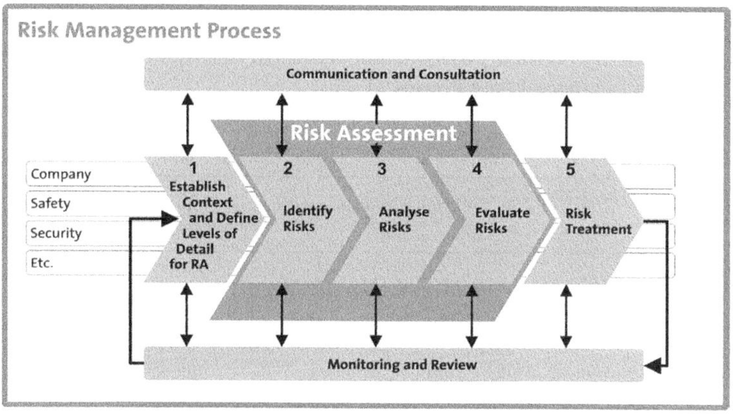

Figure 21.1. *The risk management process*

21.2.4. *Risk identification*

The first step of the risk assessment itself is the risk identification (see Figure 21.1, point 2). In this context the PTO representatives need to precisely identify the potential threats and their connected risk scenarios that can have an impact on the PT system. Risks shall be categorized according to their nature, and grouped into the relevant risk families – e.g. security risks, safety risks, environmental risks, etc. (see Figure 21.2). The amount of security levels to be studied depends directly on the level of detail the risk assessment team would like to reach.

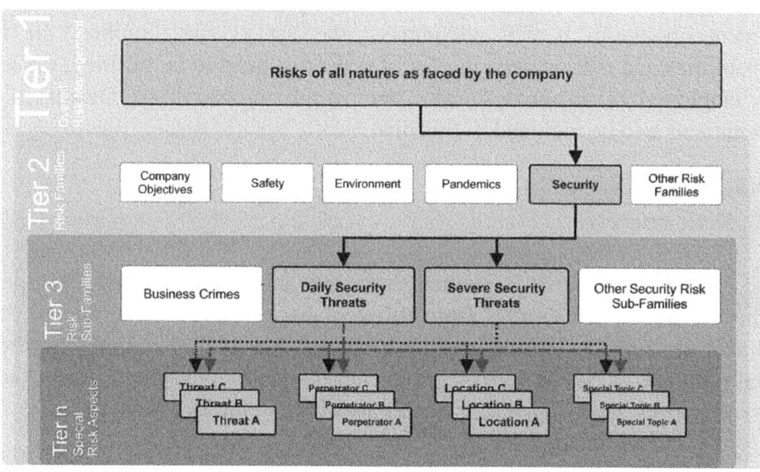

Figure 21.2. *Categorizing risks into risk families: example of security threats*

The risk families shall subsequently be assigned to the risk owner, namely to the PTO staff, depending on their positions and skills. In this manner each of the threats can be adequately managed (see Figure 21.3).

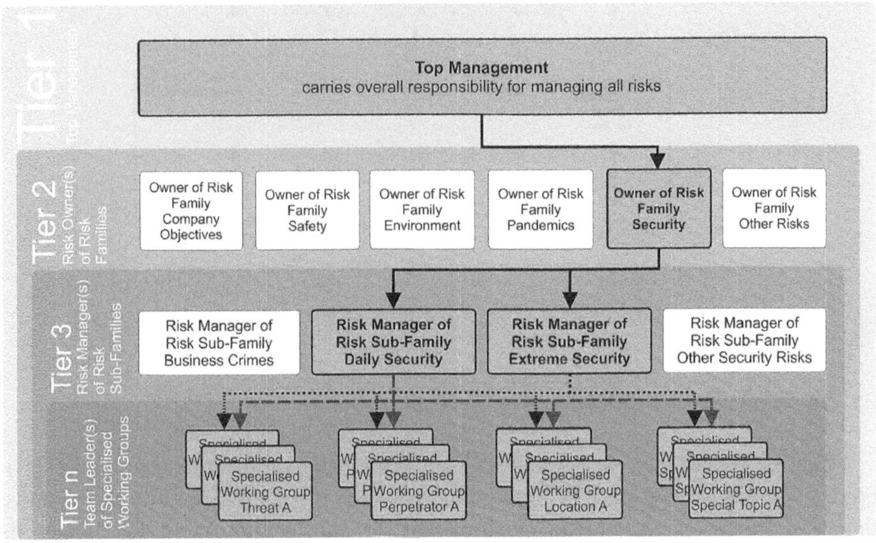

Figure 21.3. *Assigning risk families to risk owners and managers*

Once the categorization has been finished, a working plan for the risk assessment needs to be drawn up. The plan will practically create the structure for the risk assessment operation. It will encompass: the list of risk families, the list of responsibilities and risk owners, the list of tasks that need to be fulfilled, the level of details employed in the analysis and the overall methodology. Deadlines and a timeframe will also be included in this plan.

21.2.5. *Risk analysis*

The second pillar of risk assessment is the risk analysis (see Figure 21.1, point 3). It is the moment when experts from different backgrounds – operations, security department, technical departments and others, but also external stakeholders, especially police and law enforcement agencies – will analyze and assess the risks and risk scenarios. Risk analysis activities are divided into two main steps.

The first step is a detailed identification of the list of assets and operations that need protection and the list of possible threats. The list of threats must encompass both types of threats that the PTOs face: daily security threats and severe security threats. The risk assessment team must therefore:

– gain an understanding of the risks that the PTO realistically faces – what is the cause of potential threats and which may be the perpetrators;

– identify and evaluate the strengths and weaknesses of the PTO assets – both tangible and intangible;

– identify and assess the critical parts and functions within the PTO operational diagram, so as to identify the importance of the PTO functions that need to be protected;

– identify and assess the already existing safeguards for both safety and security, as there is a certain overlap between the two types of measures. These safeguards are divided between: the human factors (the trained front-line and security staff), existing emergency plans and procedures and the relevant equipment in place.

The goal is to come up with a clear and simplified structure of the PTO system.

The second step consists of defining several key main parameters (and their levels) against which the threats and assets are subsequently measured. As said, definitions are based on a qualitative approach; consequently, they can be adapted by the different risk assessment teams according to their needs.

One of the main parameters used is the likelihood of a security incident to happen. Within the risk assessment methodology, likelihood is measured by combining the probability of occurrence of an event (almost sure, very likely, possibly or unlikely) and the definition criteria of these probabilities (how often do these threats actually occur in a given time). This definition of likelihood must be applied to both types of threats PTOs face, as shown below.

The second key parameter within the risk assessment methodology is the consequence (or impact), measured on a qualitative scale of 1–4 (sometimes 5): strong, medium, weak, no consequence. It too has to be applied to daily and severe security threats.

A third key parameter is vulnerability. The first understanding of vulnerability is in the sense of "weakness", namely how potential perpetrators would consider an attack on a given target in terms of "ease of execution" and "feasibility". A second understanding meaning is the "attractiveness" of the target to potential perpetrators. The proposed levels for vulnerability are: likely (+), possible (0) and unlikely (–).

After having set the levels of the aforementioned parameters, the risks need to be categorized according to their own levels – high, medium or low. Once the definitions have been refined and implemented, the subsequent step is to measure the risk categories based on their likelihood and consequences.

Within this pillar the risk assessment matrix is compiled, based on the previous definitions and, advisably, on the organizational structure of the PTO. The aim of this step is two-fold:

– to assess the identified threats and risk scenarios into risk categories according to their likelihood, consequence and vulnerability;

– to arrange risks in a hierarchical ranking within each risk category.

Practically, the identified list of assets of the PTO will be measured against the likelihood and impact of the list of possible threats detected.

21.2.6. *Risk evaluation*

The next pillar is risk evaluation (see Figure 21.1, point 4), where the risk analysis results are obtained and analyzed. Based on the findings, the risk assessment team can thus define the threats and/or the risk scenarios which need to be assessed in more detail. It is strongly advised that these threats are dealt with by the more specialized working groups within the risk assessment team.

The risk evaluation methodology is based on the ALARP (As Low As Reasonably Practicable/Possible) principle. Here, the first measure is to assess the adequacy and effectiveness of the existing safeguards. Potential additional safeguards – using additional security technologies, new/upgraded security plans, or deploying additional security staff – which can further prevent or mitigate the risks, must also be assessed. In the case where additional safeguard measures will be installed, the main security-related aspects concerning their implementation need to be taken into account: the (negative) effects on operations, the acceptability factor (both internal and external), possible ethical issues at stake, etc. The financial aspect also plays a significant role, since all decisions have to be made with regards to the available funds.

It is advisable that a risk register should be compiled in order to provide an overview of these matters. Such a register should mainly comprise: the identified risk and a number to assign to it, the risk owner, the risk manager, the description and cause of the risk, its category, the safeguards in place and what other measures could be employed so as to mitigate the threat.

21.2.7. *Risk mitigation*

The final step of the risk assessment procedure is the risk mitigation/treatment (see Figure 21.1, point 5). The matrix resulting from the risk assessment is presented to the PTO top management, which will assess the results of the treatment of identified, analyzed and evaluated risks. An approval of the proposed measure signifies that the inherent residual risk levels are tolerable. In the case where these residual levels are not considered as tolerable, the management can either decide on the implementation and execution of additional safeguards, or they can generate a new risk mitigation measure. Once these measures have been agreed upon and implemented, the PTO management is due to monitor and review the security status of the organization in order to ensure the smooth running of the measures in place.

It is important to notice that any risk assessment is a living process; it has to be regularly repeated to ensure that all changes, at the threat level or within the PT system itself (e.g. construction or refurbishment measures), are sufficiently reflected.

In case the results are deemed insufficient or inconclusive, the risk assessment procedure – identification, analysis and evaluation – should be restarted.

It has to be repeated if the review and monitoring process reveals that the previous risk assessment has failed to take into consideration certain aspects, or when the security or even the organizational paradigms change to the extent that the previous risk assessment results simply become obsolete.

21.2.8. *Support tools*

The SECUR-ED project partners have also designed a software dedicated to the risk assessment process. It will enable a clearer and easier manner of conducting the risk assessment process. In addition to its time-saving characteristics the software is also designed to quickly signal eventual errors, thus providing a high degree of reliability.

21.3. Conclusions

In conclusion, risk assessment helps to identify security gaps, supports security operation planning and management and helps to draw up adequate preventive actions and immediate responses. Furthermore, a risk assessment process can provide input to the improvement of the security policies and guidelines, as well as the future selection of security equipment with the purpose of enhancing the protection levels of the entire public transport service.

Another crucial aspect is the fact that it builds and strengthens bridges with law enforcement and first responders (LEFR). This aspect is very important, not just due to the highly effective cooperation that needs to be ensured between the PTO staff and the relevant authorities, but also because the risk assessment offers the LEFR the chance to familiarize themselves with the particular environment of the PT networks – especially in the case of metro infrastructures. These mutually supportive sessions of information and analysis offer the guarantee of an improved reaction and response by all stakeholders in the case of a security incident.

One important aspect that needs to be taken into account when conducting a risk assessment is the fact that sometimes the priorities indicated by the risk assessment matrix are not necessarily the same as those outlined by the PTO Management Board; consequently, the ultimate implementation of the risk assessment result is based on striking a balance between the overall needs and resources of the organization.

21.4. Acknowledgments

The authors would like to thank Hamburg Consult Gmbh (HCO) and personally Dr Matthiass Mueth for his valuable input and support.

21.5. Bibliography

[HAM 13] HAMBURG CONSULT GMBH, SECUR-ED D31.2 Best practices for conducting risk assessments, ISO 31000:2009(E), 2013.

[MAT 09] MATTHIASS M., "Generic Guidelines for Conducting Risk Assessment in Public Transport Networks", COUNTERACT D3a-n, SSP4/2005/TREN/05/FP6/SO7.48891, March, 2009.

22

The GETAWAY Project – Improving Passenger Evacuation Techniques in Railway Stations (and Other Transport Hubs)

As there is a continuous move toward larger and more complex surfaces and sub-surface railway station environments, the safe evacuation of increasing numbers of travelers has become a priority. The complexity of station geometries, often incorporating interconnectivity between routes, means that safe evacuation routes will become less obvious. The GETAWAY project was conceived to provide additional clarification and guidance for the evacuation of large numbers of persons within a railway station during an emergency, such as that caused by a fire. It utilizes existing monitoring systems, such as fire detectors and CCTV cameras, combined with specially developed decision-making software and active dynamic signage systems. The objective is to assist the station management with systems to safely evacuate persons away from the incident and to a place of safety as quickly and as reliably as possible. Initial trials have already demonstrated the potential benefits of the system, and further trials were planned in 2014.

22.1. Introduction

Modern rail and metro station layouts are complex due to the interconnections between multiple rail services utilizing the station, the mix of modes of transport such as above ground rail and underground rail, and the mix of retail and transport related functions. Evacuation from large and complex public buildings, such as transport terminals, is usually hindered by a lack of detailed knowledge of the

---

Chapter written by Paul BRYANT and Christos GIACHRITSIS.

internal connectivity of the building space. In such premises, occupants are usually unaware of their most suitable means of escape.

It is a well-known problem [LON 07] that building occupants usually elect to make use of familiar exits, typically the exit through which they entered the building. Meanwhile emergency exits or exits not used for normal circulation are often ignored, and their actual usage heavily rely on staff for directing the evacuation appropriately. In fire situations, where smoke may also obscure vision, the problem is often fatally compounded. Large-scale fires involving fatalities, such as King's Cross Underground Station [FEN 98] and Düsseldorf airport [WEI 97], are examples of situations where the inability to locate efficient means of escape contributed to loss of life. Other situations, where rapid evacuation of rail/underground systems is essential, could be due to the result of terrorist actions such as the Madrid [BBC 04] and London [BAR 06] bombings.

The presence of signage within an enclosure has traditionally been an essential aid in reducing the amount of time spent on wayfinding [GWY 99], particularly in large, complex or unfamiliar structures such as transport terminal buildings. Emergency signage provides occupants with options, suggestions and the opportunity to decide on the best possible route for evacuating an enclosure. Furthermore, a successful signage system can effectively reduce the apparent complexity of an enclosure by increasing the wayfinding efficiency. However, to be effective, the occupants must first be able to see the signage.

Research carried out by the University of Greenwich [XIE 09] suggests that emergency signage systems can be extremely effective: since people follow them if they see them. However, only 38% of people see the present standard passive signage (Figure 22.1) in emergency situations. Therefore, there is a need to update the present (passive) emergency signage with one that is more detectable by passengers. Moreover, a new signage system should be intelligent and adaptable to the changing environment in order to assess and direct occupants to an optimal egress route, based on insight into the shortest route to safety coupled with the provision of minimum exposure to incident hazards such as heat, smoke and toxic gases.

Figure 22.1. *Standard evacuation sign*

22.2. External factors

This project was conceived against a back drop of continuing growth in the transport sector and the need to ensure enhanced levels of safety for travelers and those working within station environments. The European Commission Transport White Paper (2002) states that rail safety should increase regardless of the volume of traffic. In other words, the aim should be to reduce the real numbers of rail mishaps even as safety and security risks and traffic volume increase. Despite efforts to decouple transport demand from economic growth, passenger traffic is still expected to increase by 40% and freight transportation by 70% by 2020.

Given the importance of this task, the GETAWAY project is one of the current EC funded FP7 projects covering safety and security for transport terminals. Two other projects covered similar scopes, each having a defined set of tasks and deliverables:

1) The "SAVE-ME" project (http://www.save-me.eu.) scope was to investigate the setting up of a "holistic system architecture within station environments that will allow inclusion of different elements and modules, as well as a common ontological architecture for hazards recognition, cause, severity and mitigation". The intention is to make use of personal mobile devices to guide passengers in the event of events such as fire and terrorism.

2) The "SECURESTATION" project (http://www.securestation.eu) aimed "to improve passenger station and terminal resilience to terrorist attacks and safety incidents through technologies and methodologies enabling design to reduce the impact of blast, fire and the dispersion of toxic agents on passengers, staff and infrastructure". One of the deliverables for this project is to deliver a Constructive Design Handbook addressing new build and refurbishment cases for owners and operators to increase station security and safety. It also aims to create harmonization and the standardization of risk assessment methodologies, technologies and design solutions thereby supporting wide application by the numerous EC public transport organizations and associated key stakeholders.

In order to ensure continued improvement in both railway safety and security, the GETAWAY project is based upon the idea that an overall systems approach is clearly necessary, including a full analysis of the interrelated elements and determination of risks. The main targets for research activities in this area include hazard reduction. Among other things, hazard reduction is directly linked to improvements in the quality of the input and output processes. Thus, improving the overall quality management system is also a key priority.

Although much of the focus is on reducing railway accidents involving trains (collisions etc.), another important area is the need to prepare for the possible evacuation of major crowded rail stations. Some recent serious incidents have highlighted the importance of this aspect of rail transport safety, and have demonstrated the risks of a lack of preparation. Perhaps the most striking was the disaster in 2003, in which at least 198 people were killed, and at least 147 injured, by a fire that was started by an arsonist in the Jungangno Station of the Daegu Metropolitan Subway in Daegu, South Korea. Reports suggest that various additional factors compounded the disaster, including communication errors and inadequate emergency equipment.

Another incident occurred in 2006, after a subway fire started in stored wood ties (wooden sleepers for rail tracks) in Brooklyn, NY, requiring 4,000 passengers to be evacuated, injuring at least 25 people, and causing delays to thousands of peak hour commuters. Earlier events, such as the fire at King's Cross Underground station on 18 November 1987 which killed 31 people, also demonstrate the problems that can occur when large crowds need to escape from confined locations. Other high-profile incidents around the world of death and injury resulting from problems in evacuating large numbers of people from burning buildings, have underlined the challenges involved in creating safety policies for areas such as rail terminals, that will ensure the safe evacuation of large groups of people from complex and sometimes confined spaces.

Designing and preparing for such eventualities also needs to take account of human behavior in emergency situations. The public's behavior, which may not always be rational and/or predictable, in a fire has to be accommodated by engineers devising evacuation strategies. When automatic fire detection triggers an alarm, occupants' response is often anything but automatic. Time is often squandered in non-evacuation activities – a phase known as pre-movement time, during which occupants are trying to figure out what is happening and what, if anything, they should do.

Education is the key to making better-informed decisions about a fire and to appreciate the importance of taking appropriate action rapidly. In turn, clear signage has an important part to play in guiding people to safety, but this also needs to have been devised to reflect the station layout and with the best possible understanding of how people are likely to react in that environment.

22.3. Objectives of the GETAWAY project

The main objective of GETAWAY is to improve safety in transportation terminals through an innovative signage system which will:

– significantly improve the detectability of signage systems by 50%;

– achieve this without increasing the size of the sign or changing the visual content of the sign, thereby making it acceptable to owners, architects and regulatory authorities;

– make use of environmental information relating to the developing hazardous situation in real time;

– make use of video analysis techniques that can determine the number of people within given regions within the station environment to an accuracy of 90%;

– make use of evacuation simulations involving hundreds of people which can be performed at least three times faster than real-time.

In order to achieve this, GETAWAY will develop and trial an Intelligent Active Dynamic Signage Systems (IADSS) for real-time direction of transport terminal passengers during the evacuation of a transport terminal in the event of an emergency. The GETAWAY system is intended to be included as part of the design of new transport terminals or fitted into existing transport terminals to optimize evacuation from these complex structures.

22.4. The GETAWAY system concept

Before commencing to develop the GETAWAY system, a review for the relevant aspects of transport passenger emergency procedures at railway stations in line with the project objectives was carried out. The review has been used in the determination of system specifications by identifying factors and constraints for the development of the IADSS. The review will also be used in the development of the guidelines for the use of the GETAWAY system within existing stations and when designing new terminals.

The GETAWAY-IADSS concept consists of five subsystems (Figure 22.2) each performing distinct functions:

1) The Fire Detection and Alarm System (FDAS) which monitors the evolution of the fire incident at the transportation terminal through continuous readings of temperature and smoke sensors.

2) The Image Analysis Engine (IAE) which estimates the number of passengers in the affected areas based on footage from CCTV cameras.

3) Evacuation Simulation Engine (ESE) which performs faster than real-time evacuation simulations based on predefined evacuation scenarios. The simulated scenarios are passed to the Decision Engine for assessment.

4) The Decision Engine (DE) which controls the data flow between the GETAWAY components. It passes environmental and crowd data to ESE to perform the evacuation simulations. It also assesses and ranks the evacuation scenarios received from ESE on the basis of a cost function. The DE User Interface (UI) will display the optimal evacuation scenario together with other important information about the fire incident. Based on this information, the Incident Manager will be able to select an evacuation scenario and activate the DSS in order to guide passengers to safety.

5) The Dynamic Signage System (DSS) which will guide passengers to safety.

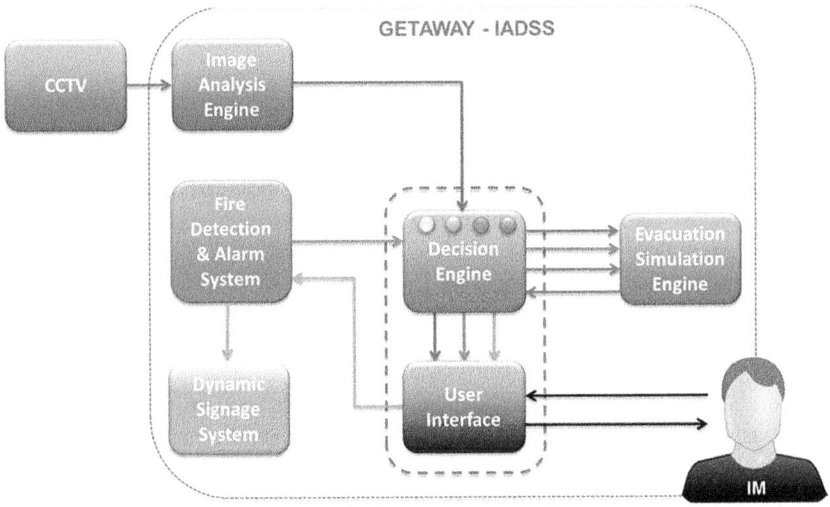

Figure 22.2. *The GETAWAY-IADSS concept and its components*

Furthermore, new image analysis algorithms have been developed based on perspective normalization and extraction of holistic features. These algorithms have been implemented and tested using footage obtained from CCTV cameras. Results have shown that these techniques can result in more than 90% accuracy when estimating crowd numbers.

A prototype version of the UI (Figure 22.3) has also been designed which will provide the Incident Manager with information about station geometry, evolution of the event (including environmental and crowd data) and all available evacuation scenarios including the optimum scenario. The UI will also allow the Incident Manager to override the evacuation scenarios suggested by the Decision Engine and implement his/her own evacuation plan.

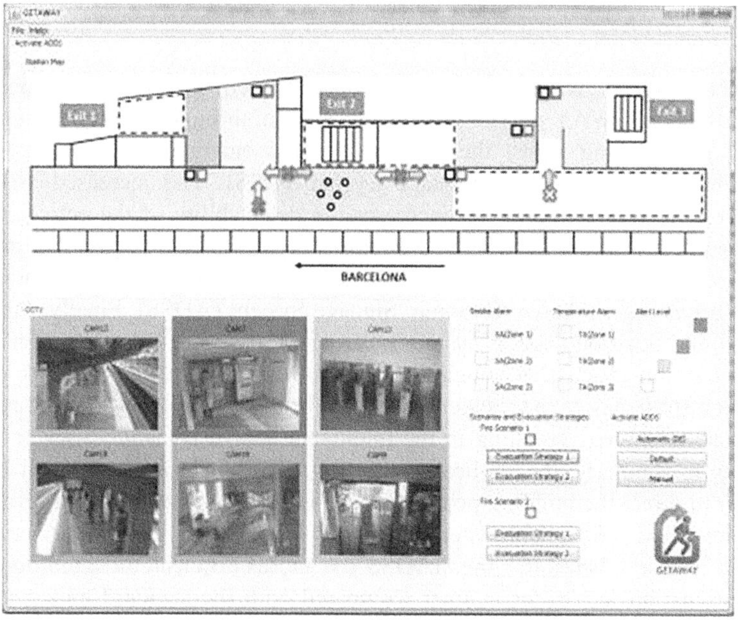

Figure 22.3. *The decision engine user interface*

One of the key elements of the project is the development of the active signage system. This signage was developed to address current detectability and decision-making issues that people face when evacuating complex structures. While different designs were considered, the most effective ones were the progressively illuminated green arrow indicating a viable route (Figure 22.4(a)) and the illuminated red cross over the "emergency exit" sign indicating a non-viable route (Figure 22.4(b)).

Figure 22.4. *The dynamic signage system indicating a viable a) and non-viable b) exit. For a color version of the figure, see www.iste.co.uk/jacob/safety.zip*

22.5. The GETAWAY-IADSS development

The Intelligent Active Dynamic Signage System (IADSS) was developed through the GETAWAY project in three steps. First, through the introduction of lit, flashing and running signs, the traditional static emergency signage system is transformed into a Dynamic Signage System (DSS). The increased affordance offered by the DSS significantly increases the delectability of the emergency exit sign. Second, by linking the DSS to the alarm system, the DSS will only be activated when needed, thereby overcoming the "learnt irrelevance" of emergency signs, thereby becoming an Active Dynamic Signage System (ADSS). Finally, the ADSS is linked to a decision support system utilizing CCTV feed, Fire Detection System information and evacuation simulation to automatically and rapidly rank existing evacuation strategies as the incident develops, bringing intelligence to the selection of evacuation strategy resulting in the Intelligent Active Dynamic Signage System (IADSS). In summary, the application of the IADSS enables the terminal Incident Manager to direct the building population to follow an intended optimal evacuation procedure based on available situational information by activating the appropriate dynamic signs. Therefore, the IADSS is more efficient in guiding people in an evacuation in a safer manner, compared with the standard emergency exit signage systems. The following benefits of the IADSS have been demonstrated through actual live trials using the IADSS in a realistic application within a rail terminal:

1) A significant increase in the numbers of people using the indicated exit (i.e. following the intended evacuation procedure) instead of their nearest exit (which may become unsafe or unviable due to the incident).

2) A positive influence on evacuating persons in that they did not find the new signage design confusing.

3) Increased confidence for evacuating persons that they were moving in the correct direction and that they had selected the correct exit.

The GETAWAY-IADSS is a direct upgrade of standard emergency signage systems to improve the efficiency of signage conveying information to people in an emergency evacuation. Subject to the requirements, the upgrade can range from one-to-one replacement of the existing emergency exit signs to a substantial enhancement to the entire signage system, which integrates signage, detection and CCTV systems with added intelligent features to provide a comprehensive solution. The full IADSS comprises of five subsystems: Active Dynamic Signage System (ADSS), Fire Detection System (FDS), CCTV Analysis Engine (CAE), Decision Engine (DE) and Evacuation Simulation Engine (ESE) (see Figure 22.5).

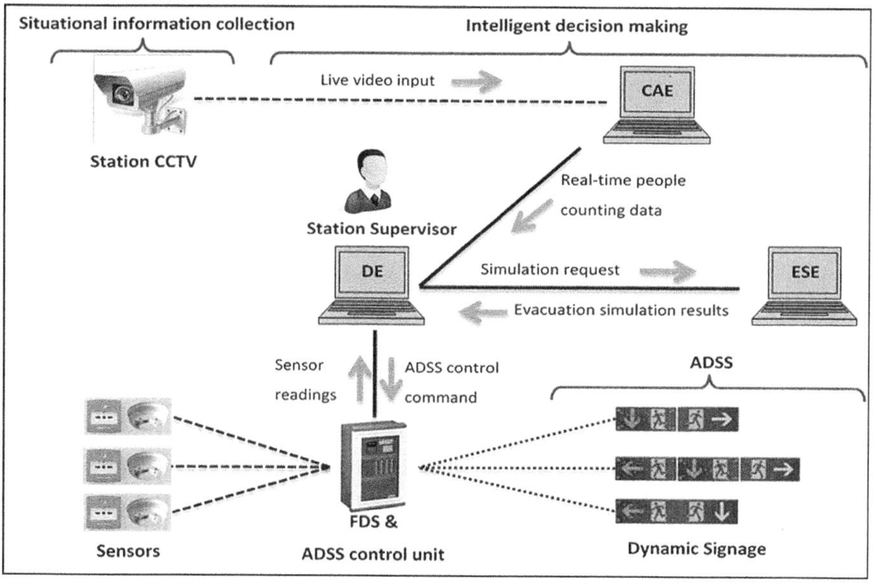

Figure 22.5. *The developed IADSS (GETAWAY system)*

22.6. The Active Dynamic Signage System (ADSS)

The dynamic signage design is designed to enhance the signage affordance while maintaining the maximum compliance with existing signage regulations and practice (e.g. Directive 92/58/EEC, EN ISO 7010). This design increases the detectability of the signs through the introduction of lit, flashing and running signage component (see Figure 22.4 above) to the existing standard signage design. The conventional static signage system is then turned into a dynamic signage system (DSS), whereas the size of the sign and the format of the signage information remain unchanged. Moreover, the dynamic nature of the sign (i.e. the flashing cycle) is only activated during an emergency situation when the alarm is tripped, thereby overcoming the "learnt irrelevance" of emergency signs.

The new dynamic signage design was tested through a series of experimental trials as part of the GETAWAY project to quantify the effectiveness of the design and compare that achieved by the conventional signage. The results of the laboratory-based trials show that 77% of people "see" the dynamic sign and 100% of them go on to follow the sign. This is a significant improvement compared with the conventional signage which is only "seen" by 38% of the testing population under almost identical experimental settings and conditions. In addition, the results show the dynamic signage reduces the time people spend on choosing the direction of evacuation at places where doubt exists about the choice of escape route or exit. The increase in detection rate and the reduction in decision time mean the dynamic signage had a more significant impact on participant exit selection than the conventional static signage.

Furthermore, it is considered important to convey to the population that a normally viable exit route is no longer considered safe, for instance, if information from the Fire Detection System suggests so. Thus, exit directions which are considered potentially hazardous are effectively shut down by negating emergency exit signs pointing in those directions. As with the flashing light concept, it is desirable to develop a negated sign which could replace existing signage on a like for like basis, is easily interpreted by the population, is relatively inexpensive to install and which would fail safe should it fail to operate. Several new negated design concepts to indicate that an exit route highlighted by an emergency sign is no longer viable were tested via international survey to gauge understanding of the new signage concept. The results demonstrate that a simple red cross passing through the entire sign (see Figure 22.3) is correctly interpreted by over 90% of the sample as indicating that the exit route is no longer considered viable and so should not be used. Thus in addition to highlighting the desirable emergency exit route, dynamic signs may also be able to highlight which routes should not be taken.

Both dynamic signage design concepts were tested through full-scale evacuation trials that demonstrate, quantify and evaluate the performance of the IADSS as a whole in a realistic application within a rail terminal. The results show that the ADSS concepts succeeded in directing 66% of the test population to the desired target exit, while only 34% of the population made use of their nearest exit, compared with 100% in the baseline trial using conventional passive signs.

22.7. Fire Detection System (FDS) development

The Fire Detection System (FDS) has two functions within the IADSS. First, the FDS manages sensor readings and sends the data to the core component of the IADSS, where the information is used to form situational awareness concerning the environment. The IADSS can assess and suggest evacuation strategies that best avoid areas considered potentially hazardous. Second, considering the capability of the FDS in connecting and controlling devices across the building, it can be adapted to receive instructions from the core component of the IADSS to configure the signs and active them accordingly (turn flashing signs on or off and switch between the green arrow and red-cross symbols).

22.8. CCTV Analysis Engine (CAE)

The CCTV Analysis Engine (CAE) connects to an existing CCTV and use the live video input to identify the location and number of occupants in the building. This information is then sent to core component of the IADSS, where the information is used to form situational awareness concerning the occupancy status. Based on the distribution of the occupants, the IADSS can assess and suggest evacuation strategies that direct people to use routes that are less hazardous and more efficient.

22.9. Decision Engine (DE) and Evacuation Simulation Engine (ESE)

The Decision Engine (DE) and Evacuation Simulation Engine (ESE) are the core components of the IADSS which are responsible for identifying the optimal evacuation strategy (see Figure 22.6). Within the IADSS, the ESE runs faster than real-time evacuation simulations for pre-defined evacuation strategies, taking into account the situational information collected by the FDS and CAE. The DE utilizes the simulation results to identify the optimal evacuation strategy, which is considered to be not only more efficient, but also safer than the others.

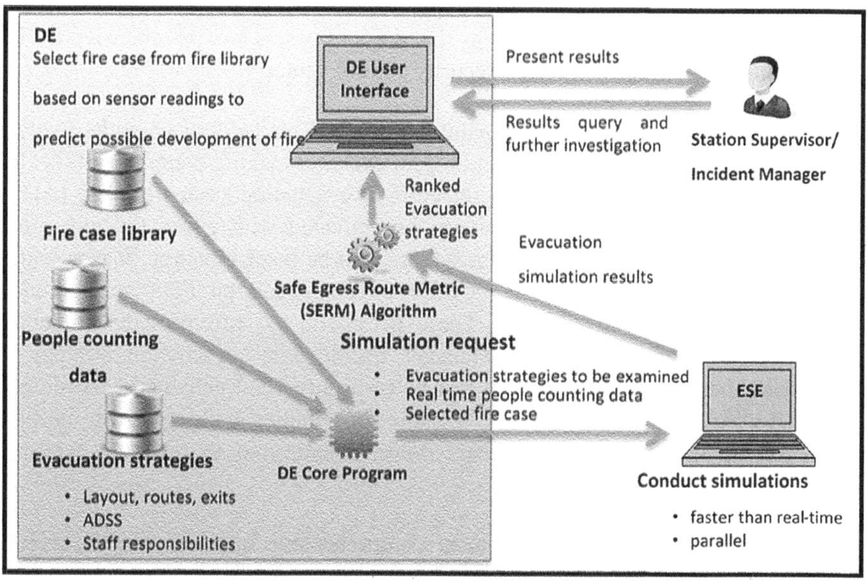

Figure 22.6. *Intelligent decision-making within the IADSS*

22.10. The level of IADSS application

The application of the IADSS is scalable and therefore subject to the upgrade requirement and the features of existing peripheral systems, the upgrade of an existing signage system to IADSS can have several levels. The upgrade requirement concerns the size of the building and the complexity of wayfinding within the building. Small buildings or buildings with simple routes may require a modest IADSS configuration, while bigger buildings and buildings with complex routes may require a comprehensive IADSS configuration. Furthermore, the IADSS will make use of existing peripheral systems such as FDS and CCTV. The features of these peripheral systems will influence the extent of the IADSS configuration.

In general, there are three levels of IADSS configuration.

Level 1: the first level is the simplest upgrade which involves a one-to-one replacement of the emergency exit signs with the dynamic exit signs. The dynamic signs will have the flashing green arrow activated when triggered during an emergency evacuation. The increased affordance offered by the dynamic signs will increase the detectability of these signs in an emergency, hence reinforcing the compliance with the direction indicated by the signs. This level of IADSS configuration is suitable for small buildings with simple routes, where the enhancement of the compliance with a single evacuation strategy is required. It can also be used in large complex buildings that do not possess a comprehensive CCTV coverage and a continuously staffed security control center.

Level 2: the second level is the medium upgrade. This level of upgrade involves upgrading to an IADSS without the intelligent decision-making components (i.e. the CAE, DE and ESE) that support the selection of evacuation strategies. The IADSS will include dynamic exit signs that can be activated with either flashing green arrow or flashing red cross. For this installation to be most effective, it is essential for the building to have extensive CCTV coverage and an FDS as well as a continuously staffed control center. These peripheral systems provide situational information to help the Incident Manager decide what evacuation strategy to implement. Then the signage system can be configured and activated accordingly. This level of IADSS configuration is suitable for complex buildings with multiple manageable evacuation strategies that could be subjected to different types of threats.

Level 3: the third level is the most comprehensive upgrade which involves upgrading to a full IADSS as illustrated in Figure 22.5. The IADSS will include FDC to detect the early sign of an incident and CCTV and CAE to monitor the occupancy status of the building. The information will be used by the intelligent decision-making components, i.e. the DE and ESE, to rank pre-defined evacuation strategies in order to support the Incident Manager in making a decision on the selection of evacuation strategy. Finally, the signage system can be configured and activated according to the selected evacuation strategy. This level of IADSS configuration is suitable for complex buildings with multiple manageable evacuation strategies that could be subjected to different types of threats and which require support in deciding the optimal evacuation strategy.

22.11. Evaluation of the GETAWAY system

The GETAWAY system has been progressively tested and evaluated in order to assess its effectiveness in improving the evacuation of large crowds from rail terminals. The evaluation includes three trial series: Trial Series 1 was designed to assess the effectiveness of the DSS in improving the affordance (e.g. detectability and clarity of directions) of the evacuation sign while Trial Series 2 and 3 were designed to evaluate the effectiveness of the ADSS and IADSS, respectively, in evacuating large crowds from train stations using volunteers in real settings as well as simulations using the building EXODUS and SMARTFIRE developed by the Fire Safety Engineering Group at the University of Greenwich (http://fseg.gre.ac.uk).

Trial Series 1 was held at Greenwich University in London. The trials were conducted over three days in August 2012 on the Maritime campus of the University of Greenwich. A total of 58 participants took part in the trials. The subjects were selected from some 200 registered volunteers.

The trials involved putting each of the participants individually through the evacuation test, videoing the progress of the participant and having the participants complete a questionnaire. These results were used to determine the likelihood of uninformed and unfamiliar people perceiving and following the DSS while moving within the built environment.

Data analysis from Trial Series 1 showed that 77% of the participants saw the DSS resulting in a 103% increase in the detection rate compared with the conventional sign. In addition, all of the participants that saw the DSS followed the direction indicated by it. Questionnaires completed by these volunteers directly after the test indicated that the DSS instructions were clearly perceived by the participants.

As most of the participants chose the direction indicated by the sign upon detecting the sign in both sets of trials, the increase in detection rate means the DSS had a more significant impact on participant exit selection than the conventional passive signage.

A similar trend was also observed for the other dynamic signs positioned in the locations where there were other influencing factors such as the presents of additional doors. The DSS also helped people to reduce their decision times. The 41 (out of 53) participants who detected the sign required 1.8 s on average to decide on a route (i.e. to follow the sign), while the other 12 participants who did not see the sign spent on average 5.7 s in deciding upon a route. The difference in the decision times between those who detected the dynamic sign and those who did not was statistically significant [XIE 13].

Trial Series 2, held in late spring and early summer of 2013, was to demonstrate the benefit of the GETAWAY ADSS compared with the conventional static signage system. This was based on the assessment of the effectiveness of the two signage systems at influencing passenger route selection in a similar evacuation scenario. Trial Series 2 included two sub-trials: the first used the conventional static signage system and the second, the ADSS.

The initial requirement was to determine a railway station that provides appropriate layout and geography that can adequately test the system. In order to properly test the GETAWAY concept, a fire scenario should present the evacuees with at least two choices for evacuation. Ideally at least one of the choices will be a route that will not be normally used. The station should also be of a proper size to allow for on-site trials using large number of participants. There are other practical issues that would need to have been considered such as efficacy of the monitoring systems, in particular, the impact of the local environment on the CCTV monitoring systems and the practical issues for the location of illuminated signage systems.

The precise fire scenarios and evacuation strategies will, by necessity, be unique to the station in question. For the real-world trial, a station on the outskirts of Barcelona, Sant Cugat, was chosen. This station has a choice of exit routes yet contains a relatively straightforward layout. The station has two platforms: platform 1 and platform 2. In order to allow easy data collection and reduce the scale of the disruption to passengers, only platform 2 and the associated three exits were used during the trials (see Figure 22.7).

Exit 1 on the left is a main exit which leads to the street. Exit 2 in the middle is an entrance to the underpass which leads to the main entrance located in platform 1 (not shown). Exit 3 on the right is a secondary exit which also leads to the street. The publicly accessible area forms three circulation/escape routes as indicated by the arrows in Figure 22.7. One hundred and ninety volunteers took part in the normal evacuation trial and 150 in the ADSS trial. The volunteers were instructed and monitored by a number of persons from the GETAWAY project teams. The evacuation scenario was based upon the assumption that an incident has occurred (e.g. on platform) that requires the evacuation of the station via the (routine and emergency) routes available.

Although the volunteers were aware of the exercise, they were given minimal information regarding the evacuation procedure. This would allow participants to make their choice of escape route based on the information available to them (i.e. normal signage in first trial and ADSS in the second trial), especially given that they were not familiar with the station and will therefore be more dependent on the guidance provided.

Station staff were briefed of the nature of the trial to be conducted to avoid unwanted staff intervention during the trial, encouraging or directing participants to use a particular exit or exit route, unless conditions developed during the trial are deemed to pose a risk to the safety of the participants.

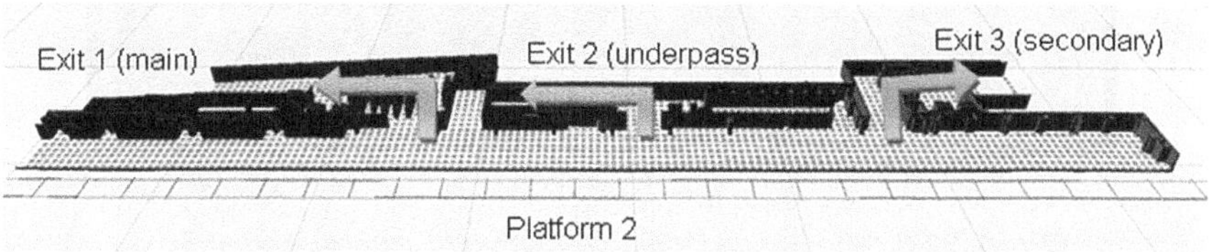

Figure 22.7. *The test area of the Sant Cugat station for trials*

The results showed that with the standard signage system, 99% of the participants used their nearer exit to evaluate the station which may not necessarily be a safe exit. However, when the ADSS was installed and activated only 57% of participants located near the non-viable exits evacuated the station from these exits. The rest of the participants (43%) evacuated through the safe exit indicated by the ADSS.

In total, 63% of the participants used the safe exit indicated by the ADSS. This figure includes those participants initially located close to the safe exit and those that redirected to the safe exit and therefore represents the proportion of the population (63%) that followed the intended procedural intervention posed by the ADSS in Trial Series 2.2. In addition, analysis of questionnaires indicated that, when only the standard signs were available, most of the participants were aware of them as a source of information. However, the standard signs played a much less important role in their exit selection than the proximity of the exits. On the contrary, when the ADSS was installed, the participants perceived both the no entrance signs and flashing arrow sign and they treated signage as the single important factor in their selection of the evacuation exit.

In the second half of the project the developments focused on the refinement of the components to meet the requirements of the Sant Cugat station in Barcelona which was used for the final full scale trials. The fire libraries in the DE were updated to match the requirements of the Sant Cugat station. The fire libraries included fire models for different fire scenarios at Platform 2. In addition the ranking algorithms for the different evacuation scenarios generated by the ESE were also updated to take into account the refined fire models. Moreover, the CAE was also refined using footage from Trial Series 2. This was necessary to ensure best possible performance during the final Trial Series 3. Finally, a new signage system was designed and developed to indicate viable (flashing green arrow) and non-viable (flashing red cross) (Figure 22.8).

Figure 22.8. *The Active Dynamic Signage System used in Trial Series 3 conveying both positive (viable route) and negative (non-viable route) information. For a color version of the figure, see www.iste.co.uk/jacob/safety.zip*

The evaluation of the GETAWAY-IADSS prototype took place during the Trial Series 3 at the Sant Cugat station in Barcelona in June 2014, with the participation of 375 volunteers. The trials demonstrated that the use of the IADSS led to a significant reduction in the numbers of people using their nearest exit to evacuate the station (as it was found when the standard sign was installed). For example, video analysis showed that, overall, 66.4% of the participants evacuated the station using the safe exit following the guidance of the IADSS. In addition, 57.3% of the participants that were not near the safe exit followed the IADSS and evacuated the station from the safe exit instead of choosing to evacuate using their nearest exit as they did with the standard signs.

Furthermore, analysis of questionnaires showed that overall there was a strong indication that the ADSS had a positive influence on the participants and that they did not find the signs confusing. For example, 90% of the participants agreed/strongly agreed that they did not need assistance in interpreting the meaning of the signs and 81% agreed/strongly agreed that the signs assisted them in finding an appropriate exit. In addition, there was a strong indication that the ADSS reinforced the feeling that the participants were moving in the correct direction and that the participants had selected the correct exit. For example, 79% of the participants agreed/strongly agreed that the ADSS signs above the exits confirmed that they were going in the right direction and 80% agreed/strongly agreed that the signs made it clear which exit to use.

Analysis of the DE performance showed that the ranking achieved by the emulated and actual test fire were very similar suggesting that using the fire library approach adopted by the IADSS rather than actually simulating the fire has the potential to correctly identify the best and worst exit strategies. In addition, the DE produced a similar exit strategy ranking with an engineer indicating that automatic assessment of complex computer-based evacuation simulation output is possible and reliable. The human operator chose a strategy which was ranked second best by the DE and therefore less effective than the optimal evacuation strategy (84.4s vs. 124.9s evacuation time). This indicates that the DE is capable of automatically selecting an evacuation strategy similar to that of a human operator producing a similar outcome to that of a human operator and, therefore, can be a reliable aid to emergency managers during a crisis.

22.12. Conclusion

The GETAWAY project developed and demonstrated an integrated active and intelligent directional emergency signage system, which guides terminal passengers to safety taking into account the evolution of the emergency incident. The active nature of the signage system means that more people are likely to notice the directional signs than is currently the case with the present passive signs. In addition, the intelligent nature of the signage system means that the signs provide situation-specific directional information. This means that the signage system identifies different routes as the incident develops and congestion, fire and its products, dictate alternatives. The emergency signage system can be integrated to work in harmony with the terminal's existing fire safety systems.

The GETAWAY-IADSS was validated using three full scale tests at an overground station in Barcelona (Sant Cugat Station) and simulations of large crowd evacuations at a London Underground station (Gloucester Road Station) with more complex geometry. Results showed that participants evacuated the station following instructions of the GETAWAY-IADSS. In addition, participants found the IADSS instructions to be clear and easy to follow allowing them to make fast decisions about the viable exit that would lead them to safety.

The long-term aim is for the rail terminal designers to be able to trial new terminal designs on the computer, modeling the efficiency of evacuation procedures with or without the new emergency signage system in place. This will provide not only a demonstration of the potential benefits of the emergency signage system, but also a mechanism by which terminal design can be verified quickly, cheaply and effectively. Furthermore, the proposed system will have application in other complex environments such as shopping malls, airport terminals, sports stadia, etc.

22.13. Acknowledgments

This work was carried out in the context of the GETAWAY FP7 project funded by the EC (Grant Agreement Number 265717). The authors would like to thank the GETAWAY consortium for their work, which was used to prepare this chapter. The list of the partners involved are as follows: BMT Group Ltd, University of Greenwich, Evaclite Ltd, Ferrocarrils de la Generalitat de Catalunya, Hochiki Europe (U.K.) Ltd, Vision Semantics Ltd, London Underground Ltd, Kingfell Bulgaria and BMT Techmar NV.

22.14. Bibliography

[BAR 06] BARNES R., Report of the 7 July Review Committee, Greater London Authority, Chairman, Richard Barnes, June 2006.

[BBC] BBC News, Available at: http://news.bbc.co.uk/1/hi/in_depth/europe/2004/madrid_train_attacks/default.stm

[CHE 13] CHEN K., GONG S., XIANG T. *et al.*, "Cumulative attribute space for age and crowd density estimation", *Proceedings of IEEE Conference on Computer Vision and Pattern Recognition*, Portland, Oregon, USA, June 2013.

[FEN 88] FENNELL D., Investigation into the King's Cross Underground Fire, The Stationery Office Books, 1988.

[GWY 99] GWYNNE S., GALEA E.R., OWEN M., *Escape as a Social Response*, Published by the Society of Fire Protection Engineers, 1999.

[LON 07] LONDON UNDERGROUND, *Station Planning Standards and Guidelines – Good Practice Guide*, G-371A, A2, May 2007.

[LOY 13] LOY C.C., CHEN K., GONG S. *et al.*, "Crowd counting and profiling: methodology and evaluation", in NISHINO A., MANOCHA AND SHAH (eds), *Modeling, Simulation and Visual Analysis of Crowds*, Springer, December 2013.

[LOY 13] LOY C.C., GONG S., XIANG T., "From semi-supervised to transfer counting of crowds", *Proceedings of IEEE International Conference on Computer Vision*, Sydney, Australia, December 2013.

[WEI 97] Weinspach P.M., Gundlach J., Klingelhofer H.G. *et al.*, Analysis of The Fire on April 11th, 1996, Recommendations And Consequences For Düsseldorf Rhein-Ruhr-Airport, Staatskanzlei Nordrhein-Wstfalen, Mannesmannufer 1 A, 40190 Düsseldorf, Germany, 1997.

[XIE 09] XIE H., FILIPPIDIS L., GALEA E.R., "Experimental study of the effectiveness of emergency signage", *Proceedings of the 4th International Symposium on Human Behaviour in Fire*, Robinson College, Cambridge, UK, 13–15 July 2009, pp. 289–300, 2009.

[XIE 13] XIE H., GALEA E.R., LAWRENCE P.J., "Experimental study of the effectiveness of Dynamic Signage System", *Proceedings of the 13th International Fire Science & Engineering Conference*, Interflam 2013 24–26th June 2013, Royal Holloway College, London, UK, vol. 2, pp. 967–978, 2013.

23

Interpretive Structural Modeling of Security Systems for Better Security Management in Railways

The inherent reason lying at the core of the issue of designing an effective security policy in railways is its complexity, leading to a poorly articulated structure of the security system. While there have been attempts to study the complex nature of security issues, scientific research in this direction has left much to be desired. A systems' approach can provide the required solution to the existing problem. In this chapter, the authors have discussed the application of Interpretive Structural Modeling (ISM) to the security system in Indian Railways. ISM is a soft systems methodology, based on paired comparison of structural elements with respect to a given relationship, to structure complex systems. The results presented in the chapter are based on research work that involved security experts drawn from the railways along with other stakeholders as domain experts, but do not reflect any official version of Indian Railways.

23.1. Introduction

The service providers in public transport systems for passengers and goods are expected to ensure security during the entire process of such transportation. The air and water transport systems have fairly well controlled access control mechanisms. Therefore, they suffer far less from security hazards than the road and rail transport systems.

The security agencies entrusted with the responsibility of securing railways have been constantly upgrading their skills, processes and technology to meet security challenges, including terrorism and insurgency. This is being ensured by better

Chapter written by Anoop SRIVASTAVA, Sanjeev SWAMI and Devender BANWET.

planning, organizing suitable training, procuring the latest arms, ammunitions, gadgetry, etc. In some railways, the strength of the security personnel has also been augmented from time to time.

However, with improvements in the professional standards of service providers in sectors, like telecoms, banking, aviation, etc., the expectations of similar standards from the railway system have also risen over time. Therefore, securing an open system like a traditionally designed complex railway network continues to be a difficult challenge for railway administrations throughout the world.

In this chapter, the authors have discussed the application of a systems analysis technique, i.e. Interpretive Structural Modeling (ISM), for the security system of the Indian Railways. The results presented in the chapter are based on a research work that involved security experts from Indian Railways along with other stakeholders such as domain experts. However, these results do not reflect any official version of the Indian Railways.

23.2. Complexity of railway systems

The inherent reason lying at the core of the problem of security in railways is the complexity of the railway system itself. Like in most of the large-scale systems, the complexity exists not only in the organizational structure and operational functions of railways but also in the relations and interactions between different elements of the system. It is for this reason that the problem of security in railways cannot be tackled without duly considering such complexity.

Modern-day problems of security management in railways are being dealt with by applying the available technology to the greatest possible extent. However, quite often, this is being done at the symptomatic level. "Technological fixes" so developed often create an illusion that such solutions merely require huge investment of public funds [SAG 77]. There is no denying that budgetary inputs are necessary for adopting new technology. Yet, funds are not the only thing necessary because the issues concerned with railways' security are quite intimately entwined in the complexity of the railway system.

In a highly networked set up like that of the Indian Railways, which is spread over almost the entire length and breadth of the country, the extent of the complexity can be gauged by the sheer size of its infrastructure and operations. With 64,600 kilometers of route length connecting 7,146 stations, the Indian railway system is the largest railway network under a single administration. Also, with 12,335 passenger trains carrying 23 million passengers per day, apart from hauling 2.67 million tons of freight traffic on a daily basis, it is one of the most heavily loaded transportation systems in the world [IND 12].

While there have been attempts to study the complex nature of security issues, scientific research in this direction has left much to be desired. A systems approach can provide the required solution to the existing problem. Systems Science, including its four divisions mentioned by Warfield [WAR 06] – Science of Description, Science of Design, Science of Complexity and Science of Action – can serve the purpose of understanding the structural design associated with the security scenario in the railway system. It can also enable the security experts to choose the best suited method for tackling the security threat in any given situation.

Two basically neutral processes of systems science – the Nominal Group Technique and Interpretive Structural Modeling – can be applied to reduce or organize the complexity of a system. Once organized, the complexity may not remain anymore in the system [WAR 06]. Clarity of structural relationships leads to better understanding of the system and enables the concerned authorities find appropriate solutions to perceived problems.

23.3. Nominal Group Technique (NGT)

The Nominal (meaning name only) Group Technique is a structured group dynamic exercise, where the participants meet and work in a group but do not interact as they would naturally do in a conventional group set up for brainstorming [DEL 71]. Research has proved that NGT is significantly superior to interactive group exercises like brainstorming in generating ideas, solutions to problems, etc. [VAN 71].

When NGT is applied, a facilitator/coordinator presents the issue in question to the participants (nominal group) for the generation of ideas, solutions, etc., as the case may be. Initially, the participants list their opinions without interacting with each other. Their opinions are pooled together and displayed by canceling out the repetitions. Finally, the most important points are selected after voting.

The process of identifying elements for the Indian Railway Security System underwent two workshop sessions of 4 hours each, in which 15 domain experts participated. A working paper prepared by the research group was presented to the participants at the beginning of the workshop. Based on the content analysis of relevant literature, policy elements concerned with National Security (NS), Security in Transport and Aviation System (NTAS) and Security in Railway System (NRS) were listed. These elements were coded and key-worded. A nominal group of domain experts was asked to go through the lists of elements, edit them and add or delete elements as per their expert knowledge. An initial list of 131 elements was agreed upon by experts in the three groups of elements of security, namely NS, NTAS and NRS. In subsequent sessions of the workshop, the domain experts identified 10 dimensions of security policy and placed the elements in decreasing order of importance within the relevant dimensions. The dimensions are presented in Table 23.1.

Dimension	No. of elements
DI (Social and Cultural Security)	13
D2 (Collection and Dissemination of Intelligence)	14
D3 (Modernization)	18
D4 (Inter-State Security)	14
D5 (Inter-Border Security)	06
D6 (Economy and Finance)	12
D7 (Training, R&D and Education)	17
D8 (Human Resources and Infrastructure Development)	20
D9 (Legal Aspects)	8
D10 (Administrative and Office Procedure)	8
Total	131

Table 23.1. *Dimensions of railway security policy elements*

An additional round of discussion was held and the initial list of 131 elements was reduced to 56 on the basis of weighted average for each element. The paper titled "Systems Approach to Security Management in Indian Railways" written by the authors of this chapter may be referred for the complete list of 56 elements [SRI 09, SWA 09, BAN 09].

23.4. Interpretive Structural Modeling (ISM)

ISM is a soft systems methodology, which is based on paired comparison of structural elements with respect to a given relationship to structure complex systems. It has been successfully used to provide significant inputs for policy-making in many complex human systems. The ISM process transforms unclear, poorly articulated, complex mental models of systems into visible and well-defined simpler models, which could be useful for many purposes. It deals with the interpretation of the embedded object by systematic iterative application of graph theory, resulting in a directed graph for the concerned complex system. Each stage of methodology may be viewed as transforming a model from one form to another, referred to as model exchange isomorphism.

ISM requires active participation of the stakeholders of the concerned system as domain experts. The experts' responses help in preparing structural self-interaction matrix (SSIM), which studies the existence of the contextual relation between any two elements. Based on the data of various matrices, a graphical depiction ("digraph") is constructed, which provides information about hierarchy between the elements. A structural model provides a clearer picture and an understanding of contextual relationships. It is derived from the contextual information connected in the digraph.

[WAR 90a] mentions that the ISM process, since its inception in the early 1970s, has been used in different fields like forestry, fishery, corporate planning, administration, etc., in the settings of various countries, including Brazil, India, Japan, UK and USA. The hub of methodological development activity in this area has been the Centre of Interactive Management, first at University of Virginia, and later at George Mason University, USA.

In India, ISM was introduced in 1998, especially at IIT Delhi. It has been a part of the methodology base of the company Tata Consultancy Services (TCS). [WAR 90b] lists a major governmental application of ISM in Doordarshan, the TV Broadcasting organization of the Indian Government.

[ANG 93a, ANG93b] discuss the development of a systems engineering framework for policy design for the Government of India in the handloom sector of the Indian textile industry. In developing this policy, they specifically used the ISM as a tool, along with other consensus methodologies such as Brain Writing, Nominal Group Technique, Option Field and Option Profile Methodologies and Fuzzy Set Evaluation Procedures. Similarly, [GAU 95, MIS 95 SAT 95] have used ISM to arrive at alternatives of integrated policy designs for Indian industries.

More recently, the methodology of ISM to analyze complex problems has been applied to as diverse applications as literature [DAY 08], information risk management in supply chains [FAI 07, BAN 07, SHA 07], creation of flexible and lean value chain using outsourcing [MOH 08, SHA 08, BAN 08], the cement industry [SAX 06], electrical/electronics industry's green supply chain [GOV 13], etc. Shahabadkar *et al.* [SHA 12] provide a taxonomical overview of the literature on the use of ISM in supply chain management. Although the abovementioned examples are only a representative subset of the application of the ISM in various contexts, they are sufficient to indicate the potent use of this technique in diverse settings and applications.

23.4.1. *Interpretive Structural Modeling for Indian Railway Security System*

The elements identified through NGT were further fine-tuned in terms of the language and the keywords used and fed into the ISM software, made available by Prof. Benjamin Broom of George Mason University, USA on the advice of Prof. John A. Warfield. After introducing the concept of ISM and outlining the requirement of its application to railway security, an ISM exercise was held in a workshop setting by projecting a series of software generated questions before the participant group consisting of security professionals and other stakeholders. The questions were in the format, "Does element A depend on element B?" (pair wise comparison, with the transitive contextual relation being "dependence" of one element on another).

Each time a question flashed on the screen, the participants carefully read it and responded as *yes* or *no* answers. In the case of a difference of opinion, discussions were held and a consensus arrived at. In the absence of a consensus, the majority view was adopted as the group answer. The resultant matrix was presented to the participants. Due to the limitation of the software, the computer could not draw the required structure. However, the ISM structure was created on the basis of the data generated by the software. It was presented to another set of experts in a workshop and adopted after minor changes suggested by them.

The elements finally got structured in three main condensed groups as depicted in Figure 23.1.

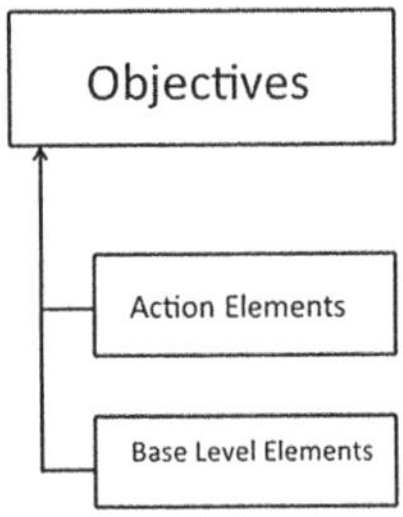

Figure 23.1. *Condensed ISM of Railway Security*

The following five objectives (Table 23.2) emerged in cycles (Figure 23.2) at the top-most two levels.

S. No.	Code	Brief Description	Key Term	Dimension
1)	NRS10	To protect Railway Property, Passengers and Passenger areas.	Passenger-Property Protection	D1
2)	NS4	To protect Human Rights of all citizens.	Human Rights Protection	D1
3)	NTAS17	To modernize security system, especially in transport sector.	Security Modernization	D6
4)	NS10	To provide secure environment for development & economic growth.	Secure Environment for Development	D6
5)	NTAS7	To initiate measures for safety and security of life and property at sea.	Sea-life Security	D9

Table 23.2. *Top-most objectives appearing in cycles*

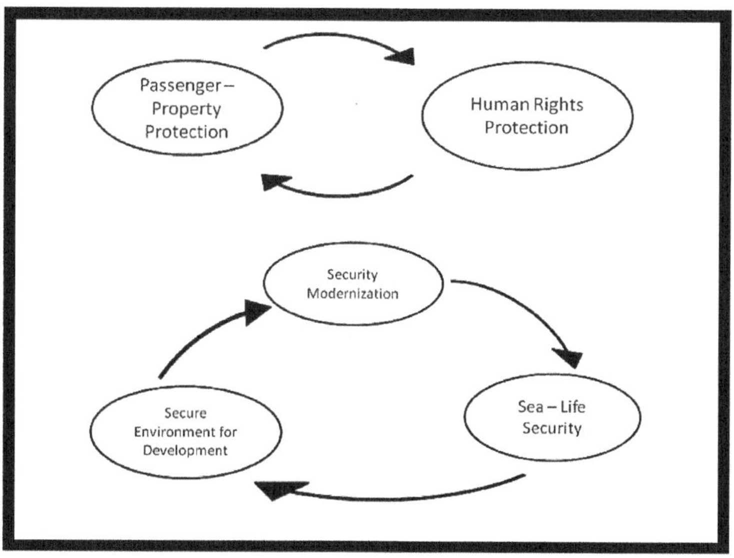

Figure 23.2. *Cyclic representation of top level elements (objectives) for railway security*

Appearance of elements in a cycle indicates that they are interdependent. For railways, element NRS10 (*Passenger-Property Protection*) is of great importance as an objective. But it is unthinkable to ensure protection of passengers and their property/railway property without ensuring *protection of human rights*. The opinion of the domain experts and system science experts regarding the ISM structure for security in railways is that the prime objective of any policy for security in railways has to be the *protection of life and property* of all stakeholders in the railway system while *ensuring human rights* in general. Therefore, the two objectives can be clubbed together as one for the sake of simplicity.

The other three "objective level" elements appearing in cycle, i.e. *Security Modernization*, *Secure Environment for Development* and *Sea-life Security*, are important policy features on which achievement of the prime objective is heavily dependent. This is due to the fact that when criminals have access to sophisticated technology to aid them in commission of crime, neglecting *modernization of security infrastructure* would prove to be a great hindrance in protection of passengers and property in the Railway premises. *Sea-life security* is important for a country like India. For instance, the terrorist attack at Chhattrapati Shivaji Terminus, Mumbai on 26 November 2008, that resulted in the killing of 166 passengers as well as injury to over 300 passengers, was carried out by highly skilled terrorists, who had entered India by the sea route [DAY 08]. Again, maintaining a *secure environment for development* in a country is essential for smooth functioning of a public transport system like Railways.

At the very basic level, the following seven elements emerged to be a prerequisite for successfully designing and implementing the security policy.

Unlike in the case of objective-defining elements, the bottom-most elements of the structural framework are not in any cycle and may be said to be independent of other elements, but affect the achievement of the policy objectives. There was a consensus amongst the experts that the ISM has yielded significant elements at the basic level. While the *requirement of funds* emerged as the most significant requirement for designing any effective policy, another element, i.e. *empowerment of Railway Protection Force (RPF)*, a security force which is expected to play a vital role in ensuring security in Indian Railways, also surfaced as an important requirement. Similarly, *public awareness about causes and consequences of crime* was adjudged by the experts as an important requirement to ensure their participation in law enforcement. This is quite relevant, as, for instance, after a countrywide response to the criminal victimization of a young girl while traveling in a bus in New Delhi late last year, the awareness amongst the traveling female passengers increased. Accordingly, increased reporting by them has led to more and more arrests in the Indian Railway system in the context of crime against female passengers.

S.No.	Code	Brief description	Key term	Dimension
1.	NRS29	To create awareness amongst the public about the causes and consequences of crime against railways.	Public awareness about causes and consequences of crime in Railways.	D7
2.	NS2	To avoid red tape and malpractices at every step.	Avoid red tape	D10
3.	NRS1	To empower RPF in terms of adequate legal powers to deal with offences against passengers and other rail users, railway employees and railway administration.	Legal empowerment of RPF	D9
4.	NS28	To ensure adequate provision of funds for equipment, weaponry, mobility and communication of security forces, especially in the naxal/militant/terrorist-affected States.	Funds for infrastructural/ operational requirements of police forces	D6
5.	NRS22	To include cost on Security in the Railway Projects.	Security cost in railway projects	D6
6.	NTAS21	To utilize services of retired personnel, scouts and guides, etc.	Services of retired personnel, etc.	D4
7.	NS60	To decentralize powers of different security agencies.	Power decentralization	D9

It has been noticed that there is always a gap between the available strength of manpower and the sanctioned strength. Even if the gap is made good by recruitment, the gestation period to train fresh manpower properly and place them in position consumes a lot of time. The gap remains even after the new recruits get trained because by that time some fresh vacancies arise. This gap can be filled if the *services of retired personnel* are utilized, at least in less strenuous duty beats. Experts also agreed that there is a need for a *decentralization of power* and reforms to do away with *red tape*.

Between the elements appearing at the "objective level" and those appearing at the "base level", a number of significant elements have figured at the "action level." However, some of the higher placed action level elements could also qualify as intermediate objective level elements and some of those that are close to the bottom of the structure as semi-base level objectives. The experts were of the opinion that the policy formulation for enhancing the security arrangements in the railway system should not lose sight of some of the most significant action elements like *strengthening and ensuring coordination between the intelligence agencies, training and research, augmentation of manpower, modernization, monitoring internal Security, attempting Holistic Security in terrorism/insurgency affected areas, etc.* Apart from ensuring due attention to these elements at the time of policy-making and implementation, the linkages would have to be kept in mind, as an erroneous sequence in taking action may not yield the desired results.

23.5. Policy implications

It is clear from the ISM structure arrived at that the policy objectives depend upon certain actions. These actions, in turn, depend upon some basic inputs. Therefore, a policy for enhancing security arrangements in the railways will have to address at least the following as basic requirements:

1) Funds for Infrastructural/Operational requirements of Police Forces.

2) Inclusion of Security Cost in Railway Projects.

3) Legal Empowerment of Security Forces.

4) Decentralization of Power.

Some of the important actions required to be taken should include strengthening and ensuring coordination between the intelligence agencies, training and research, augmentation of manpower, modernization, monitoring of internal security, attempts to provide holistic security in terrorism/insurgency affected areas, and the like.

23.6. Conclusions and avenues for future research

The workshops on NGT yielded a comprehensive set of elements that were categorized into ten dimensions. Miller's Index, which is computed by taking the total number of interactions of elements and dividing it by 7, before the NGT was $131^2/7$ =2451.56. It got reduced to 448 when the list of elements was pruned down to 56 items. If the complexity is dealt with at the dimension level, Miller's Index would be 10X10/7= 14.29. Ideally, it should be 1 or less than 1. A Miller's index of value one is the border value between the domain of complexity and the domain of normality [DAY 08].

The ISM workshops provided a structural framework, connecting the elements of security policy on the basis of a contextual relationship. This structure affords a clearer systemic picture to the policy-makers and can enable them to choose from a given set of action elements, depending upon the available inputs at the level of basic elements. The extent to which the outcome of achieving the objectives is likely to be realized, in view of the extent to which the basic inputs are available, can be predicted reasonably well on the basis of the ISM of Security System of Railways.

However, the complexity in the security system of railways can be reduced further, by employing other tools of systems science. There is, therefore, need for further research in that direction. Drawing option profiles based on relevant option fields and evaluating them using Analytical Hierarchy Process to arrive at the best option can be one such way of moving ahead. The authors are already engaged in such research and expect to develop the same in the near future.

23.7. Acknowledgments

The authors are highly indebted to the Most Revered Professor P.S. Satsangi, Chairman, Advisory Committee on Education in Dayalbagh for Invaluable Guidance received from Him. Profuse thanks are also due to Mr. Ranjit Sinha, the then Director General, Indian Railway Protection Force and now Director, Central Bureau of Investigation for the support provided by him. The authors specially thank Prof. D.S. Mishra, Prof. S.K. Gaur, and Prof. C. Patvardhan of Faculty of Engineering, DEI for their valuable suggestions and academic support during the course of this work, and Prof. J. Warfield and Prof. B. Broom for providing the software for conducting the ISM workshop.

23.8. Bibliography

[AGN 93] AGNIHOTRI V.K., SATSANGI P.S., "Policy analysis through structural modeling: Indian Handloom Policy", *Paritantra: A Journal of Systems Sciences and Engineering*, vol. 1, no. 1, pp. 53–65, 1993.

[DAY 08] DAYAL B., SATSANGI P.S., "Managing complexity in systems through neutral processes of nominal group technique and interpretive structural modeling: a novel application from literary field", *Proceedings of XXXII National Systems Conference 2008*, December 17–19, 2008.

[SAG 77] SAGE A.P., *Methodology for Large-Scale Systems*, McGraw-Hill Book Company, New York, NY, 1977.

[WAR 06] WARFIELD J.N., *An Introduction to Systems Science*, World Scientific Publishing Company Pte. Ltd, Hackensack, NJ, 2006.

List of Authors

Letty AARTS
Institute for Road Safety Research
The Netherlands

Ladji ADIAVIAKOYE
ESEO
Angers
France

Michael ALEKSA
AIT Austrian Institute of Technology
Mobility Department
Transportation Infrastructure
Technologies
Vienna
Austria

Giuseppina AMATO
Queens' University Belfast
UK

Miha AMBROŽ
University of Ljubljana
Faculty of Mechanical Engineering
Department of Modelling in Engineering
Sciences and Medicine
Ljubljana
Slovenia

Emmanuelle AMOROS
University of Lyon
and
IFSTTAR
Epidemiological Research and
Surveillance Unit in Transport,
Occupation and Environment
(UMRESTTE)
Bron
France

Maurice ARON
University Paris-Est
IFSTTAR
France

Jean-Pascal ASSAILLY
IFSTTAR
Versailles
France

Jean-Michel AUBERLET
ESEO
Angers
France

Georgia AYFANDOPOULOU
Center for Research and
Technology Hellas
Hellenic Institue of Transport
(CERTH-HIT)
Thessaloniki
Greece

Jean Emmanuel BAKABA
German Insurers Accident Research (UDV)
German Insurance Association (GDV)
Berlin
Germany

Devender BANWET
Indian Institute of Technology (IIT)
New Delhi
India

Cécile BARBIER
IFSTTAR
Versailles
France

Mihai BARCANESCU
UITP
Brussels
Belgium

Charlotte BAX
SWOV
The Netherlands

Evangelos BEKIARIS
Center for Research and Technology Hellas
Hellenic Institue of Transport (CERTH-HIT)
Thessaloniki
Greece

Daniel BELL
FACTUM Chaloupka & Risser OHG
Vienna
Austria

Abdelkrim BENSAID
Laboratoire IDEES-Caen
UMR CNRS 6266
University of Caen Basse-Normandie
France

Ruth BERGEL-HAYAT
UPE IFSTTAR GRETTIA
Marne la Vallée
France

Stéphanie BLAIZOT
University of Lyon
and
IFSTTAR
Epidemiological Research and Surveillance Unit in Transport, Occupation and Environment (UMRESTTE)
Bron
France

Attila BORSOS
Department of Transport Infrastructure
Széchenyi István University
Győr
Hungary

Marc BOURCERIE
ESEO
Angers
France

Maroua BOUZID
Laboratoire GREYC
UMR CNRS 6072
University of Caen Basse-Normandie
France

Chris BRITTON
Parsons Brinckerhoff
UK

Paul BRYANT
Kingfell
Plovdiv
Bulgaria

Ilona BUTLER
ITS
Warsaw
Poland

Salvatore CAFISO
Department of Civil and Environmental Engineering
University of Catania
Italy

Grazia LA CAVA
Department of Civil and Environmental Engineering
University of Catania
Italy

Julien CESTAC
IFSTTAR
Versailles
France

Eleni CHALKIA
Center for Research and Technology Hellas
Hellenic Institue of Transport (CERTH-HIT)
Thessaloniki
Greece

Simon COHEN
Génie des Réseaux de Transport Terrestres et Informatique Avancée (GRETTIA)
IFSTTAR
Marne-la-Vallée
France

Saskia DE CRAEN
SWOV
The Netherlands

François DEFOSSEZ
MERMEC Group S.p.A.
France

Hélène DE SOLÈRE
CERTU
France

Alessandro DI GRAZIANO
Department of Civil and Environmental Engineering
University of Catania
Italy

Etti DOVEH
Technion
Haifa
Israel

Emmanuelle DUPONT
IRSE
Brussels
Belgium

Ceki ERGINBAS
TRL Ltd
UK

Petros EVGENIKOS
National Technical University of Athens
Greece

Chiara FERARINI
University of Modena and Reggio Emilia (UNIMORE)
Reggio Emilia
Italy

Veronique FEYPELL-DE LA BEAUMELLE
OECD/ITF
Paris
France

Dominique FLEURY
Laboratoire Mécanismes d'accident
IFSTTAR
Marne-La-Vallée
France

Åsa FORSMAN
Swedish National Road and Research Institute
Linköping
Sweden

Richard FREEMAN
UCL Institute of Education
London
UK

Alejandra Beatriz GARCÍA MELÉNDEZ
CIDAUT Foundation
Valladolid
Spain

Christos GIACHRITSIS
BMT Group Ltd
Teddington
UK

Victoria GITELMAN
Technion
Haifa
Israel

Gabriele GIUSTINIANI
CTL
Rome
Italy

Anita GRASER
AIT Austrian Institute of Technology
Mobility Department
Dynamic Transportation Systems
Vienna
Austria

Per-Olov GRUMMAS GRANSTRÖM
Swedish Transport Administration
Sweden

Mohamed MOULOUD HADDAK
University of Lyon
and
IFSTTAR
Epidemiological Research and
Surveillance Unit in Transport,
Occupation and Environment
(UMRESTTE)
Bron
France

Lars INGE HASLIE
Road Directorate
Norway

James HOLGATE
VicRoads
Australia

John N. IVAN
Civil & Environmental Engineering
University of Connecticut
USA

Bernard JACOB
IFSTTAR
France

Jernej KORINŠEK
University of Ljubljana
Faculty of Mechanical Engineering
Department of Modelling in Engineering
Sciences and Medicine
Ljubljana
Slovenia

Jean-Bernard KOVARIK
Ministry of Infrastructure, Transport
and the Sea
Paris La Défense
France

Sami KRAÏEM
FSTTAR
Versailles
France

Francesca LA TORRE
University of Florence
Italy

Houda LABIOD
Telecom ParisTech
Paris
France

Marcello LAGUNA
KTH, Stockholm
Sweden

Gernot LENZ
AIT Austrian Institute of Technology
Mobility Department
Dynamic Transportation Systems
Vienna
Austria

Brigitte LONC
Renault
Guyancourt
France

Klaus MACHATA
KfV
Vienna
Austria

Dimitris MARGARITIS
Center for Research and Technology Hellas
Hellenic Institute of Transport (CERTH-HIT)
Thessaloniki
Greece

Heike MARTENSEN
Belgian Road Safety Institute
Belgium

Mohand MEDJKANE
Laboratoire IDEES-Caen
UMR CNRS 6266
Université de Caen Basse-Normandie
France

Evangelos MITSAKIS
Center for Research and Technology Hellas
Hellenic Institue of Transport (CERTH-HIT)
Thessaloniki
Greece

Rim MOALLA
Renault
Guyancourt
France

Andrew MORRIS
Loughborough University
Loughborough
UK

Juan MUGUIRO
ATOS Consulting
Spain

Nicole MUHLRAD
IFSTTAR
Lyon-Bron
France

Gyula OROSZ
Hungarian Transport Administration
Budapest
Hungary

Jörg ORTLEPP
German Insurers Accident Research (UDV) German Insurance Association (GDV)
Berlin
Germany

Panagiotis PAPADIMITRATOS
KTH
Stockholm
Sweden

Eleonora PAPADIMITRIOU
NTUA
Athens
Greece

Francis PAPON
Université Paris-Est
IFSTTAR
Department of Transport Economics and Sociology (DEST)
Noisy-le-Grand
France

Giuseppina PAPPALARDO
Department of Civil and Environmental Engineering
University of Catania
Italy

Bine PENGAL
Slovenian National Building and Civil Engineering Institute
Slovenia

Oscar Martin PEREZ
CIDAUT Foundation
Valladolid
Spain

Luca PERSIA
CTL
Rome
Italy

Jonathan PETIT
UT
Enschede
The Netherlands

Jean-François PEYTAVIN
Laboratoire Mécanismes d'accident
IFSTTAR
Marne-La-Vallée
France

Patrick PLAINCHAULT
ESEO
Angers
France

Ivan PREBIL
University of Ljubljana
Faculty of Mechanical Engineering
Department of Modelling in Engineering Sciences and Medicine
Ljubljana
Slovenia

Eliane PROPECK-ZIMMERMANN
Laboratoire Image, Ville, Environnement
UMR 7362 CNRS
University of Strasbourg
France

Kris REDANT
Belgian Road Research Centre (CRR)
Belgium

Mark ROBINSON
NewRail Centre for Railway Research
Newcastle University
Newcastle-upon-Tyne
UK

Thierry SAINT-GÉRAND
Laboratoire IDEES-Caen
UMR CNRS 6266
Université de Caen Basse-Normandie
France

Josep Maria SALANOVA GRAU
Center for Research and Technology Hellas
Hellenic Institue of Transport (CERTH-HIT)
Thessaloniki
Greece

Peter SALEH
AIT Austrian Institute of Technology
Mobility Department
Transportation Infrastructure Technologies
Vienna
Austria

Gian-Marco SARDI
S.I.P.Si.Vi
Cuneo
Italy

Johan SCHOLLIERS
VTT Technical Research Centre of Finland
Tampere
Finland

Gerard SEGARRA
Renault
Guyancourt
France

Régine SEIDOWSKY
University Paris-Est
IFSTTAR
France

Andrea SOEHNCHEN
UITP
Brussels
Belgium

Anoop SRIVASTAVA
Railway Protection Force (RPF)
New Delhi
and
Dayalbagh Educational Institute (DEI)
Agra
India

Markus STRAUB
AIT Austrian Institute of Technology
Mobility Department
Dynamic Transportation Systems
Vienna
Austria

Sanjeev SWAMI
Dayalbagh Educational Institute (DEI)
Agra
India

Rachel TALBOT
TSRC
Loughborough
UK

Pete THOMAS
Transport Safety Research Centre
Loughborough University
UK

Robert THOMSON
Swedish National Road and Transport
Research Institute
Sweden

Cristian ULIANOV
NewRail Centre for Railway Research
Newcastle University
Newcastle-upon-Tyne
UK

Anna VADEBY
Swedish National Road and
Research Institute
Linköping
Sweden

Pierre VAN ELSLANDE
IFSTTAR
Salon-de-Provence
France

Gilles VALLET
IFSTTAR
Lyon-Bron
France

Gordana VASIĆ FRANKLIN
NewRail Centre for Railway Research
Newcastle University
Newcastle-upon-Tyne
UK

Stephan WITTMANN
AIT Austrian Institute of Technology
Mobility Department
Transportation Infrastructure
Technologies
Vienna
Austria

George YANNIS
School of Engineering
National Technical
University of Athens
Greece

Joanna ZUKOWSKA
Gdansk University of Technology
Gdansk
Poland

Index

R, S, T

U, V, W

Printed and bound by CPI Group (UK) Ltd, Croydon, CR0 4YY
15/09/2026
14962356-0005